Lecture Notes in Mechanical Engineering

Lecture Notes in Mechanical Engineering (LNME) publishes the latest developments in Mechanical Engineering—quickly, informally and with high quality. Original research reported in proceedings and post-proceedings represents the core of LNME. Volumes published in LNME embrace all aspects, subfields and new challenges of mechanical engineering. Topics in the series include:

- Engineering Design
- Machinery and Machine Elements
- Mechanical Structures and Stress Analysis
- Automotive Engineering
- Engine Technology
- Aerospace Technology and Astronautics
- Nanotechnology and Microengineering
- Control, Robotics, Mechatronics
- MEMS
- Theoretical and Applied Mechanics
- Dynamical Systems, Control
- Fluid Mechanics
- Engineering Thermodynamics, Heat and Mass Transfer
- Manufacturing
- Precision Engineering, Instrumentation, Measurement
- Materials Engineering
- Tribology and Surface Technology

To submit a proposal or request further information, please contact the Springer Editor of your location:

China: Dr. Mengchu Huang at mengchu.huang@springer.com
India: Priya Vyas at priya.vyas@springer.com
Rest of Asia, Australia, New Zealand: Swati Meherishi at swati.meherishi@springer.com
All other countries: Dr. Leontina Di Cecco at Leontina.dicecco@springer.com

To submit a proposal for a monograph, please check our Springer Tracts in Mechanical Engineering at http://www.springer.com/series/11693 or contact Leontina.dicecco@springer.com

Indexed by SCOPUS. The books of the series are submitted for indexing to Web of Science.

More information about this series at http://www.springer.com/series/11236

Maddali Ramgopal · Sachindra Kumar Rout ·
Sunil Kr Sarangi
Editors

Advances in Air Conditioning and Refrigeration

Select Proceedings of RAAR 2019

Springer

Editors
Maddali Ramgopal
Mechanical Engineering Department
Indian Institute of Technology Kharagpur
Kharagpur, India

Sachindra Kumar Rout
Mechanical Engineering Department
C. V. Raman College of Engineering
Bhubaneswar, India

Sunil Kr Sarangi
School of Mechanical Sciences
Indian Institute of Technology
Bhubaneswar, India

ISSN 2195-4356 ISSN 2195-4364 (electronic)
Lecture Notes in Mechanical Engineering
ISBN 978-981-15-6359-1 ISBN 978-981-15-6360-7 (eBook)
https://doi.org/10.1007/978-981-15-6360-7

This Springer imprint is published by the registered company Springer Nature Singapore Pte Ltd.
The registered company address is: 152 Beach Road, #21-01/04 Gateway East, Singapore 189721, Singapore

Preface

The importance of refrigeration and air conditioning in modern economies has been well recognized. It is becoming even more important as more and more countries located in tropical and subtropical regions of the globe are growing in industrial activity and social prosperity. The International Energy Agency has observed that the sale of home air conditioners will grow by about 10% each year till 2030, and that, it will lead to increase in power demand to (as yet) unthinkable levels. Unless the efficiency of R&AC machinery is increased to much higher levels, a cold crunch will be inevitable. Thus, the primary target before the research community is to invent technologies that offer far higher overall thermodynamic efficiency than is possible today. Air conditioners are energy-intensive equipment. They not only consume scarce energy of our world, but also cause environmental pollution through the power stations. Efficient cooling machines, better building insulation and smarter building management will not only save on electricity, but will also lead to a proportionate reduction in environmental pollution. The key to such enhancement of energy efficiency and accompanying environmental benefits is in training, education and research.

While domestic, institutional, commercial and industrial air conditioning constitutes the most important application of moderately low temperature, the mobile sector of the economy is no less important. Automobile air conditioning, trains, planes, ships and submarines not only consume a sizable fraction of our air-conditioning effort, but also pose different challenges to the designer and the manufacturer. Once again, trained human resource—the thinking minds—holds the key to success. Air conditioning means control of three major quality parameters—temperature, humidity and air quality. While the former two have been studied extensively over a full century, appreciation for the last item, i.e. indoor air quality, has come during the past one or two decades. Instrumentation is expensive, standardization is difficult, and health risks are hard to quantify. The topic is of much current interest.

Apart from air conditioning, refrigeration at temperatures close to ice temperature, higher and lower, offers immense industrial and commercial benefits. Processing of meat, poultry, fish and shell fish, milk and milk products require an unbroken cold chain from the factory/diary/ocean to the kitchen. Vegetables need cold stores, particularly in poor tropical countries. Many industrial processes are assisted by specified temperature and humidity conditions. The appropriate technology of such refrigeration is varied, product and process dependent, and often quite different from those most appropriate to air conditioning.

The R&AC industry went through a major epoch at the end of the last century, when scientists identified the ozone-depleting potential of the then popular CFC refrigerants. Massive global research efforts have gone to find alternatives to CFC refrigerants, evolving through HCFCs, HFCs, olefins, plain hydrocarbons, inorganic fluids and their mixtures. Besides ozone-depleting potential, scientists are now examining global warming potential of refrigerants. In addition, normal engineering concerns such as operating pressures under various ambient and target temperatures, chemical stability, miscibility with lubricating oils, toxicity, flammability, hygroscopicity, stability and cost are being freshly examined. Generation of such massive amount of data needs experimental facilities, computing capabilities, and the most important of all, human resource.

While climate control (temperature, humidity and air quality)—domestic, institutional, industrial, and commercial—continues to be the dominant application of low-temperature technology, other applications of refrigeration are also increasing at a rapid pace. Food cooling and transportation—cold stores, supermarkets, transport vehicles, shipping, and export—are becoming increasingly important in every developing country. In addition, several important industrial processes, e.g. cryogenics and superconductivity, pharmaceuticals, and fine chemicals, often need refrigeration as an intermediate process step.

Equipment for air conditioning and refrigeration has been evolving over several decades. New technologies like scroll compressors, plate heat exchangers, electronic controls, variable frequency drives, intelligent building management, and IoT have been major disruptive technologies in the field of R&AC. The use of additive manufacturing and precision engineering, integration with renewable energy sources, remote operation through use of information and communication technology, and optimization of operating conditions through soft computing techniques are some examples of product and performance improvement through the application of new knowledge. Once again, proper human resource is the key to success.

Human resource is important for executing new projects and mass manufacturing existing products. It is even more important for discovering new phenomena and inventing new products and new processes. This new knowledge is created not only in college classrooms and university laboratories, but also through formal and informal sharing of knowledge in seminars and conferences, in oral and poster presentations. The International Conference on Recent Advancement in Air Conditioning and Refrigeration (RAAR-2019) hosted by the C. V. Raman College of Engineering, Bhubaneswar, from 28 to 31 November 2019 was one such event, now in its second edition. The conference acted as a trigger for discussion on various topics among scientists, professors, engineers and students distributed across age and generation bands as well as across varied experience levels. The proceedings archives the accumulated knowledge and will help share it among those who did not attend the conference for various reasons.

It has been our singular pleasure to organize the administration of the paper collection and review process, to edit the final contents and put up to the publishers for bringing out the book. We hope that this volume will find a place among the leading university and college libraries of the world. Before closing, we would like to record our gratitude to the Biju Patnaik University of Technology, Rourkela, for supporting the conference under its TEQIP-III project and to M/s Springer Publications for bringing out the proceedings.

Kharagpur, India — Maddali Ramgopal
Bhubaneswar, India — Sachindra Kumar Rout
Bhubaneswar, India — Sunil Kr Sarangi
February 2020

Contents

About the Editors

Dr. Maddali Ramgopal is currently a Professor of Mechanical Engineering at Indian Institute of Technology Kharagpur. He obtained his Ph.D. degree from the Indian Institute of Technology, Madras, in 1996. His major research interests are in the fields of refrigeration, air conditioning, thermal comfort and sorption-based energy conversion systems with more than 50 papers in respected international journals. Currently, he is working towards the development of a practical CO2-based refrigeration system. Prof. Ramgopal serves as a member of the Editorial Advisory Board of the journal Buildings & Environment, an Elsevier publication. He is a member of the International Institute of Refrigeration (IIR), Commission B2, and also of the Technical Committee, Indian Society for Heating, Refrigeration and Air conditioning Engineers (ISHRAE).

Dr. Sachindra Kumar Rout a Ph.D. from the National Institute of Technology, Rourkela, India, is currently an Assistant professor at the Department of Mechanical Engineering, C. V. Raman College of Engineering, Bhubaneswar, and Head of the Centre of Excellence on Heating, Ventilation & Air Conditioning. His research interests include refrigeration & air conditioning, application of solar energy in heating and cooling, cryogenic refrigeration and liquefaction systems and closed cycle cryocoolers. He is also interested in the development of laboratory equipment for teaching or HVAC&R. He has published 14 papers in respected international journals. Dr. Rout also served as an editor of an issue of the Energy Procedia series.

Dr. Sunil Kr Sarangi, Ph.D. from the State University of New York, Stony Brook, is currently Professorial Fellow, School of Mechanical Sciences, Indian Institute of Technology Bhubaneswar, and Academic Advisor to the C. V. Raman College of Engineering, Bhubaneswar. His research interests include cryogenic refrigeration and liquefaction systems including process design, helium compressor, purifier, heat exchanger, expansion turbine and system integration; air separation; closed cycle cryocoolers; cryogenic applications in industry, biology and medicine; refrigeration and air conditioning. He has published more than 100 papers in respected international journals. Prof. Sarangi, a Fellow of the Indian

National Science Academy and a Life Fellow of the Indian Cryogenics Council, has been bestowed with the Distinguished Teacher Award of INAE – 2016, Lifetime Achievement Award of the Indian Cryogenics Council, 2017, and Odisha Citizen Award 2015 for contributing and providing iconic leadership in the field of education.

Modelling and Simulation of Photovoltaic Thermal Cooling System Using Different Types of Nanofluids

Sanjeev Jakhar, Mukul Kant Paliwal, and Atul Kumar

Abstract Solar technologies like flat plate solar collectors are being widely used for low-grade thermal energy for household purposes. These days, photovoltaic thermal (PV/T) collectors are also gaining momentum as source of combined heat and electric power. Commonly used base fluid in PV/T collector is water which have low thermal conductance, and thus, addition of nanoparticles in base fluid will lead to the enhancement in overall thermal conductance. Keeping this as main focus, a research has been carried out to evaluate the performance of PV/T system with different nanoparticles. For that, the simulation was carried out by performing grid test and then simulated on ANSYS to obtain results. For the same, nanofluids with 20 nm particle dimensions and 299 K inlet temperature were loaded with 0.5, 1 and 1.5% particle volume fraction with different Reynolds numbers varying from 250 to 1500. The simulated model was validated with the literature, and obtained results showed that the heat transfer coefficient (HTC) without any nanoparticles ranges from 245.5 to 519.8 W/m^2 K for Reynolds number of 250–1500, respectively. On other hand, with nanoparticles, the HTC increases and ranges between 250.6–529.20 W/m^2 K, 255.42–539.8 W/m^2 K and 261.1–550.8 W/m^2 K for 0.5%, 1.0% and 1.5% volume fraction, respectively, for Reynolds number of 250–1500. In the end, it is concluded that the simulation results are in good agreement with the literature.

Keywords Laminar flow · Photovoltaic thermal (PV/T) · Heat transfer coefficient (HTC) · Solar energy

Nomenclature

C Specific heat (J kg^{-1}K^{-1})
ρ Density kg m^{-3}
μ Dynamic viscosity (Pa s)

S. Jakhar (✉) · M. K. Paliwal · A. Kumar
Department of Mechanical Engineering, School of Engineering and Technology, Mody University of Science and Technology, Lakshmangarh, Sikar 332311, Rajasthan, India
e-mail: sanjeevj450@gmail.com

M. Ramgopal et al. (eds.), *Advances in Air Conditioning and Refrigeration*, Lecture Notes in Mechanical Engineering, https://doi.org/10.1007/978-981-15-6360-7_1

K Thermal conductivity (W/m K)
$\emptyset$ Particle volume fraction
ΔP Pressure drop (Pa)
bf Basefluid
nf Nanofluid

1 Introduction

As we embark this twenty-first century, the new technology and new age also bring many concerns with it. One such concern is the rapidly growing population of the world, which is becoming a grave concern day by day. Population is over shadowing many features and that includes one of the primary sources—"energy consumption". The world's energy consumption is increasing day by day, and with increase in demand, the generation is also increasing. The majority of the energy generated is from conservative sources of energy like fossil fuels, etc. With high energy demand throughout the world and rapid consumption rate of fossil fuels, it is estimated that the supply of fossil fuels would be exhausted within next 70–100 years. A report by International energy agency in 2001 estimated that with the existing rate of consumption, the oil reserves will last up to the year 2100 and the coal reserves will exhaust in the next 170–200 years [1]. These fossil fuels releases stored greenhouse gases on combustion which eventually leads to global warming. This disturbs the global carbon cycle and leads to an increase in atmospheric CO_2 levels. Energy load transfer from conventional to non-conventional is one of the smart steps to meet the increasing energy demand. The depletion of the conventional fuels will occur sooner or later, and the concerns related to it are also clouding the decision whether to continue them on accounts of their environmental impacts. These problems can be overcome with the use of renewable energy sources. One of them is solar energy which can be harnessed with the use of photovoltaic (PV) systems [2]. PV systems are commercially proven tools and are being implemented throughout the country on a large scale. However, the main issue with PV panel is that the operating temperature of panels increases due high solar radiation, which further results in the degradation of its efficiency. The lifespan of PV panels is also reduced rapidly when exposed to upper temperatures for lengthy duration. Therefore, surface temperature of the PV panels plays a very important role in determining the system efficiency and life span of the panels. The optimum efficiency of PV panels is achieved at 25 °C. However, for every degree rise in temperature, the efficiency of panel decreases by 0.45%. Thus, to maintain optimum efficiency of PV panels, the temperature needs to be control using appropriate cooling technologies.

To control the high operating temperature of PV panels during peak load, cooling technologies may be used. However, their choice depends on a variety of parameters like heat exchanger area, mass flow rate of fluid and heat transfer coefficient (HTC) [3]. In the past few years, a significant amount of work has been done on

PV panel cooling techniques by many researchers. One of the cooling techniques uses nanoparticles for improving system performance [4]. The cooling was achieved by designing a photovoltaic/thermal (PV/T) system. A PV/T system is a cogeneration technology which generates both electrical and thermal output. These days, commercial PV/T systems are also available to provide electricity and hot water from a single unit. Colloidal suspensions which are finished by mixing nanoparticles in a base fluid with different volume fractions are known as nanofluids (NFs). After performing several experiments and investigations on different nanoparticles in Argonne National Laboratory (USA), the idea of nanofluids was initially materialized by Choi [5]. Nanoparticles are basically solids which have a great influence on properties of water such as dynamic viscosity, specific heat of fluid, thermal conductivity and density [6]. With the addition of nanoparticles, HTC increases, but on the other hand, pumping power also increases due to increase in viscosity. The balance between these two opposing thermo-physical properties is imperative while considering heat transfer of nanofluids.

Meibodi et al. [7] studied the presentation of silicon oxide-based nanofluids for heat transfer medium in the PV/T system. They reported a highest of 8% increase in the thermal efficiency. In other related case studies, sedimentation was examined as a major restriction of adopting NFs in solar plate collector as stated by Colangeo et al. [8]. Li and Xuan [9] examined the performance of Cu/water NF in a tube having length 800 mm and a diameter of 10 mm while maintaining same Reynolds number (Re) and stable heat flux. They observed a maximum of 60% enhancement in the performance for 2% volume fraction of nanoparticles. In the above reported investigations, the comparison is made between performances of the system employing nanofluid with the system using base fluids keeping Reynolds number equal. However, this criteria of evaluation can be misleading because with increment in density and apparent viscosity of NFs implies that flow rate also needs to be increased. This leads to increase in pumping power, which offset a few of the benefits of improvement in thermal conductivity. Sadegi et al. [10] accomplished that Brownian diffusion play a significant role on heat transfer system. Mohammed et al. [11] concluded that Brownian diffusion can be abandoned for nanofluids flow with Re greater than 100. Various research amount of research work has been done on nanofluids to enhance the heat transfer rate.

As discussed above in the literature review, the research also discussed use of nanofluids to enhance the heat transfer rate. The use of nanofluids was also reported by different researchers for various applications. But the use of nanofluids for PV/T systems to increase heat transfer rate is not discussed widely. Especially for arid and semi-arid regions of India, the work related to PV/T and uses of nanofluids are at nascent stage. The PV/T systems play a quite important role in arid and semi-arid regions due to availability of high solar radiation. Keeping in view of gaps in the literature, the main objective of the current work is to enhance heat transfer in PV/T system using nanofluids for climate conditions of semi-arid regions of Rajasthan. The main objective of this paper is to maintain the operating temperature of PV panels with the help of cooling technique in which base fluid water will be loaded with different volume percentage particle volume fraction of nanoparticles and their effect on thermal performance. In this paper, Al_2O_3 and zirconium nanoparticles were used with volume fraction of 0.5, 1 and 1.5% at different Reynolds numbers.

2 Mathematical Modelling

Nanofluids are either single-phase homogenous mixture of nanosized particulate material with base fluid or heterogeneous mixture having two phases [12]. Ahmed et al. [13] found that if Reynolds numbers of certain nanofluid exceeds 100, then Brownian diffusion and thermophoresis would be negligible in that fluid. In present study, the value of Reynolds number varies between 250 and 1500 so that nanofluid consisting of single phase can be analysed using conventional approach for its heat transfer behaviour.

2.1 *Computational Domain*

In this study, the work is carried out using a flat plate solar PV/T circular-shaped channel (2 m length and 12 mm diameter) and the nanofluid for investigating its thermal and pressure drop behaviour in a 2-D computational domain. The schematic figure of the PV/T system is shown in Fig. 1.

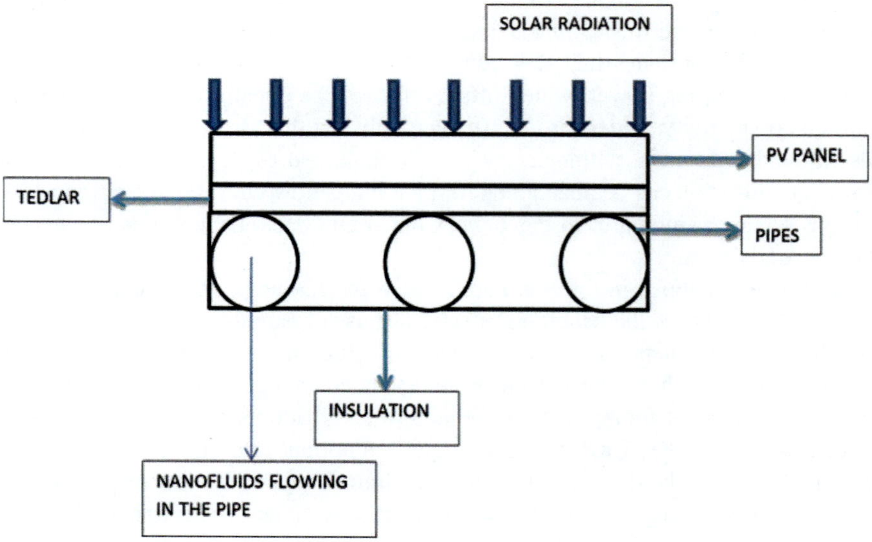

Fig. 1 Schematic diagram of PV/T system

2.2 Governing Equations

ANSYS Fluent 15.0 was used for analysis of governing equations. Second-order upwind scheme was adopted for solving energy and momentum terms, whereas standard process was applied for pressure discretization. Simple algorithm was preferred for the interaction of pressure and velocity factors [14]. Various equations viz. energy, momentum and continuity equations were iterated till the residuals were reduced lower than 10^{-6}. The subsequent governing equations are used for mathematical formulation of single-phase method.

Conservation of mass

$$\mathrm{div}(\rho\vec{V}) = 0 \tag{1}$$

Conservation of momentum

$$\mathrm{div}(\rho\vec{V}\vec{V}) = -\mathrm{grad}\,\mathrm{P} + \nabla.(\nabla\vec{V}) \tag{2}$$

Conservation of Energy

$$\mathrm{div}(\rho\vec{V}\mathrm{cT}) = \mathrm{div}(k\,\mathrm{grad}) \tag{3}$$

2.3 Boundary Conditions

For the analysis, the governing equations are taken as mentioned in Eq (1) and Eq (3) having boundary limits of fluid temperature of 299 K at the inlet of the PV/T system. The 2-D simulation was carried out instead of 3-D. For a circular tube, the side cross section comes out to be rectangular view and we have taken assumptions on side bottom wall for the simulations. The inlet velocity of the nanofluid was also estimated using governing equations. For the sidewalls, no slip boundary was assumed, while bottom was assumed to be adiabatic. For the sidewalls, in 2-D planar computational domain, simulated solar radiation was taken as 1000 W/m^2. Pressure outlet boundary condition was applied at the channel outlet.

2.4 Nanofluids Thermo-Physical Property Models

Density, specific heat, viscosity, thermal conductivity, etc., affect the heat transfer and pressure drop characteristic to a larger extent. These properties either require measurement through sophisticated instruments or through various mathematical

Table 1 Thermo-physical properties of nanoparticles and base fluid [15]

Nanoparticle	Density (kg m^{-3})	Specific heat (J kg^{-1} K^{-1})	Thermal conductivity (W m^{-1} K^{-1})
Water	998.2	4182	0.6
Alumina	3970	765	40
Zirconium Oxide	5600	418	2.8

correlations with appropriate allowances. The values of these properties are taken from empirical data as listed in Table 1.

Various properties of nanofluids can be determined by using Newton's mixture equation [15], as specified in the following equations

$$\rho_{nf} = \rho_{bf}(1 - \emptyset) + \rho_p\emptyset \tag{4}$$

$$C_{nf} = \frac{(1-\emptyset)(\rho c)_{bf} + \emptyset(\rho c)_p}{(\rho)_{nf}} \tag{5}$$

$$K_{nf} = \frac{K_p + 2K_{bf} + 2(K_p - K_{bf})\emptyset}{K_p + 2K_{bf} - 2(K_p - K_{bf})\emptyset} \tag{6}$$

$$\mu_{nf} = \mu_{bf}\left(1 + 7.3\emptyset + 123\emptyset^2\right) \tag{7}$$

2.5 *Validation of Theoretical Results*

Utomo et al. [16] revealed that Nusselt number of a nanofluid can be derived using heat transfer correlations of a liquid with ±10% variation. The values of parameters obtained in the present work were validated with the help of Shah's correlation [17]. As Shah's correlation was carried for the uniform heat flux, in the present study, modified boundary conditions would be applicable. For the confirmation trails, the value of Reynolds number is taken as 250 and 1500 with 1% particle fraction of nanoparticles with uniform heat flux of 100 W/m^2 through top and bottom parts of the tube of PV/T system. It was found out that the numerical values of the findings agree excellently with results of confirmation trials.

3 Results and Discussions

In this section, we have discussed simulations results obtained by using different types of nanofluids as HTF in PV/T collector. For the same, the Reynolds numbers were varied from 250 to 1500 with 0.5, 1 and 1.5% particle volume fraction of nanoparticles. Figure 2 shows the HTC for water and alumina nanoparticles with different Reynolds numbers. As shown in Fig. 2, the HTC ranges between 250.6–529.20 W/m^2 K, 255.42–539.8 W/m^2 K and 261.1–550.8 W/m^2 K for 0.5%, 1.0% and 1.5% volume fraction, respectively, for Reynolds number in the range of 250–1500. Without any nanoparticles, for the same Reynolds number, the HTC ranged between 245.5 and 519.8 W/m^2 K. This shows that HTC increases with use of nanoparticles and is directly proportional to its volume fraction. This may be due to the fact that with addition of nanoparticles in the water, the thermal conductivity and specific heat of nanofluid increases which increases the HTC.

The graphical representation of water pressure and alumina nanoparticles, at different Reynolds numbers, is shown in Fig. 3. It shows that with increase in nanoparticles, the pressure drop within PV/T increases. This may be due to the increase in viscosity of the nanofluid due to addition of nanoparticles. For Reynolds numbers ranging from 250 to 1500, the pressure drop ranged between 31.44 and 235.70 Pa when no nanoparticles were used. On the other hand, the pressure drop ranged between 34.5–249.9 Pa, 37.01–269.12 Pa and 40.10–291.4 Pa for 0.5, 1.0 and 1.5% volume fraction of nanoparticles having Reynolds number of 250–1500. Thus, with increase in volume fraction of nanoparticles, the pressure drop also increases. However, the increase of pressure drop is not desirable as it also increases the pumping power of the system. Thus, for the current study, maximum volume fraction of 1.5% was taken.

The variation of HTC for different shapes of PV/T pipes with 0.5% alumina nanoparticle is shown in Fig. 4. When investigation was carried out for different

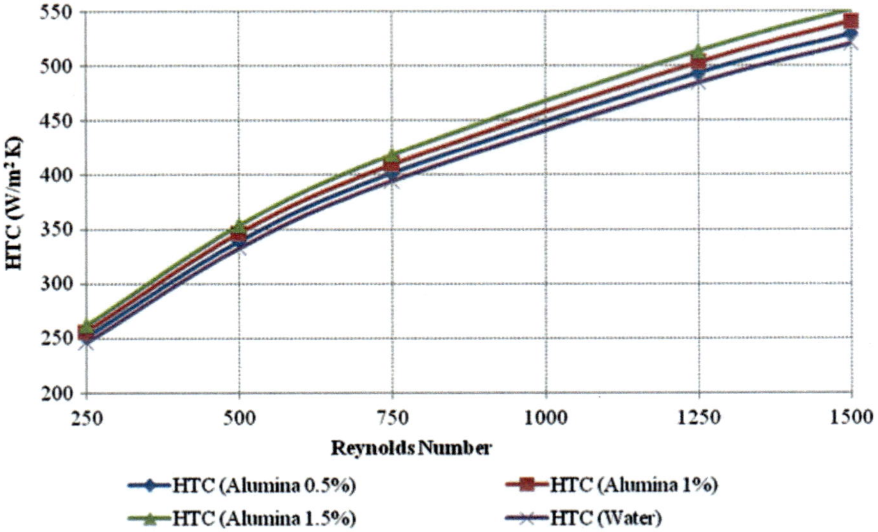

Fig. 2 Comparison of HTC of basefluid and nanofluids with different particle volume fractions for various Reynold numbers

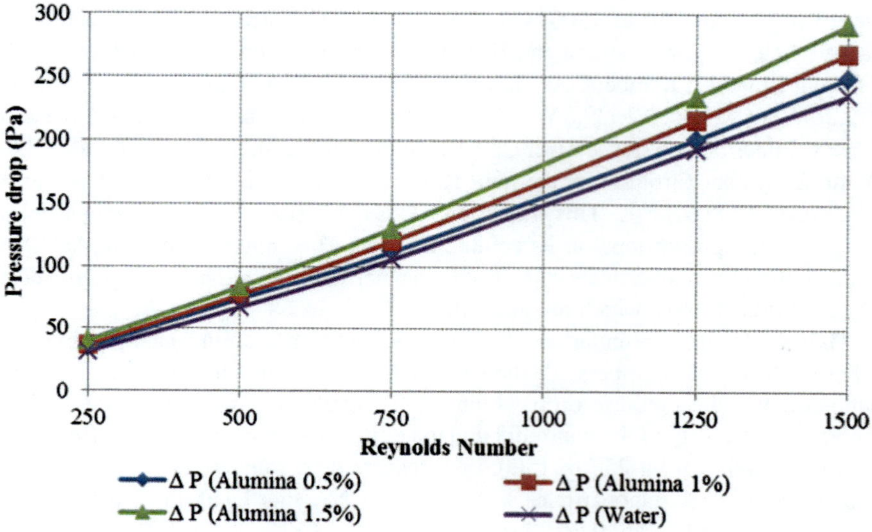

Fig. 3 Comparison of pressure drop of basefluid and NF

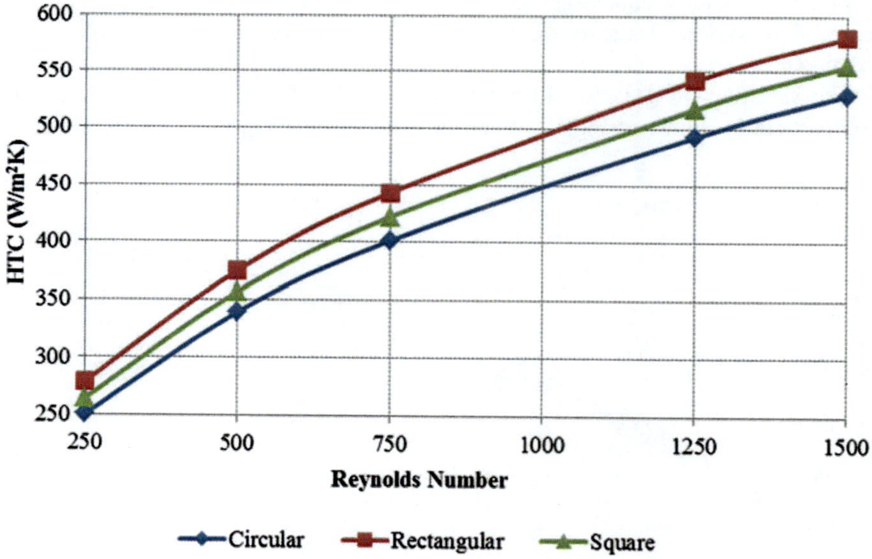

Fig. 4 Variation of HTC for different shapes of PV/T pipes with 0.5% alumina nanoparticle

shapes of PV/T pipes with alumina/water at 0.5% particle loading and Reynolds number at 1500, it was observed that rectangular duct exhibits highest HTC as 581.30 W/m^2 K. For same Reynolds number, square duct and circular duct gives HTC of 554.95 and 529.30 W/m^2 K, respectively. The maximum HTC was obtained for rectangular duct was may be due to its large hydraulic diameter. Figure 5 shows the pressure drop for the various shapes of PV/T pipes with 0.5% alumina nanoparticle. The observed pressure drop ranges between 34.5–249.9 Pa, 44.5–336.1 Pa and 39.1–289.6 Pa for circular, rectangular and square duct, respectively, with Reynolds number of 250–1500. It was found out that the maximum pressure drop occurs for rectangular duct.

Figure 6 shows the HTC and pressure drop different nanoparticles with 1% volume fraction for different Reynolds numbers. It was found that the HTC varies between 251.6–532.4 W/m^{2}K and 255.42–539.8 W/m^{2}K for zirconium oxide and aluminium

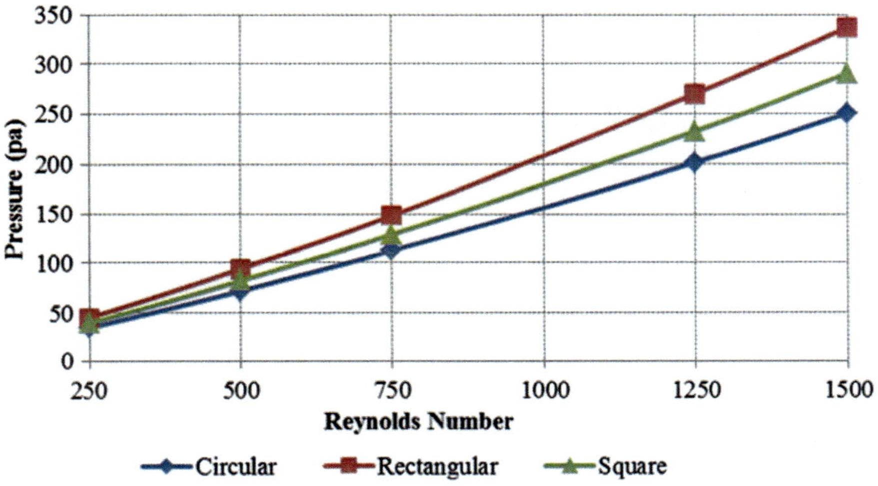

Fig. 5 Variation of pressure for different shapes of PV/T pipes with 0.5% alumina nanoparticle

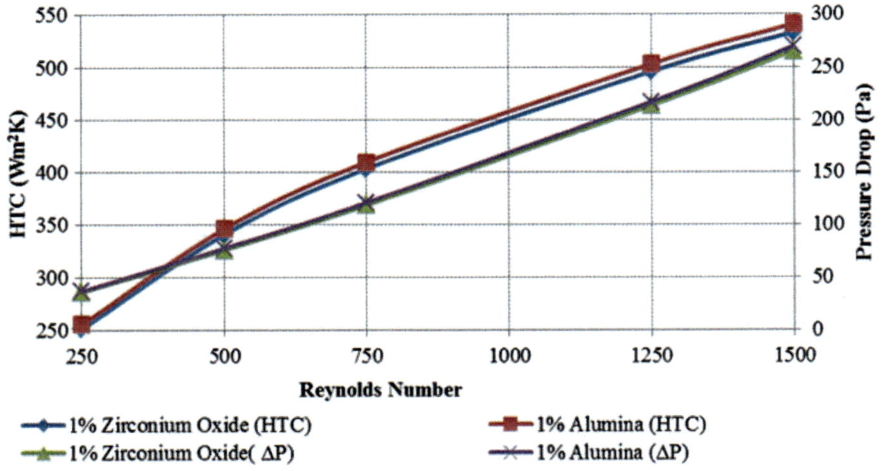

Fig. 6 HTC and pressure variation for different nanoparticles with 1% volume fraction

oxide nanoparticles, respectively. It was observed that there is not much difference in performance of both nanoparticles, thus the cost-effective nanoparticle could be used for PV/T system.

4 Conclusion

In this study, steady-state laminar convective heat transfer of Al_2O_3/water and ZrO_2/water nanofluids are numerically examined in a circular tube of PV/T system. The outcome of the simulation and major conclusions are summarized below:

- Nanofluids could be used to improve the performance of PV/T systems. Their superior thermo-physical properties result in major improvement in heat transfer process which in turn would lead to reduction in the size of solar devices.
- On the basis of HTC, the results obtained are better for 1.5% alumina/water at Reynolds number of 1500. For higher performance of PV/T cooling system, high value of HTC is desirable.
- In case of pressure drop, water has lower pressure drop in comparison with different volume fractions of nanoparticles. With increment in pressure drop, pumping power also increases.
- In comparison to alumina and zirconium oxide at same percentage of particle loading, alumina is preferred as: HTC was higher for alumina than the other. For both nanoparticles, there is no much difference observed in pressure drops.
- On the contrary, the use of nanofluids increases the heat transfer with in PV/T, but also increases the pumping power, due to high viscosity of nanofluids. Thus, a trade-off is required between the two. It is recommended that nanofluids are more beneficial for large solar PV/T collectors than the smaller systems.

Acknowledgements The authors gratefully acknowledge the support from the Department of Mechanical Engineering of Birla Institute of Technology and Science—Pilani, Rajasthan, and Ms. Priya Gupta for this research.

References

1. Reddy VS, Kaushik SC, Panwar NL (2013) Review on power generation scenario of India. Renew Sustain Energy Rev 18:43–48
2. Jakhar S, Soni MS, Gakkhar N (2016) Historical and recent development of concentrating photovoltaic cooling technologies. Renew Sustain Energy Rev 60:41–59
3. Jakhar Sanjeev, Soni Manoj S, Gakkhar Nikhil (2017) An integrated photovoltaic thermal solar (IPVTS) system with earth water heat exchanger cooling: Energy and exergy analysis. Sol Energy 157:81–93

4. Purohit N, J Sanjeev, Gullo P, Dasgupta MS (2018) Heat transfer and entropy generation analysis of alumina/water nanofluid in a flat plate PV/T collector under equal pumping power comparison criterion. Renew Energy 120:14–22
5. Choi SUS, Eastman JA. Enhancing thermal conductivity of fluids with nanoparticles. No. ANL/MSD/CP-84938; CONF-951135–29. Argonne National Lab., IL (United States), p 199
6. Das SK, Choi U, Yu W, Pradee T (2007) Nanofluids: science and technology. Wiley, USA
7. Meibodi SS, Kianifar A, Niazmand H, Mahian O, Wongwises S (2015) Experimental investigation on the thermal efficiency and performance characteristics of a flat plate solar collector using SiO2/EG ewater nanofluids. Int Commun Heat Mass Tran 65:71–75
8. Colangelo G, Favale E, De Risi A, Laforgia D (2013) A new solution for reduced sedimentation flat panel solar thermal collector using nanofluids. Appl Energy 111:80–93
9. Li Q, Xuan Y (2002) Convective heat transfer and flow characteristics of Cu-water nanofluid. Sci China Ser E: Technol Sci 45(4):408–416
10. Sadeghi R, Haghshenasfard M, Etemad SG, Keshavarzi E (2016) Theoretical investigation of nanoparticles aggregation effect on Water-alumina laminar convective heat transfer. Int Commun Heat Mass Transfer 72:57–63
11. Ali FM, Yunus WMM, Talib ZA (2013) Study of the effect of particles size and volume fraction concentration on the thermal conductivity and thermal diffusivity of Al2O3 nanofluids. Int J Phys Sci 8(28):1442–1457
12. Kamyar A, Saidur R, Hasanuzzaman M (2012) Application of computational fluid dynamics (CFD) for nanofluids. Int J Heat Mass Transf 55(15-16):4104–4115
13. Ahmed M, Eslamian M (2015) Laminar forced convection of a nanofluid in a microchannel: effect of flow inertia and external forces on heat transfer and fluid flow characteristics. Appl Therm Eng 78:326–338
14. Versteeg HK, Malalasekera W (2007) An introduction to computational fluid dynamics: the finite volume method. Pearson Education
15. Helvaci HU, Khan ZA (2017) Heat transfer and entropy generation analysis of HFE 7000 based nanorefrigerants. Int J Heat and Mass Transfer 104:318–327
16. Utomo AT, Haghighi EB, Zavareh AI, Ghanbarpourgeravi M, Poth H, Khodabandeh R, Pacek AW et al (2014) The effect of nanoparticles on laminar heat transfer in a horizontal tube. Int J Heat Mass Transf 69:77–91
17. Shah RK, London AL (1978) Laminar flow forced convection in ducts. Suppl Adv Heat Transf

Prospect of a Fully Solar Energy-Driven Compact Cold Store for Low Income Farming Communities

Sachindra Kumar Rout, Madhu Kalyan Reddy Pulagam, and Sunil Kr Sarangi

Abstract Unlike industrially developed economies, weaker societies lack infrastructure for transportation and storage of agricultural produce. This leads to demand for many small cold store facilities distributed widely, instead of a few large and energy-efficient units. Solar energy gives a clean and environment-friendly option for meeting this need; but solar energy is available only for about a third of the day, leaving a challenge to save applications needing round the clock air conditioning. Recently (2015), a new technology has been proposed by Al-Ugla et al. [1] of Saudi Arabia to circumvent the problem by employing a modified vapour absorption refrigeration cycle. The authors have shown on the basis of theoretical modelling that it is feasible to design and build a modified vapour absorption refrigerator that will provide 24 h cooling without storing the refrigerating effect below room temperature. The proposed system is expected to be compact in volume and economical to operate. We are on the path of developing a practical cold store to store about 5 tonnes potato or equivalent quantity of other produce in rural setting at moderately low temperature. The system is based on the classical LiBr/H_2O absorption refrigeration cycle, the heat supply being given by solar power. Night operation is implemented by providing extra storage tanks to save rich mixture, lean mixture and pure water when solar heat is no more available for generating the refrigerant from the rich mixture. A vapour absorption refrigeration process is similar to the more common vapour compression system except that the mechanical compressor of the latter is replaced by an absorption-pumping-desorption process to achieve the objective of increasing the density, and consequently the pressure of the refrigerant. A common chemical system consists of LiBr and H_2O, the latter being the refrigerant and LiBr the absorbent, and the process is illustrated in Fig. 1. While the vapour absorption cycle is well established, particularly, for generating refrigerating effect using low-grade heat or solar radiation, the focus of our project is to create a system that saves potential refrigerating effect (not a cold fluid) for use during night hours. The innovation introduced by Al-Ugla et al. revolves around extending the duration of operation by adding three secondary storage tanks—one for the strong (more

S. K. Rout (✉) · M. K. R. Pulagam · S. K. Sarangi
C. V. Raman Global University, Bhubaneswar, Odisha, India
e-mail: Sachindra106@gmail.com

M. Ramgopal et al. (eds.), *Advances in Air Conditioning and Refrigeration*, Lecture Notes in Mechanical Engineering, https://doi.org/10.1007/978-981-15-6360-7_2

water) solution, one for the weak (more LiBr) solution and one for the pure refrigerant (water), connected in parallel with the respective primary containers. During the daylight hours, the normal vapour absorption process, running in solar heat, produces the required cooling effect and excess amount of poor (weak) solution and pure refrigerant (water) which is stored in separate tanks for later use. This system should be distinguished from competing refrigeration storage systems where cooling effect is stored for the night either in the form of excess chilled water or in the form of solidified phase change material. The latter materials are bulky and expensive. In contrast, the proposed system stores the strong and weak solutions in separate tanks, with very high effective cooling capacities. The resulting secondary tanks are small, light and inexpensive. The machine will operate with 70 °C as evaporator temperature and 90 °C as generator temperature. The solar collector size is around 24 m^2 for 24 h run. The major challenge is to maintain vacuum inside the system. In summary, the system proposed, herein it is characterized by the following features. (a) No external supply of electricity (grid connection) not necessary, (b) 24-hour operation (including night hours). (c) Based on vapour absorption cycle using LiBr as absorbent and water as refrigerant. (d) No expensive refrigerant or phase change material involved, nor there is a need for a large reservoir for storing refrigeration effect. (e) Small solar PV unit (200 W) with battery to power the three magnet-linked liquid pumps. The system is under development in our HVAC laboratory. Theoretical design is ongoing and results will be presented in the conference. A cost comparison between the system under development and two competing conventional systems (grid connected and solar PV + Battery) will be presented.

Keywords Cold storage · Solar · Absorption refrigeration

1 Introduction

In India, most of the people depend on cultivation and fishery as their earning source. India is the second biggest generator of vegetables (187474 MT) and fruits (96754 MT) [1] in global market. This production, in India, is contributing 11.36 and 14.04% of global fruits and vegetables production, respectively. There is a rapid progress in production of vegetables and fruits since last few years but the issue associated with postharvest management of fruits and vegetables owing to poor storage facilities and lack of infrastructure, remained extremely discouraging in India. As per an estimate by CEPHET, postharvest losses in India are estimated at 133 billion rupees per year (combining fruits and vegetables together). However, the concern is universal, as roughly one-third of the edible parts of worldwide food production gets lost or wasted, which amounts to about 1.3 billion tons annually [2]. The hot and humid climate prevailing in the tropical and subtropical countries is also seriously responsible for such enormous decay. Thus, the establishment of cold storages is the need of the hour, to reduce the wastage of perishable commodities, as well as for the economic benefit of both the growers and consumers. The desirability of using environment-friendly solar power instead of fossil fuel generated grid power need not be over emphasized. The appropriate technology for solar air conditioning is still evolving, and the current research is an innovative step in that direction.

One of the major interference towards such initiative is the considerable energy required by a cold storage for its powering and operation. This is a serious concern in the underdeveloped and developing countries, where a considerable fraction of the rural population does not have access to the grid electricity. In a developing country like India, there are about 31 million homes un-electrified. Operation of cold storage powered through alternative energy can be a perfect solution in this perspective, as that offers the dual advantage of primary energy savings and reduced environmental menace. Today, there is considerable demand for cold storage; and the demand is constantly rising. Solar energy gives an uncontaminated and environment-friendly option for fulfilling this requirement; but solar energy is available only for about a third of the day, leaving a challenge to save applications needing round the clock.

The vapour absorption cycle has been widely used across the globe for air conditioning, many of them utilizing solar heating. Several European companies have established successful business ventures by setting up solar-driven vapour absorption systems not only in Europe but also in rest of the world including Asia. In addition to LiBr-H_2O, several other material combinations have been successfully employed. Table 1 gives a summary of the various refrigerant/absorbent combinations employed by researchers across the globe. Some of them have extended the system to night hours by the use of phase change materials, which does not exploit the thermodynamic process of the vapour absorption cycle.

Table 2 gives a partial list of commercially successful air conditioning systems in the international arena which are based on vapour absorption cycle. None of them, however, makes provision for night operation by using a solar thermal system, except by adding completely alternative systems.

Table 1 Different working fluid pairs used by various researchers working with the vapour absorption cycle

SI no.	Refrigerant	Absorbent	References
1	Water	Lithium bromide	Nakahara et al. [3], Yeung et al. [4], Li and Sumathy [5] Praene et al. [6], Romero et al. [7]
2	Ammonia (NH_3)	Water	Gutiérrez [8], Jakob et al. [9]
3		Calcium chloride	Worse-Schmidt [10]
4		Strontium chloride	Erhard and Hahne [11]
5		Lithium nitrate	Rivera and Rivera [12]
5		Water–NaOH mixture	Steiu et al. [13]
6		IMPEX material	Bansal et al. [14]
7	Trifluoroethanol	Tetraethylenglycol dimethylether	Medrano et al. [15]
8	Methanol	Tetraethylenglycol dimethylether	Medrano et al. [15]

Table 2 Absorption cooling market survey

Cycle type	Working pair	Manufacturer	Heating source	Cooling source	Qe (kW)	Cooling COP
Single-effect	LiBr/H_2O	Sonnenklima (Germany)	Water (75–95 °C)	Water	10	0.78
	LiBr/H_2O	EAW (Germany)	Water (80–90 °C)	Water	15	0.71
	LiBr/H_2O	Yazaki (Japan)	Water (80–90 °C)	Water	17.6	0.70
	LiBr/H_2O	Rotartica[a] (Spain)	Water (80–100 °C)	Air (recooling)	4.5	0.67
	H_2O/NH_3	Pink (Austria)	Water (75–85 °C)	Water	10	0.64
	LiCl/H_2O	Climate-Well (Sweden)	Water (85–110 °C)	Water	10	0.70
Double-effect	LiBr/H_2O	Broad (China)	Direct-fired (150–170 °C)	Water	16	1.20
	LiBr/H_2O	Rinnai (Japan)	Direct-fired (150–170 °C)	Water	5	1.20
	LiBr/H_2O	Yazaki (Japan)	Direct-fired (150–170 °C)	Air	28	0.85
GAX	H_2O/NH_3	Robur (Italy)	Direct-fired (180–200 °C)	Air	17	0.90

[a]No longer in production

There are limited number of research article reported in open literature supporting to continuous run cold storage operated with the solar energy and other modes of renewable sources. The first research article reported by, Sethu et al. [16] described a complete design of a 10 TR solar-powered cold storage for potato, operated with aqua-ammonia vapour absorption refrigeration system (VARS). It was designed for storing 75 tons of potatoes at operating temperature 4 °C. It was connected with 152 selectively coated flat plate collectors (nominal area 304 m^2) to supply hot water at a temperature of 950 °C. Development of various alternate designs of solar-powered VARS for the climatic conditions of Saudi Arabia was presented by Said et al. [6]. Continuously operating aqua-ammonia system with refrigerant storage was found to be the most suitable design for uninterrupted cooling. A refrigerant storage VARS system was studied by Al-Ugla et al. [7], and also they analysed three alternative storage designs for a solar-powered H_2O–LiBr absorption system for 24 h operation. Yuan et al. [17] conducted theoretical analysis on OTEC-based solar-assisted for powering a fishery cold storage, considering ammonia and water as working pair and the warm/cold seawater as the heating/cooling source. They reported that the proposed cycle has lower COP than ammonia/water refrigeration cycle, but it was an energy generation system. A response strategy for active and passive cold storages was presented by Cui et al. [18], to reduce the immediate and stepped power demand through chillers shutdown. Bao et al. [19] designed a thermochemical resorption system with the working pair $MnCl_2/NH_3$ and NH_4Cl/NH for application in cold storage and for long distance refrigeration. Shi and Zang [20] made a comparative study of different methods for generation of tetra-n-butyl-ammonium bromide clathrate hydrate slurry in a cold storage air conditioning system. In recent years, the use of phase change materials (PCMs) for cold storage application has been explored quite a bit [21–23] and a comprehensive review on relevant applications of solid–liquid PCMs is available in [24].

However, despite the promise of low running cost, PCMs can be twice as expensive compared to conventional coolers [25], and hence solar-based systems seem to be a better option. A theoretical design of a grid-interactive SPV/thermal-powered cold storage was designed by Basu et al. [2] for the storage of potatoes. It was based on water–lithium bromide absorption system. Proposed system utilizes both solar thermal and photovoltaic generated electrical energy for its operation. The proposed system is found to provide a net surplus of about 36 MWh energy over a calendar year, after meeting the in-house requirements.

On summarizing both international and national scenarios, it can be concluded that while the vapour absorption-based air conditioning is a mature technology, well integrated with solar thermal input, there has been very limited work in implementing a 24 h air conditioning system except by use of alternative technologies, thus almost doubling the capital expenditure. The present work aims at conquering this handicap in a convenient and cost-effective manner.

The greatest obstacle on the path of extensive adoption of solar power for air conditioning is its unavailability in night hours. In order to provide uninterrupted refrigeration/air conditioning, the cooling system needs to be integrated with a storage system in the form of either electrical batteries or storage of refrigerating effect in phase change materials. Both routes, though feasible, are expensive, bulky and, on eventual disposal, harmful to the environment.

The proposed system is based on the classical $LiBr/H_2O$ absorption refrigeration cycle, the heat supply being given by solar power. Night operation is implemented by providing a couple of extra storage tanks to save rich mixture, lean mixture and pure water when solar heat is no more available for generating the refrigerant from the rich mixture.

2 The Technology of Vapour Absorption Refrigeration

A vapour absorption refrigeration process is similar to the more common vapour compression system except that the mechanical compressor of the latter is replaced by an absorption-pumping-desorption process to achieve the objective of increasing the density, and consequently the pressure of the refrigerant. A common chemical system consists of LiBr and H_2O, the latter being the refrigerant and LiBr the absorbent, and the process is illustrated in Fig. 1. Water vapour from the evaporator at the appropriate temperature and pressure is led to the weak solution (H_2O in LiBr)

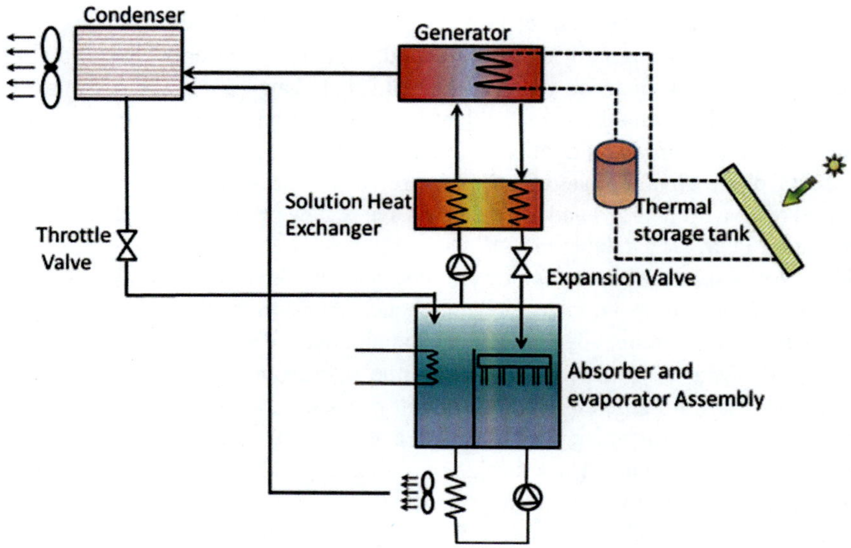

Fig. 1 Schematic diagram of a conventional solar-driven absorption system

(typically 50 °C and 10 kPa) (strong in terms of water) which absorbs the water vapour readily to become lean (in terms of LiBr). The lean solution is pumped to a higher pressure by using a small liquid pump and delivered to the generator. In the generator weak, LiBr solution is heated to liberate the refrigerant, i.e., water vapour at the elevated pressure. The stripped liquid returns to the absorber (at low pressure) through a valve to dissipate the pressure difference by fluid friction. The cycle repeats over the absorber pump-generator system that replaces the compressor of the vapour compressor cycle.

The liberated refrigerant H_2O, now at high pressure, travels to the condenser, expansion valve and the evaporator to complete the normal refrigeration cycle. Vapour from the evaporator enters the absorber enriching and cooling the lean mixture, therein the overall cycle repeats providing refrigerating effect at the evaporator, absorbing solar heat at the gas generator and dissipating heat to the ambient in the condenser. Figure 1 graphically describes a normal vapour absorption cycle driven by solar energy. The proposed system modified for generating refrigerating effect both during and in absence of solar radiation is presented in Fig. 2 in the proposal.

The innovation introduced by Al-Ugla et al. [26] revolves around extending the duration of operation by adding three secondary storage tanks—one for the strong (more water) solution, one for the weak (more LiBr) solution and one for the pure refrigerant (water), connected in parallel with the respective primary containers. During the daylight hours, the normal vapour absorption process, running in solar heat, produces the required cooling effect and excess amount of poor (weak) solution and pure refrigerant (water) which is stored in separate tanks for letter use.

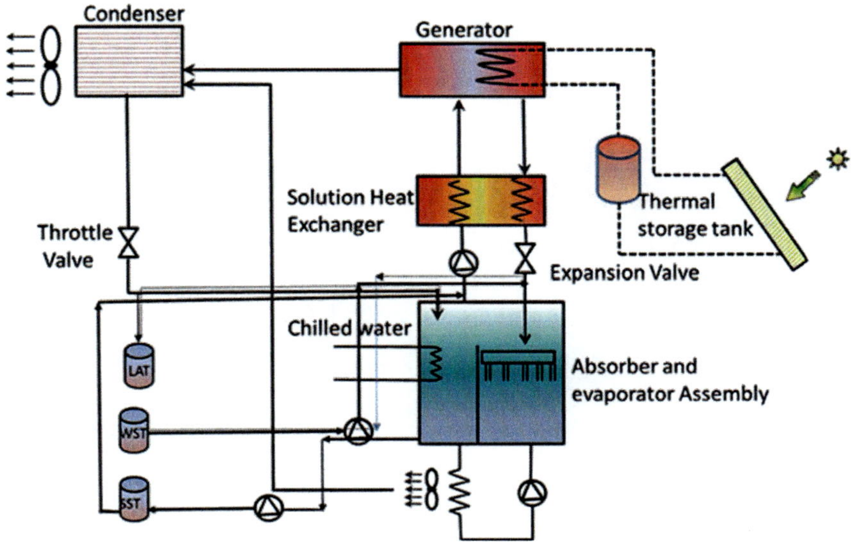

Fig. 2 Modified vapour absorption cycle providing refrigerating effect over 24 h

During the night hours, the stored weak solution and pure refrigerant are pumped and confined in a controlled manner to produce net cooling effect. By properly calculating the tank sizes and pump capacities, it is possible to fine-tune the system so that the stored liquids are fully utilized before the sun takes over the refrigeration system in the following day.

This system should be distinguished from competing refrigeration storage systems where cooling effect is stored for the night either in the form of excess chilled water or in the form of solidified phase change material. The latter materials are bulky and expensive. In contrast, the proposed system stores the strong and weak solution in separate tanks, with very high effective cooling capacities. The resulting secondary tanks are small, light and inexpensive.

In summary, we are convinced that the Al-Ugla et al. [26] scheme has the capacity of giving the required refrigeration, only with a minor enhancement of required infrastructure.

The proposed model is having capacity of 2 TR (7 KW). An EES programme has been written for calculation of the design parameter and operating condition for the proposed capacity machine with 70 °C as evaporator temperature and 90 °C as generator temperature. The solar collator size will be 24 m^2 for 24 h run design.

Application: Institutional, office, medical and cold storage.

3 Cost Comparison with Solar PV and Grid Operation

SI no.	Type	Capital cost (lakh)	Annual operating cost (electricity/battery cost) (lakh)	Total (all expenditure) cost for 10 years (lakh)	Environmental impact
1	100% grid no solar	0.50	1.50	15.5	Worst
2	Solar PV + Grid at night no battery	5.00	0.75	12.5	Bad
3	Solar PV + Battery no grid	9.00	0.6	15.0	Moderate (large battery)
4	Absorption + Grid	4.00	0.75	11.5	Bad
5	Absorption no grid proposed	9.00	0.1	10.0	Good (small battery)

References

1. National Horticulture Anual Report, India (2018–19) http://nhb.gov.in/
2. Basu DN, Ganguly A (2016) Solar thermal–photovoltaic powered potato cold storage—conceptual design and performance analyses. Appl Energy 165:308–317
3. Nakahara N, Miyakawa Y, Yamamoto M (1977) Experimental study on house cooling and heating with solar energy using flat plate collector. Sol Energy 19(6):657–662
4. Yeung MR et al (1992) Performance of a solar-powered air conditioning system in Hong Kong. Sol Energy 48(5):309–319
5. Li ZF, Sumathy K (2001) Experimental studies on a solar powered air conditioning system with partitioned hot water storage tank. Sol Energy 71(5):285–297
6. Praene JP et al (2011) Simulation and experimental investigation of solar absorption cooling system in Reunion Island. Appl Energy 88(3):831–839
7. Romero RJ et al (2001) Comparison of the modeling of a solar absorption system for simultaneous cooling and heating operating with an aqueous ternary hydroxide and with water/lithium bromide. Sol Energy Mater Sol Cells 70(3):301–308
8. Gutiérrez F (1988) Behavior of a household absorption-diffusion refrigerator adapted to autonomous solar operation. Sol Energy 40(1):17–23
9. Jakob U et al (2008) Simulation and experimental investigation into diffusion absorption cooling machines for air-conditioning applications. Appl Therm Eng 28(10):1138–1150
10. Worsøe-Schmidt P (1979) A solar-powered solid-absorption refrigeration system. Int J Refrig 2(2):75–84
11. Erhard A, Hahne E (1997) Test and simulation of a solar-powered absorption cooling machine. Sol Energy 59(4):155–162
12. Rivera CO, Rivera W (2003) Modeling of an intermittent solar absorption refrigeration system operating with ammonia–lithium nitrate mixture. Sol Energy Mater Sol Cells 76(3):417–427

13. Steiu S et al (2009) A basis for the development of new ammonia–water–sodium hydroxide absorption chillers. Int J Refrig 32(4):577–587
14. Bansal NK et al (1997) Performance testing and evaluation of solid absorption solar cooling unit. Sol Energy 61(2):127–140
15. Medrano M, Bourouis M, Coronas A (2001) Double-lift absorption refrigeration cycles driven by low–temperature heat sources using organic fluid mixtures as working pairs. Appl Energy 68(2):173–185
16. Sethu MR, Kumar A, Yardi NR (1986) Design, installation & performance of a 10 TR solar powered cold storage in India. In: Bilgen E, Hollands KGT (eds) Intersol eighty five. Pergamon, Oxford, pp 1013–1017
17. Yuan H, Zhou P, Mei N (2015) Performance analysis of a solar-assisted OTEC cycle for power generation and fishery cold storage refrigeration. Appl Therm Eng 90:809–819
18. Cui B et al (2015) Effectiveness and life-cycle cost-benefit analysis of active cold storages for building demand management for smart grid applications. Appl Energy 147:523–535
19. Bao HS et al (2012) Resorption system for cold storage and long-distance refrigeration. Appl Energy 93:479–487
20. Shi XJ, Zhang P (2013) A comparative study of different methods for the generation of tetra-n-butyl ammonium bromide clathrate hydrate slurry in a cold storage air-conditioning system. Appl Energy 112:1393–1402
21. Melone L et al (2012) Phase change material cellulosic composites for the cold storage of perishable products: From material preparation to computational evaluation. Appl Energy 89(1):339–346
22. Castell A et al (2011) Maximisation of heat transfer in a coil in tank PCM cold storage system. Appl Energy 88(11):4120–4127
23. Martin V, He B, Setterwall F (2010) Direct contact PCM–water cold storage. Appl Energy 87(8):2652–2659
24. Oró E et al (2012) Review on phase change materials (PCMs) for cold thermal energy storage applications. Appl Energy 99:513–533
25. Osterman E, Butala V, Stritih U (2015) PCM thermal storage system for 'free' heating and cooling of buildings. Energy Build 106:125–133
26. Al-Ugla AA, El-Shaarawi MAI, Said SAM (2015) Alternative designs for a 24-hours operating solar-powered LiBr–water absorption air-conditioning technology. Int J Refrig 53(Supplement C):90–100

Enhancement of Cooling Rate Using Biodegradable MgO Nanoparticles During a Cryopreservation Process

Siladitya Sukumar and Satya Prakash Kar

Abstract Cryopreservation is the method of the preservation of biological tissue samples for future usages without incurring significant changes to functional properties. A two-dimensional numerical model is developed in the current work to study the role played by MgO nanoparticles in enhancing the freezing rate of the tissue during cryopreservation. The Pennes bio-heat model is the governing equation in this case. Finite volume method is employed to discretize the governing equation while the tri-diagonal matrix algorithm is used for solving the discretized algebraic equations for obtaining temperature distribution within the tissue. The movement of solid–liquid interface during freezing is tracked using the enthalpy-porosity method. Validation of this model is first done with the result of the existing literature. Then, the effect of MgO nanoparticles in enhancing the freezing rate is studied by increasing the volume fraction of the nanoparticles in the tissue-nanoparticle system up to 30%. Finally, a comparative study is made to analyse the performances of MgO and Fe_3O_4 nanoparticles in quickening the freezing process during cryopreservation.

Keywords Cryopreservation · Finite volume method · Enthalpy-porosity method · Tri-diagonal matrix algorithm

Nomenclature

a_P	Coefficient in discretized algebraic equation
c_p	Specific heat (J/kg K)
g_l	Liquid volume fraction inside a control volume
H	Total enthalpy (J/kg)
h_{sen}	Sensible heat (J/kg)
k	Thermal conductivity (W/m K)
L	Length of tissue (m)

S. Sukumar · S. P. Kar (✉)
SME, KIIT Deemed to be University, Bhubaneswar 751024, Odisha, India
e-mail: satyapkar@gmail.com

M. Ramgopal et al. (eds.), *Advances in Air Conditioning and Refrigeration*, Lecture Notes in Mechanical Engineering, https://doi.org/10.1007/978-981-15-6360-7_3

L_f Latent heat of fusion (J/kg)
T Temperature (K)
t Time (s)
Q_m Metabolic heat generation (W/m^3)

Greek Symbols

ρ Density (kg/m^3)
ϕ Volume fraction of nanoparticles in tissue-nanoparticle system
ω Blood perfusion rate (s^{-1})
ΔH Latent heat (J/kg)
ΔT Change in temperature (K)
λ Under relaxation parameter
ΔV_P Control volume

Subscripts

e Effective
s Solid
l Liquid
f Freezing point
init Initial
b Blood
nb Neighbouring nodes
t Tissue
p Nanoparticle

Superscripts

n Iteration number
0 Previous time step

1 Introduction

The common method of preservation of biological cells, tissues and organs for further usages and references by freezing them to well below the freezing point of the biological fluids is known as cryopreservation. It should be performed without affecting their functionality, viability and mechanical properties [1]. With the growing biomedical applications like cell diagnostics, testing of drug delivery methods, cancer therapy, stem cell therapy, organ banking and transplantation, the importance of tissue cryopreservation has further enhanced [2]. The liquid content gets converted into amorphous solid or glassy state due to very high speed of freezing during cryopreservation by vitrification [3]. This ceases enzymatic activities within the tissue specimen and enables it to be preserved for a longer duration. Ahmadikia and Moradi [4] have studied about the cryopreservation method considering both Fourier and non-Fourier models. In the recent years, various researchers have worked on employing nanoparticles in enhancing the cooling rate during cryopreservation. But the lack of biocompatibility of various nanoparticles has hindered the researchers from using them in cryopreservation. The nanoparticles are required to be non-toxic, biodegradable and biocompatible to be used in cryopreservation. Yi and Zhao [5] studied about freezing of HeLa cells taking hydroxyapatite nanoparticles. Liu et al. [6] obtained greater freezing rates during cryosurgery using magnetite and maghemite nanoparticles. He et al. [7] obtained greater freezing rate during cryosurgery using MgO nanoparticles as compared to the conventional cryosurgery process. They also found that MgO nanoparticles are resulting greater cooling rates as compared to the Fe_3O_4 nanoparticles during the process. Hou et al. [8] have made a comparative study to analyse the performances of various nanoparticles in reducing temperature during tissue freezing. After understanding the utilities of various biodegradable, biocompatible and non-toxic nanoparticles in freezing of target cells and tissues during cryopreservation and cryosurgery, MgO and Fe_2O_3 nanoparticles are found out to be the most suitable to be used for cryopreservation. Therefore, a two-dimensional numerical model is developed in the present work to study the role played by MgO nanoparticles in enhancing the freezing rate during the cryopreservation process while the study of He et al. [7] was concentrated on cryosurgery. Then, a comparative study is made to find the more suitable nanoparticle between MgO and Fe_3O_4 to be used in cryopreservation.

2 Physical Model

A two-dimensional biological tissue of square shape is considered for the current study as shown in Fig. 1. Its length is 0.01 m. The thermo-physical properties of the tissue are given in Table 1. The tissue specimen is initially at 300 K temperature. Its left wall is brought in contact with liquid nitrogen at 77 K to initiate freezing while

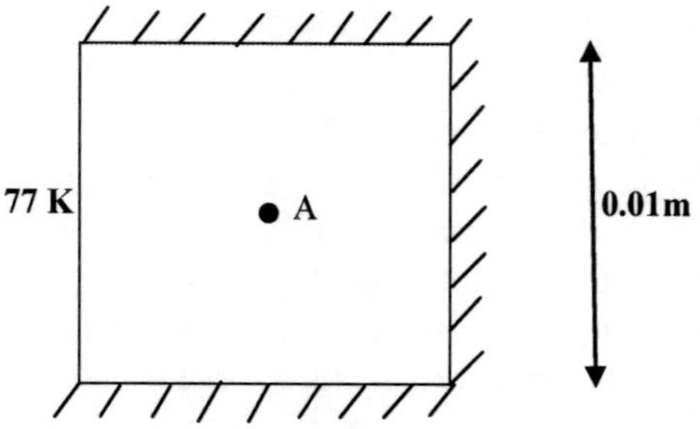

Fig. 1 Schematic diagram of the tissue specimen

Table 1 Thermo-physical properties of tissue

Properties	Value	Unit
Density	1000.0	kg/m^3
Solidus temperature	265.0	K
Liquidus temperature	272.0	K
Frozen thermal conductivity	2.0	W/m K
Unfrozen thermal conductivity	0.5	W/m K
Frozen heat capacity	3.6×10^6	J/m^3 K
Unfrozen heat capacity	2.0×10^6	J/m^3 K
Latent heat of fusion	250,000.0	J/kg

Table 2 Thermo-physical properties of nanoparticles [8]

Nanoparticles	Thermal conductivity (W/m K)	Heat capacity (J/m^3 K)
MgO	34.3	4.1×10^6
Fe_3O_4	7.1	4.1×10^6

the remaining sides are kept insulated. The thermo-physical properties of the MgO and Fe_3O_4 nanoparticles are given in Table 2.

2.1 Governing Equations

In a combined tissue-nanoparticle system, the effective thermal conductivity and heat capacity are given by the Hamilton–Crosser model.

$$k_e = k_t \frac{k_p + 2k_t - 2\phi(k_t - k_p)}{k_p + 2k_t + \phi(k_t - k_p} \tag{1a}$$

$$(\rho c_p)_e = (1 - \phi)(\rho c_p)_t + \phi(\rho c_p)_p \tag{1b}$$

The Pennes bio-heat equation in terms of enthalpy is written as

$$\frac{\partial(\rho H)}{\partial t} = k_{\mathrm{e}}\nabla^2 T + \omega_{\mathrm{b}}\rho_{\mathrm{b}}c_{\mathrm{pb}}(T_{\mathrm{b}} - T) + Q_{\mathrm{m}} \tag{2}$$

For tissue cryopreservation, $\omega_{\mathrm{b}} = 0$ and $Q_{\mathrm{m}} = 0$, so Eq. (2) becomes

$$\frac{\partial(\rho H)}{\partial t} = k_{\mathrm{e}}\nabla^2 T \tag{3}$$

$$H = h_{\mathrm{sen}} + \Delta H = c_{\mathrm{p}}T + g_{\mathrm{l}}L_{\mathrm{f}} \tag{4}$$

Now, Eq. (2) becomes

$$\frac{\partial(\rho c_{\mathrm{p}}T)}{\partial t} = k_{\mathrm{e}}\nabla^2 T - \frac{\partial(\rho L_{\mathrm{f}}g_{\mathrm{l}})}{\partial t}$$

Enthalpy Source Term,

$$S_{\mathrm{h}} = -\frac{\partial(\rho L_{\mathrm{f}}g_{\mathrm{l}})}{\partial t} = -\frac{\rho L_{\mathrm{f}}(g_{\mathrm{l}} - g_{\mathrm{l}}^0)}{\Delta t} \tag{6}$$

Initial Condition:

$$T(x, y, t = 0) = T_{\mathrm{init}} \tag{7}$$

Boundary Conditions:

$$-k\frac{\partial T}{\partial y} = 0;\ y = 0, 0 \le x \le L \tag{8a}$$

$$-k\frac{\partial T}{\partial y} = 0;\ y = L, 0 \le x \le L \tag{8b}$$

$$T = 77\,\mathrm{K}; x = 0.0 \le y \le L \tag{8c}$$

$$-k\frac{\partial T}{\partial x} = 0;\ x = L, 0 \le y \le L \tag{8d}$$

3 Numerical Method Formulation

Finite volume method by Patankar [10] is used to discretize the governing equation. The tri-diagonal matrix algorithm is employed to obtain temperature distribution inside the tissue specimen by solving the discretized algebraic equation. The volume fraction of liquid inside each control volume at all the time steps is updated using an iterative solution method. Here, the phase change process is non-isothermal in nature.

$$g_l^{n+1} = g_l^n + \frac{\lambda a_P \Delta t}{\rho L_f \Delta V_P}\left[T_P^n - g_l^n (T_l - T_s) + T_s\right] \quad (9)$$

The enthalpy-porosity method [9] is employed for capturing the moving solid–liquid interface at each time step. When $g_l = 0$, the control volume under consideration is completely solid while for $g_l = 1$, it is completely liquid. The convergence criteria are set at 10^{-6} for the iterative solution method.

4 Results and Discussion

First, the numerical model is made grid-independent as shown in Fig. 2. Temperature variation along the centreline parallel to x-axis at time, $t = 20$ s is shown in the figure. As no changes in the result are witnessed beyond the grid size of 41 × 41, and this is used to produce the further results.

Validation of this model is done with the result of the existing literature [4] as shown in Fig. 3. Variation of temperature along the width of a one-dimensional tissue at time, $t = 90$ s is shown in this figure. As a good agreement between the results is witnessed, the present model is used to produce the results in this work. After validating the model, an attempt is made to study the role played by MgO

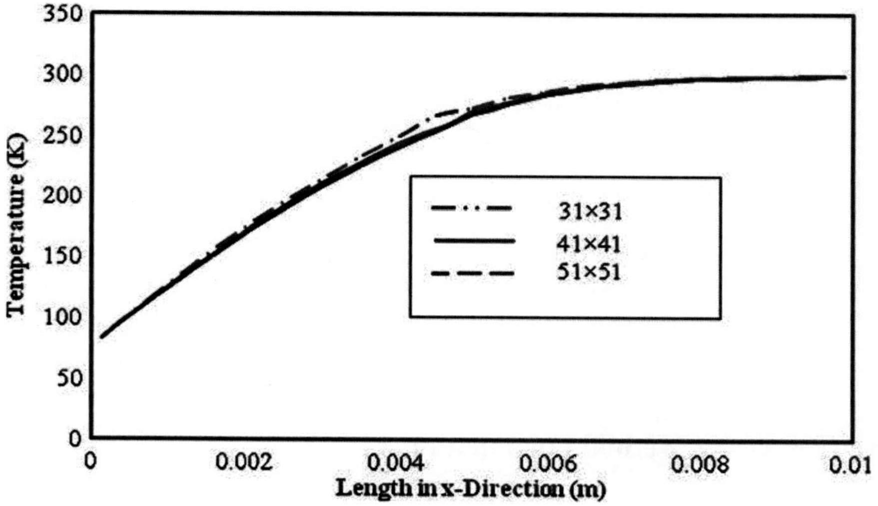

Fig. 2 Grid-independent test

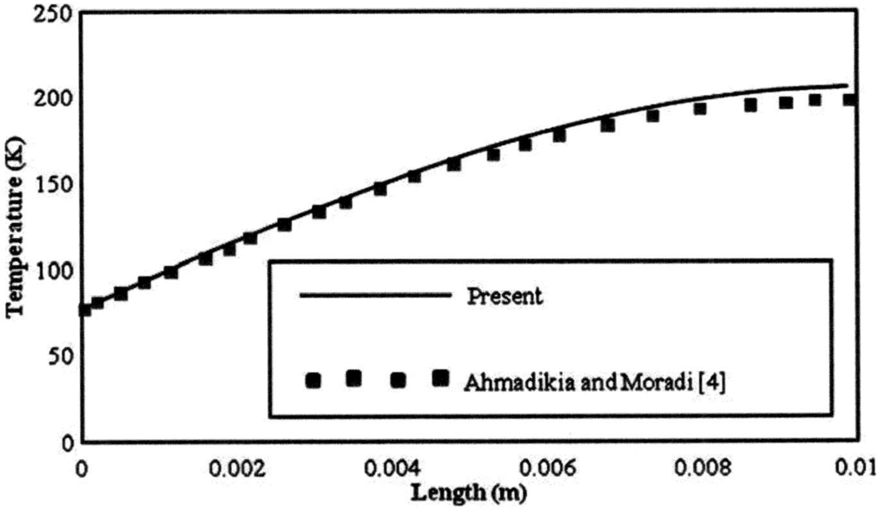

Fig. 3 Validation of the present result with the literature [4]

nanoparticles in enhancing the freezing rate of a tissue during cryopreservation. Then, a comparative study is made to assess the performances of the MgO and Fe_3O_4 nanoparticles in increasing cooling rate for the same volume fraction during cryopreservation process.

Temperature variation along the centreline in x-direction for different volume fraction of nanoparticles is shown in Fig. 4. With the increase in the value of ϕ, the thermal diffusivity of tissue-nanoparticles combined system increases significantly both in the unfrozen and frozen phases. Increase in the thermal diffusivity is mainly caused by the rise in thermal conductivity with the increase in ϕ value as change in heat capacity is quite negligible. Increase in the thermal diffusivity in unfrozen state, increases cooling rate and initiates the freezing process at a shorter time as compared to freezing without nanoparticles while increase in frozen state thermal diffusivity, enhances the heat removal rate from the vicinity of the tissue specimen following freezing. It helps in attaining a lower temperature at a certain location within the tissue specimen at a certain time. For $\phi = 30\%$, unfrozen and frozen thermal diffusivities increase by 2 times and 2.4 times, respectively, as compared to those of normal tissue specimen. So, the increase in volume fraction of MgO nanoparticles helps in the rise in cooling rate during cryopreservation of tissue. During cryopreservation, temperature of the entire preserved tissue is required to bring below 100 K in a very short time to ensure preservation for a significant duration. For $\phi = 30\%$, the entire tissue specimen is brought below 90 K temperature within 90 s time which ensures better preservation of the tissue. So, the use of MgO nanoparticles at a judicious volume fraction can ensure better preservation of biological tissue.

Figure 5 shows the variation of average cooling rate with ϕ at the centre of the tissue, A. It can be found out from the figure that for $\phi = 30\%$, the cooling rate is

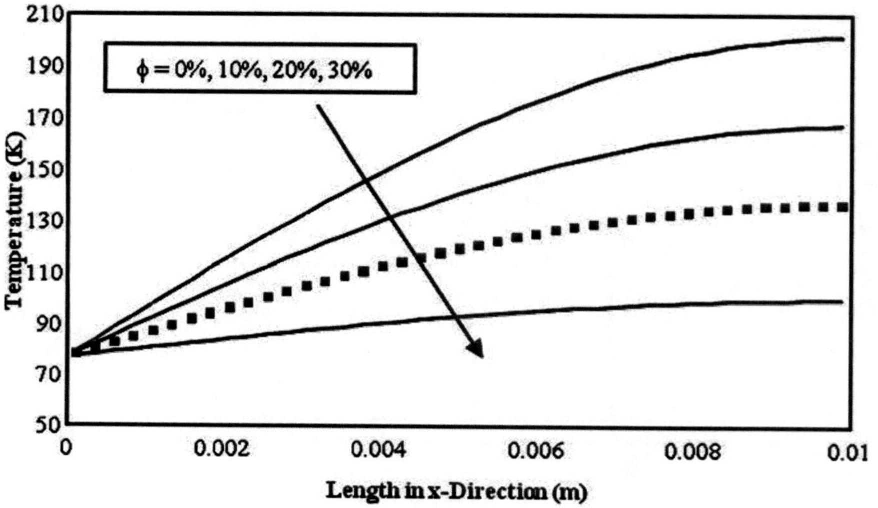

Fig. 4 Temperature variation along the centreline in x-direction for different volume fractions of MgO nanoparticles at time, $t = 90$ s

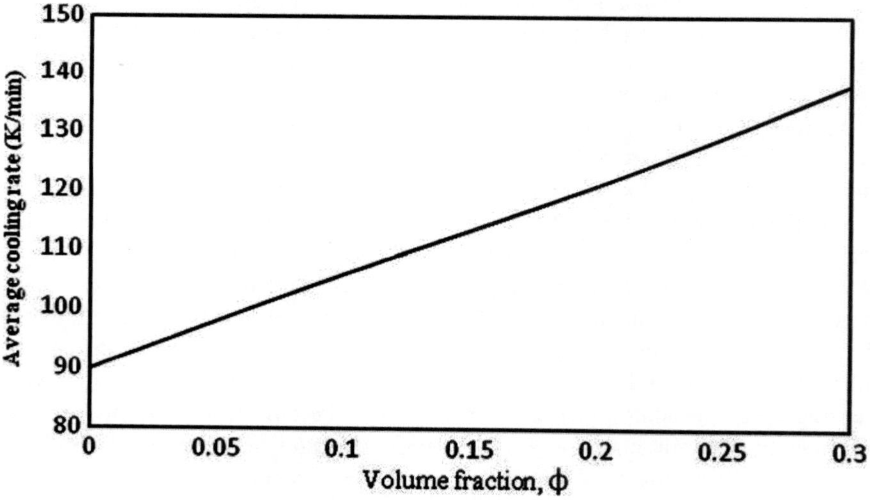

Fig. 5 Variation of average cooling rate with volume fraction of nanoparticles at the centre of the tissue

about 50 K/min more as compared to freezing of tissue without nanoparticles. This is a significant rise in cooling rate for proper cryopreservation of tissue. Figure 6 shows temperature variation along the centreline in *x*-direction for MgO and Fe_3O_4 nanoparticles. From Table 2, it is known that the heat capacity of both the nanoparticles is same, but the conductivity of MgO is more. This helps in enhancing the

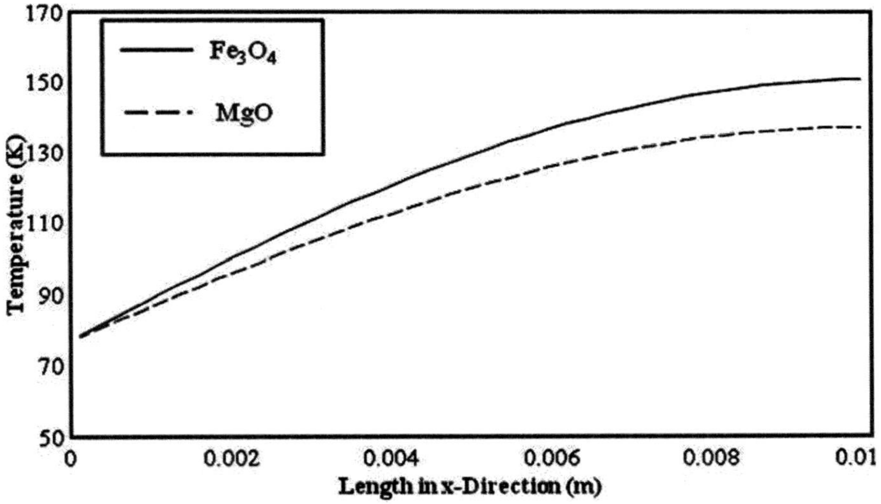

Fig. 6 Temperature variation along the centreline of the tissue parallel to x-axis for Fe_3O_4 and MgO nanoparticles ($\phi = 30\%$) at time, $t = 90$ s

effective thermal diffusivity and resulting cooling rate of the system significantly. This helps in achieving lower temperature at a certain point inside the tissue specimen for MgO nanoparticles as compared to Fe_3O_4 for the same volume fraction.

5 Conclusion

A model is developed to study the role played by the biodegradable, biocompatible and non-toxic MgO nanoparticles in enhancing cooling rate during cryopreservation of tissue. Freezing of the tissue specimen is initiated by bringing the left side in contact with liquid nitrogen at 77 K keeping other walls insulated. Then, a comparative study is made to assess the performances of MgO and Fe_3O_4 nanoparticles. The following conclusions are inferred from the current study.

1. By increasing the volume fraction of MgO nanoparticles in the tissue-nanoparticle system, the effective thermal diffusivity increases that enhance the freezing rate and help in attaining lower temperature inside the tissue specimen.
2. MgO nanoparticles are more efficient in increasing cooling rate for the same volume fraction of Fe_3O_4 nanoparticles during the cryopreservation process because of higher thermal conductivity.

References

1. Whittingham DG, Leibo SP, Mazur P (1972) Survival of mouse embryos frozen to −196 °C and −269 °C. Science 178:411–414
2. Yang J, Zhu Y, Xu T, Pan C, Cai N, Huang H, Zhang L (2016) The preservation of living cells with biocompatible micro-particles. Nanotechnology 27:265101-1-9
3. Rall WF, Fahy GM (1985) Ice-free cryopreservation of mouse embryos at −196 °C by vitrification. Nature 313(14):573–575
4. Ahmadikia H, Moradi A (2012) Non-Fourier phase change heat transfer in biological tissues during solidification. Heat Mass Transfer 4:1559–1568
5. Yi J, Zhao G (2014) Effect of hydroxyapatite nanoparticles on bio-transport phenomena in freezing HeLa cells. J Nanotechnol Eng Med 5:040904-1-7
6. Liu J, Yan JF, Deng ZS (2007) Nano-cryosurgery: a Basic way to enhance freezing treatment of tumour. In: 2007 ASME International Mechanical Engineering Congress and Exposition, IMECE 2007-43916 (2007)
7. He ZZ, Di DR, Liu J (2012) Enhancement of cryosurgery using biodegradable MgO nanoparticles. In: Proceedings of the ASME 3rd Micro/Nanoscale Heat & Mass Transfer International Conference, MNHMT 2012-75218
8. Hou Y, Sun Z, Rao W, Liu J (2018) Nanoparticle-mediated cryosurgery for tumor therapy. Nanomed Nanotechnol Biol Med 14:493–506
9. Brent AD, Voller VR, Reid KJ (1988) Enthalpy-porosity technique for modeling convection-diffusion phase change: Application to the melting of a pure metal. Numer Heat Transf 13(3):297–318
10. Patankar SV (1980) Numerical heat transfer and fluid flow, 4th edn. Hemisphere, London

Studies on Performance Improvement of an R744 Transcritical Refrigeration System Using Dedicated Mechanical Subcooling

Mihir Mouchum Hazarika, Maddali Ramgopal, and Souvik Bhattacharyya

Abstract Due to low critical temperature, the performance of a conventional, single-stage R744-based transcritical refrigeration cycle drops drastically when the heat sink temperature is high. However, the performance can be enhanced by subcooling the refrigerant at the gas cooler exit. The present study is carried out to investigate the performance of an R744-based refrigeration cycle integrated with an R744-based auxiliary cycle to subcool the refrigerant at gas cooler exit. This modified cycle is termed as the cycle with dedicated mechanical subcooling (CDMS). This modified cycle is shown to give superior performance compared to the conventional cycle and cycle with internal heat exchanger. For this proposed cycle, the important operating parameters are optimized based on thermodynamic analysis. The results reported in this study are helpful in designing R744 transcritical refrigeration system integrated with auxiliary subcooling cycle.

Keywords R744 · Transcritical · Subcooling · Optimization

Nomenclature

CDMS [–]	Cycle with dedicated mechanical subcooling
CIHX [–]	Cycle with internal heat exchanger
COP [–]	Coefficient of performance
h [J kg^{-1}]	Specific enthalpy
$\dot{m}$ [kg s^{-1}]	Mass flow rate
P [Pa]	Pressure
$\dot{Q}$ [W]	Heat transfer rate

M. M. Hazarika (✉) · M. Ramgopal
Mechanical Engineering Department, Indian Institute of Technology Kharagpur, Kharagpur 721 302, India
e-mail: mmhazarika@iitkgp.ac.in

S. Bhattacharyya
Department of Mechanical Engineering, BITS Pilani, Pilani 333 031, India

M. Ramgopal et al. (eds.), *Advances in Air Conditioning and Refrigeration*, Lecture Notes in Mechanical Engineering, https://doi.org/10.1007/978-981-15-6360-7_4

s [J kg^{-1} K^{-1}]	Specific entropy
T [K]	Temperature
$\dot{W}_{com}$ [W]	Compressor power

Special Characters

ε [–]	Effectiveness
η [–]	Efficiency

Subscripts

amb	Ambient
aux	Auxiliary
com	Compressor
ev	Evaporator
ex	Exit
gc	Gas cooler
in	Inlet
max	Maximum
min	Minimum
ref	Refrigerant
s	Suction

1 Introduction

Most of the synthetic refrigerants used in refrigeration and air-conditioning systems have high global warming potential (GWP). These synthetic refrigerants are harmful for the environment. Therefore, to protect the environment from the negative effects of synthetic refrigerants, it has become essential to use low GWP natural substances as refrigerants. Carbon dioxide (R744) is a promising refrigerant as it has very low GWP (GWP = 1). However, COP is low for conventional R744-based refrigeration and air-conditioning systems at high heat sink temperature. Several modifications have been suggested recently to improve the performance of R744-based systems. Aprea and Maiorino [1] and Torrella et al. [2] analyzed the possibilities of improving the performance of R744-based refrigeration cycle using internal heat exchanger. Robinson and Groll [3] observed that use of expander in place of throttle valve improved the COP by 25% compared to the COP of conventional cycle with internal heat exchanger. Ferrara et al. [4] and Yang et al. [5] also performed studies to investigate the use of expander in place of throttle valve in R744-based refrigeration cycles. Ahammed et al. [6] and Lee et al. [7] carried out studies on R744-based refrigeration cycles using ejector as an expansion device. Singh et al. [8] analyzed the performance of modified R744-based configurations by incorporating expansion turbine in place of throttle valve, multistage compression with intercooling, and internal heat exchanger.

Llopis et al. [9] performed theoretical studies to analyze the possibilities of improving the performance of R744-based refrigeration cycle by using an auxiliary cycle to subcool the refrigerant at gas cooler exit. They used six different refrigerants (R290, R1270, R1234yf, R161, R152a, and R134a) in the auxiliary cycle. From results, they reported comparable performance improvements with all refrigerants. However, the synthetic refrigerants with high GWP are no longer allowed in auxiliary cycle due to the implementation of European F-gas regulations. Use of hydrocarbons is also restricted due to their high flammability. Hence, the options of refrigerants suitable for auxiliary cycle are also very limited.

The present study is carried out to investigate the performance of an R744-based refrigeration cycle integrated with an R744-based auxiliary cycle to subcool the refrigerant at gas cooler exit. This modified cycle is termed as the cycle with dedicated mechanical subcooling (CDMS). The performance of this proposed cycle is compared with conventional cycle and the cycle with internal heat exchanger (CIHX). In addition, the important operating parameters are optimized for the proposed cycle with dedicated mechanical subcooling.

2 Cycles Considered in Present Analysis

The schematic diagrams of all the three cycles studied here are shown in this section. Figure 1a shows the conventional R744-based cycle. Figure 1b shows the R744-based cycle with internal heat exchanger. The modified R744-based cycle with dedicated mechanical subcooling (CDMS) is shown in Fig. 1c. This modified refrigeration cycle comprises a primary cycle and an auxiliary cycle. The auxiliary cycle is used to subcool the refrigerant at gas cooler exit of the primary cycle. R744 is used as refrigerant in the primary cycle as well as in the auxiliary cycle. Figure 1d shows the p-h plot for this modified R744 refrigeration cycle. For all the three cycles, the ambient is the heat sink.

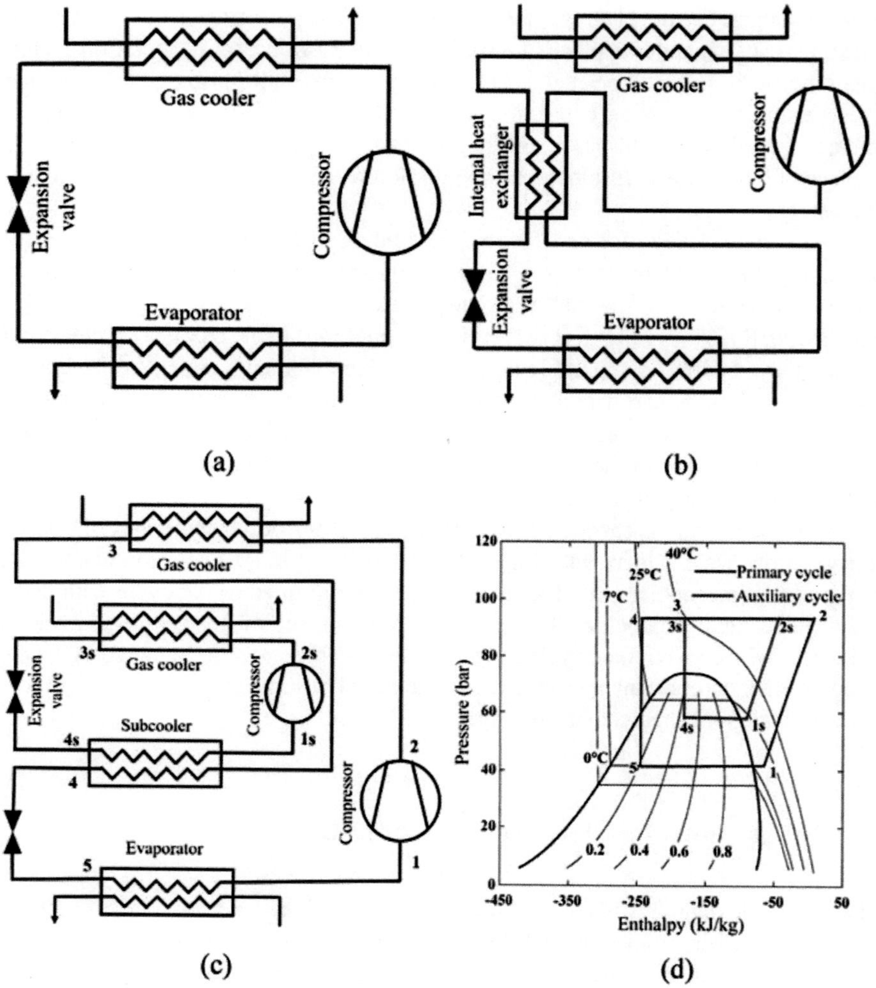

Fig. 1 **a** Schematic of conventional cycle; **b** schematic of cycle with IHX; **c** schematic of cycle with dedicated mechanical subcooling; and **d** p-h plot for cycle with dedicated mechanical subcooling

3 Mathematical Modeling

The thermodynamic models for the cycles are developed using the following assumptions [8, 9]:

(a) Steady-state operation.
(b) Pressure drop is neglected in gas cooler, evaporator as well as in the connecting pipes.
(c) Flow through the expansion valve is isenthalpic.

Based on these assumptions, mass balance and energy balance equations are applied across each component.

Compressor power is estimated from:

$$\dot{W}_{\mathrm{com}} = \dot{m}_{\mathrm{ref}} \times \left(h_{\mathrm{ex,com}} - h_{\mathrm{in,com}}\right) \tag{1}$$

$$\eta_{\mathrm{is,com}} = \frac{\left(h_{\mathrm{ex,com,s}} - h_{\mathrm{in,com}}\right)}{\left(h_{\mathrm{ex,com}} - h_{\mathrm{in,com}}\right)} \tag{2}$$

$$h_{\mathrm{in,com}} = h(P_{\mathrm{s}}, T_{\mathrm{s}}) \tag{3}$$

$$s_{\mathrm{ex,com,s}} = s(P_{\mathrm{s}}, T_{\mathrm{s}}) \tag{4}$$

$$h_{\mathrm{ex,com,s}} = h\left(P_{\mathrm{gc}}, s_{\mathrm{ex,s}}\right) \tag{5}$$

Gas cooler heat rejection rate is estimated from:

$$\dot{Q}_{\mathrm{gc}} = \dot{m}_{\mathrm{ref}} \times \left(h_{\mathrm{in,gc}} - h_{\mathrm{ex,gc}}\right) \tag{6}$$

$$h_{\mathrm{ex,gc}} = h\left(P_{\mathrm{gc}}, T_{\mathrm{ex,gc}}\right) \tag{7}$$

Heat extraction rate in evaporator is estimated from:

$$\dot{Q}_{\mathrm{ev}} = \dot{m}_{\mathrm{ref}} \times \left(h_{\mathrm{ex,ev}} - h_{\mathrm{in,ev}}\right) \tag{8}$$

$$h_{\mathrm{in,ev}} = h(P_{\mathrm{s}}, T_{\mathrm{s}}) \tag{9}$$

Heat exchange rate in the internal heat exchanger is estimated from:

$$\dot{Q}_{\mathrm{IHX}} = \varepsilon_{\mathrm{IHX}} \times \dot{Q}\mathrm{max} \tag{10}$$

$$Q\mathrm{max} = \min\begin{cases} m_{\mathrm{ref}} \times \left(h\left(P_{\mathrm{gc}}, T_{\mathrm{ex,gc}}\right) - h\left(P_{\mathrm{gc}}, T_{\mathrm{s}}\right)\right) \\ m_{\mathrm{ref}} \times \left(h\left(P_{\mathrm{s}}, T_{\mathrm{ex,gc}}\right) - h(P_{\mathrm{s}}, T_{\mathrm{s}})\right) \end{cases} \tag{11}$$

Heat exchange rate in the subcooler is estimated from:

$$\dot{Q}_{\mathrm{subcooler}} = \varepsilon_{\mathrm{subcooler}} \times \dot{Q}\mathrm{max} \tag{12}$$

$$\dot{Q}\mathrm{max} = \min\begin{cases} \dot{m}_{\mathrm{ref}} \times \left(h\left(P_{\mathrm{gc}}, T_{\mathrm{ex,gc}}\right) - h\left(P_{\mathrm{gc}}, T_{\mathrm{subcooler}}\right)\right) \\ \dot{m}_{\mathrm{ref,aux}} \times \left(h\left(P_{\mathrm{subcooler}}, T_{\mathrm{ex,gc}}\right) - h(P_{\mathrm{subcooler}}, T_{\mathrm{subcooler}})\right) \end{cases} \tag{13}$$

COP is estimated from:

$$COP_{\text{conventional}} = \frac{Q_{\text{ev}}}{\dot{W}_{\text{com}}} = \frac{\left(h_{\text{ex,ev}} - h_{\text{in,ev}}\right)}{\left(h_{\text{ex,com}} - h_{\text{in,com}}\right)} \tag{14}$$

$$\text{COP}_{\text{CIHX}} = \frac{Q_{\text{ev}}}{\dot{W}_{\text{com}}} = \frac{\left(h_{\text{ex,ev}} - h_{\text{in,ev}}\right)}{\left(h_{\text{ex,com}} - h_{\text{in,com}}\right)} \tag{15}$$

$$\begin{aligned}\text{COP}_{\text{CDMS}} &= \frac{\dot{Q}_{\text{ev}}}{\dot{W}_{\text{com}} + W_{\text{com,anx}}} \\ &= \frac{\dot{m}_{\text{ref}} \times \left(h_{\text{ex,ev}} - h_{\text{im,ev}}\right)}{\dot{m}_{\text{ref}} \times \left(h_{\text{ex,com}} - h_{\text{in,com}}\right) + \dot{m}_{\text{ref,aix}} \times \left(h_{\text{ex,com,aix}} - h_{\text{in,com,aix}}\right)}\end{aligned} \tag{16}$$

4 Results and Discussion

Based on the numerical models discussed in Sect. 3, codes are developed in MATLAB [10]. The thermophysical properties of R744 are evaluated using REFPROP [11] which is integrated with MATLAB. Table 1 shows the operating conditions maintained during simulations. The results obtained from numerical simulations are presented here. The first part of this section presents the performance of R744 refrigeration cycle with dedicated mechanical subcooling (CDMS) as compared to conventional cycle and cycle with internal heat exchanger. The second part presents the optimization of important operating parameters for the CDMS configuration.

4.1 Performance of Cycle with Dedicated Mechanical Subcooling Compared to Conventional Cycle and Cycle with IHX

To investigate the effect of gas cooler pressure, thermodynamic cycle simulations are carried out for the three proposed configurations: (a) cycle with dedicated mechanical subcooling (CDMS), (b) cycle with internal heat exchanger (CIHX), and (c) conventional cycle. Figure 2 shows the variation in COP with changes in gas cooler

Table 1 Operating conditions for simulation

Ambient temperature (°C)	Approach temperature of gas cooler (K)	Evaporation temperature (°C)	Superheat (K)	Effectiveness of IHX and subcooler (%)
30–50	3	5	7	80

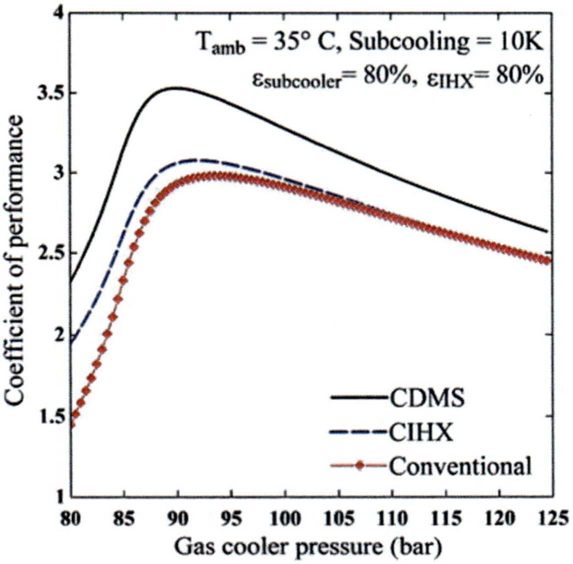

Fig. 2 Effect of gas cooler pressure on COP for all three cycles

pressure. It is observed that for all three configurations, maximum COP is obtained for a specific gas cooler pressure which is treated as the optimum gas cooler pressure for the respective configuration under specific operating conditions. At supercritical pressure, as the isotherm forms a peculiar s-shape, there exists an optimum gas cooler pressure for which COP is maximum at a specific gas cooler exit temperature. Results also show that there is significant improvement in COP for the CDMS configuration compared to the other two configurations. Maximum COP obtained for CDMS configuration increases by 14 and 19% compared to the maximum COP obtained for CIHX and conventional cycle, respectively. In addition, it is observed that the optimum gas cooler pressure shifts toward lower pressure for CDMS configuration compared to the other two configurations. Optimum gas cooler pressure obtained for CDMS configuration decreases by 2 and 4% compared to the optimum gas cooler pressure obtained for CIHX and conventional cycle, respectively. Hence, it can be concluded that the configuration integrated with dedicated mechanical subcooling cycle performs better compared to the conventional cycle and the cycle with internal heat exchanger.

4.2 Optimum Intermediate Pressure and Degree of Subcooling for Cycle with Dedicated Mechanical Subcooling

4.2.1 Selection of Intermediate Pressure

For the CDMS configuration, the dedicated subcooling cycle is used to subcool the refrigerant at the exit of gas cooler. To get the desired degree of subcooling, the subcooler of the dedicated subcooling cycle should be maintained at suitable intermediate pressure. The intermediate pressure should be selected to achieve maximum COP. Figure 3 shows cycle COP with changes in intermediate pressure. It is observed that COP increases as the intermediate pressure approaches toward gas cooler pressure. However, with an increase in the intermediate pressure, the difference in refrigerant temperature at gas cooler exit and subcooler inlet decreases. If this temperature difference is too low, then it is not possible to achieve the desired degree of subcooling. That is, there exists a maximum intermediate pressure that should be maintained to achieve the desired degree of subcooling as well as maximum COP. This intermediate pressure should be treated as optimum intermediate pressure under specific operating condition. Figure 4 shows optimum gas cooler pressure, optimum intermediate pressure, and maximum COP with changes in ambient temperature.

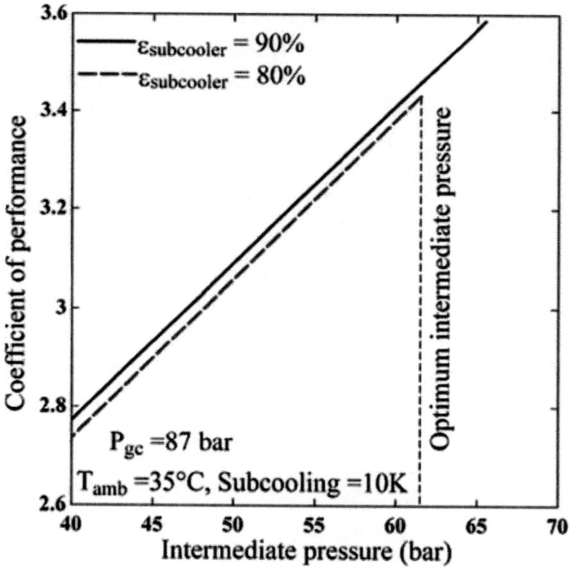

Fig. 3 Effect of intermediate pressure on COP

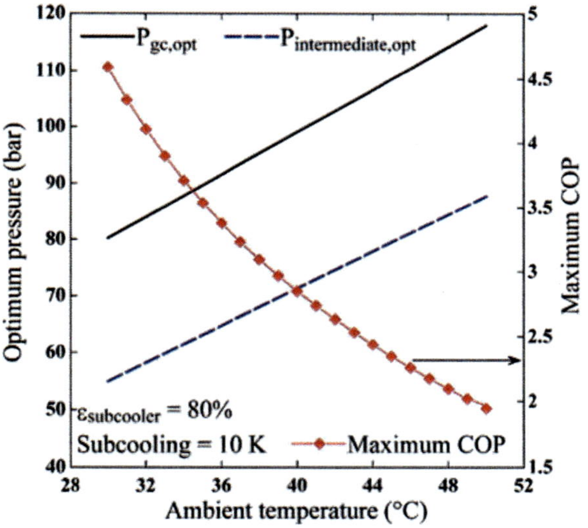

Fig. 4 Optimum gas cooler pressure and intermediate pressure to maintain 10 K subcooling

4.2.2 Selection of Degree of Subcooling

To investigate the effect of degree of subcooling, cycle simulations are carried out and the results are plotted in Fig. 5. Gas cooler pressure and intermediate pressure plotted in Fig. 5 are optimum pressures required to maintain the specific degree of subcooling under specific operating condition. It is observed that as the requirement of degree of subcooling increases, the difference between gas cooler pressure and intermediate pressure increases. In addition, it is observed that with an increase in

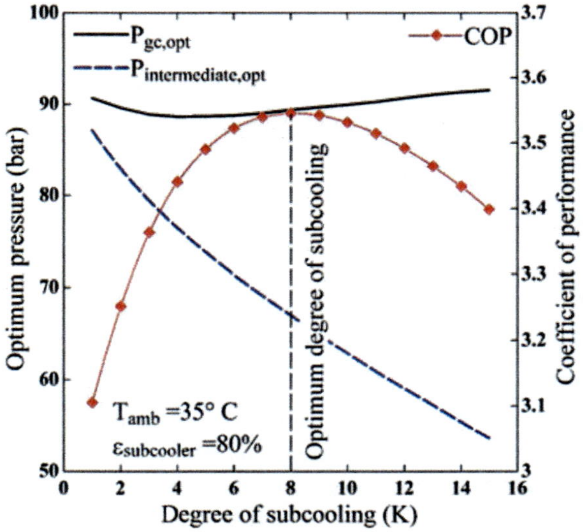

Fig. 5 Optimum gas cooler pressure and subcooler pressure and COP with change in degree of subcooling

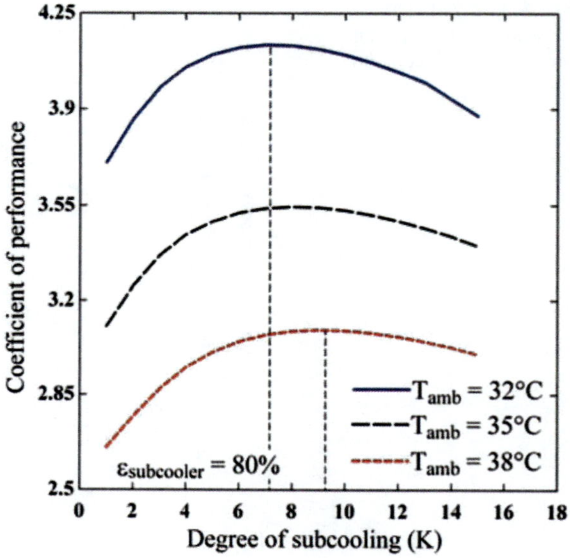

Fig. 6 COP with change in degree of subcooling for different ambient temperatures

degree of subcooling, COP first increases, reaches maximum, and then decreases. As the degree of subcooling increases, cooling capacity as well as compressor power increases. However, as long as the rate of increment in cooling capacity is higher than the rate of increment in compressor power, COP increases. COP starts decreasing when increment in cooling capacity becomes lower than the increment in compressor power. Figure 6 shows cycle COP with changes in degree of subcooling for different ambient temperatures. It is observed that higher the ambient temperature, higher is the optimum degree of subcooling for which maximum COP is obtained.

It is observed that with changes in the ambient temperature, the optimum degree of subcooling changes. Cycle simulations are carried out to estimate the optimum degree of subcooling with changes in ambient temperature. Figure 7 shows the optimum degree of subcooling and maximum COP with changes in ambient temperature. Optimum gas cooler pressure and intermediate pressure required to be maintained to achieve the optimum degree of subcooling and the maximum COP are shown in Fig. 8.

5 Conclusions

This paper presents theoretical studies on an R744-based refrigeration cycle with dedicated mechanical subcooling (CDMS). Results show that this proposed cycle gives superior performance as compared to conventional cycle and cycle with internal heat exchanger. The gas cooler pressure, the intermediate pressure at subcooler, and

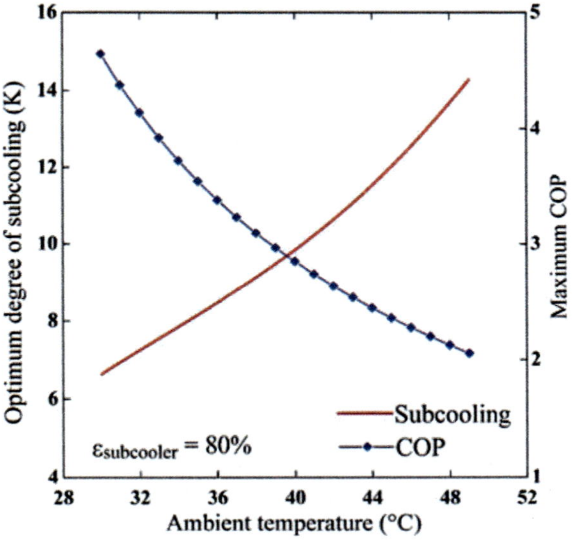

Fig. 7 Optimum degree of subcooling and maximum COP with change in ambient temperature

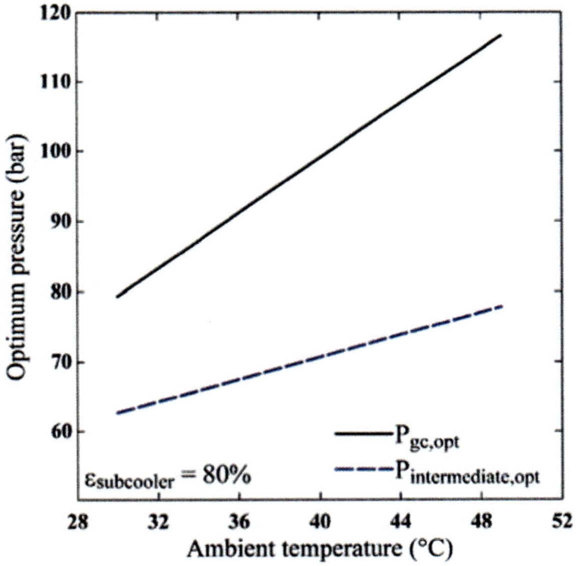

Fig. 8 Optimum gas cooler pressure and intermediate pressure with change in ambient temperature

the degree of subcooling are the three parameters that need to be optimized to extract maximum COP from this cycle.

(a) It is observed that maximum COPs are 3.53, 3.08, and 2.98 for CDMS, CIHX, and conventional cycle, respectively, at ambient temperature of 35 °C. In addition, the optimum gas cooler pressure for CDMS configuration is found to be 90 bar with subcooling of 10 K at an ambient temperature of 35 °C. However, for

CIHX and conventional cycle, the optimum pressures are 92 bar and 93.5 bar, respectively.

(b) Intermediate pressure at subcooler should be as high as possible to achieve maximum COP for this proposed CDMS configuration. Hence, the highest possible intermediate pressure required to maintain the desired degree of subcooling is treated as the optimum intermediate pressure under specific operating condition. This study presents the optimum intermediate pressure for a wide range of ambient temperature.

(c) Degree of subcooling is the third parameter that needs to be optimized for this proposed CDMS configuration. It is observed that there exists an optimum degree of subcooling at which COP is maximum under specific operating condition. This study presents the optimum degree of subcooling for a wide range of ambient temperature. At an ambient temperature of 35 °C, the optimum degree of subcooling, intermediate pressure, and gas cooler pressure are 8 K, 89.3 bar, and 66.7 bar, respectively, for the CDMS configuration. For these specified conditions, the maximum COP is 3.55.

References

1. Aprea C, Maiorino A (2008) An experimental evaluation of the transcritical CO_2 refrigerator performances using an internal heat exchanger. Int J Refrig 31:1006–1011
2. Torrella E, Sanchez D, Llopis R, Cabello R (2011) Energetic evaluation of an internal heat exchanger in a CO_2 transcritical refrigeration plant using experimental data. Int J Refrig 34:40–49
3. Robinson DM, Groll EA (1998) Efficiencies of transcritical CO_2 cycles with and without an expansion turbine. Int J Refrig 21:577–589
4. Ferrara G, Ferrari L, Fiaschi D, Galoppi G, Karellas S, Secchi R, Tempesti D (2016) Energy recovery by means of a radial piston expander in a CO_2 refrigeration system. Int J Refrig 72:147–155
5. Yang JL, Ma YT, Liu SC (2007) Performance investigation of transcritical carbon dioxide two-stage compression cycle with expander. Energy 32:237–245
6. Ahammed E, Bhattacharyya S, Ramgopal M (2014) Thermodynamic design and simulation of a CO_2 based transcritical vapour compression refrigeration system with an ejector. Int J Refrig 45:177–188
7. Lee JS, Kim MS, Kim MS (2014) Studies on the performance of a CO_2 air conditioning system using an ejector as an expansion device. Int J Refrig 38:140–152
8. Singh S, Purohit N, Dasgupta MS (2016) Comparative study of cycle modification strategies for trans-critical CO_2 refrigeration cycle for warm climatic conditions. Case Stud Therm Eng 7:78–91
9. Llopis R, Cabello R, Sanchez D, Torrella E (2015) Energy improvements of CO_2 transcritical refrigeration cycles using dedicated mechanical subcooling. Int J Refrig 55:129–141
10. MATLAB (2008a) The MathWorks, Inc., Version 7.6.0. Natick, MA
11. REFPROP (2010). Thermodynamic properties of refrigerants and refrigerant mixures, Version 9.0. National Institute of Standards and Technology, Gaithersberg

Experimental Investigation of Parabolic Trough-Type Solar Collector Integrated with Storage Tank Under the Northern Indian Climatic Conditions

Devander Kumar and Sudhir Kumar

Abstract Currently, the trough-type solar collector incorporated with storage unit is receiving significant attention because of their capacity to retain excess heat during non-availability of insolation. Thus, the objective of present study is to investigate the trough collector thermal performance which is built-in with storage container experimentally. The study is performed for south-facing and tracking modes of working in the end of November month. The performance is estimated in terms of collector thermal efficiency, gain in useful heat, storage tank water temperature rise, system overall and charging efficiency by fabricating and testing the simple structure of parabolic trough collector (PTC) system. Results show that PTC is capable to heat the water stored in the storage tank during both modes of working. The highest collector thermal efficiency is obtained as 54.4 and 53.04% during south-facing and tracking modes, respectively. The maximum system charging efficiency is found to be 87.98% in south-facing and 89.8% in tracking modes.

Keywords Trough-type collector · Solar beam radiation · PTC · Thermal efficiency · Overall and charging efficiency · Storage tank · South-facing and tracking modes

Nomenclature

A_a	aperture area (m^2)
C_{pf}	working fluid specific heat (J/kg K)
$C_{p,st}$	stored fluid specific heat (J/kg K)
D_{ai}	receiver internal diameter (m)
D_{ao}	receiver external diameter (m)

D. Kumar (✉)
AEE(M), ONGC, Ahmedabad, India
e-mail: lambadev1@rediffmail.com

S. Kumar
National Institute of Technology, Kurukshetra 136119, Haryana, India
e-mail: mail2sudhir@rediffmail.com

M. Ramgopal et al. (eds.), *Advances in Air Conditioning and Refrigeration*, Lecture Notes in Mechanical Engineering, https://doi.org/10.1007/978-981-15-6360-7_5

D_{ci}	glass cover internal diameter (m)
D_{co}	glass cover external diameter (m)
E_{co}	collected energy (J)
E_{st}	stored energy (J)
F	focal length (m)
H	inner storage tank height (m)
I	solar beam irradiance (W/m^2)
L	length of the collector (m)
L_{sr}	length of supporting rod (m)
$\dot{m}$	working fluid mass flow rate (kg/s)
m	stored fluid mass in storage container (kg)
Q_u	gain in useful heat (W)
S_a	arc length of parabola (m)
T_{fi}	temperature of fluid at inlet (°C)
T_{fo}	fluid exit temperature (°C)
$\Delta\mathrm{T}$	receiver inlet and exit temperature difference (°C)
T_{st}	storage tank fluid temperature (°C)
w_{I}	solar beam irradiance error (%)
w_{m}	mass flow rate error (%)
$w_{\Delta T}$	temperature rise in the receiver error (%)
w_{η}	collector thermal efficiency error (%)
W_a	aperture width (m)
x	half of aperture width (m)
X	thermocouple height from bottom of inside storage tank (m)
y	parabola depth from focal axis (m)

Subscripts

k	at any instant time
$k+1$	one hour period after kth time
i	inner
o	outer

Greek symbols

ϕ_r	chosen rim angle in (°) as 90°
η_{I}	instant thermal efficiency
η_{ch}	charging efficiency
η_{ov}	overall efficiency
Φ_{br}	ball bearing diameter (m)
Φ_{ic}	inside diameter of collar (m)

Φ_{oc} outside diameter of collar (m)
Φ_{sr} supporting rod diameter (m)

1 Introduction

The utilization of fossil fuel excessively has created various serious problems like emission of carbon dioxide into the atmosphere, global warming which ultimately affects our climate and environment. Today, the climate change and global warming are the most serious threats which the entire world community is facing. Such difficulties can be decreased only if our belief is increased on the alternative renewable sources, sustainable and green energy instead of fossil fuel [1]. For achieving global justice and satisfying the rising requirement of energy all over the world, plentiful accessible sun energy throughout the world can be taken as one of efficient, sustainable and effective choices to make of energy contribution for wide industrial uses, reduce green house gases and maintain ecological balance [2]. Recently, the trend has been increased in different types of solar energy extracting systems and technologies, i.e., concentrating and non-concentrating. For obtaining high temperature, trough collector amongst the different categories of concentrating collectors is the highly accepted system and practiced technology. The major drawback that imposes the limitation on utilization of solar insolation for different types of applications is that it is an energy source of variable nature depending upon the time. Thus, the solar systems need the energy storage to deliver energy during off-sunshine hours and inadequate amount of insolation. Keeping the above, the current research seeks to predict the PTC performance incorporated with a storage tank throughout the charging process under the northern Indian climatic conditions. PTCs are typically providing temperatures varying around 100–450 °C. Trough-type collectors are efficiently used in a variety of applications such as energy production, space cooling and heating, steam production, electricity generation by using produced steam, water distillation, refrigeration and air-conditioning, hot water production, cooking, industrial process heating [3, 4]. In this section, a brief discussion has been illustrated on PTC development along with choices of storage is also discussed.

Numerous attempts have been made to predict and improve the performance of parabolic trough solar collector in the literature such as Refs. [5–13]. Heiti and Thodos [5] predicted the PTC thermal behavior experimentally utilizing copper absorber tube covered by a borosilicate glass tube having aperture uncovered and covered by a polycarbonate plate. It has been found that the performance of collector is higher when the receiver is coated and collector is uncovered. Hamad [6] has carried out an investigation experimentally for predicting the thermal behavior of a parabolic concentrator having the receiver of new designs with various values of water mass flow rate. The author has concluded that collector performance is mainly dependent on the working fluid mass flow rate. The variation in collector efficiency is found from 26 to 62% when flow rate of water is varied from 0.00056 to 0.0094 kg/s.

It is also reported that no considerable variation has been observed in efficiency when the mass flow rate is >0.0028 kg/s. Dudley et al. [7] evaluated the thermal performance of PTC utilizing various configurations of receiver and a different coating on the receiver surface experimentally. It is reported that performance of collector varies with each variation of receiver configuration. They have also developed a 1-D model of collector receiver, compared analytical model results with the experimental outcomes and found well agreement. Zarza et al. [8] have presented the Direct Solar Steam (DISS) project outcomes and conclusions. The authors concluded that there is viability of direct steam generation (DSG) with horizontal PTCs and recirculation is better option among various DSG processes. It is also noticed that power plant efficiency can be improved to supply steam by a DSG solar field at 550 °C/100 bar, rather than at 400 °C/100 bar. Testing and performance characterization of PTC with tracking system have been performed by Brooks [9]. Two cases of receiver are tested as per ASHRAE-93-1986 (RA91) standard; in first, receiver is uncovered, and second is with evacuated glass envelope. Reduced heat losses and superior performance are found with the evacuated glass unit. Reddy et al. [11] have investigated and compared the yield of PTC with a variety of porous disk receiver configurations experimentally. It is stated that such receiver can be used to enhance the PTC efficiency and decrease the receiver angular thermal gradient.

A detailed 1-D heat transfer investigation has been carried out for optimizing the PTC, and its performance is estimated under the different operating conditions [11]. The obtained numerical model results have been compared with the Sandia National Laboratories (SNL) experimental results and other studies. The results are in good agreement in comparison with other models. Li and Wang [12] measured PTC performance using two types of evacuated tube with a various heat transfer fluid (HTF) experimentally. Heating efficiency of 70–80% and less than 40% is found with water and N_2 gas, respectively. Kumar and Kumar [13] carried out the analytical study for the year-round performance evaluation of PTC and found well agreement of results with Sandia National Laboratories (SNL). Additionally, it is shown that the method and model utilized are feasible and reliable.

Saini et al. [14] estimated the solar cooker performance based on the PTC making use of acetanilide as heat storage material and water and engine oil as the heat transfer fluids. The authors' outcomes show that the system is capable to cook the food two times in a day. The storage material has stored more quantity of heat with thermal oil as compared to the water.

Active and passive schemes of thermal storage have been designed and investigated for solar energy applications [15]. Results showed that water heat storage works adequately with variable solar energy applications. The number of issues has been studied about the sensible heat storage (SHS) system [16]. SHS is found correct solution and cheaper storage technology between demand and supply.

The preceding literature studies show that a small number of investigations have been carried out on thermal performance assessment of trough-type collector incorporated with thermal energy storage tank experimentally. Presently, there is no comprehensive study present on experiment-based performance estimation and comparison of a mini-scale PTC incorporated with storage unit during south-facing and one-axis tracking modes underneath the northern Indian climatic conditions where the solar irradiance is available in magnificence. In a step ahead, the aim of this research is to construct and investigate experimentally the thermal behavior of PTC integrated with storage tank during tracking and south-facing modes and also present the working experience with PTC system. Important features of this system are its ease of construction, mechanical stability, reasonable cost and weight.

2 Experimental Setup Description and Procedure

2.1 Experimental Setup

Figure 1 shows the experimental setup photograph of PTC built-in with storage tank. It comprises a PTC, a heat storage tank, fluid circulating pump, connecting pipes and valve. The main components are described as follows.

2.1.1 Parabolic Trough Collector

The trough-type collector consists of a reflecting parabolic surface mounted on the parabolic shape supporting frame. The anodized aluminum sheet having specular reflectivity of 86% is utilized for making PTC reflector. Initial dimensions of reflector are 1.255 m width and 1.22 m length. MS flat plates and mild steel (MS) square pipe are used for constructing the supporting structure of collector. Aluminum sheet and flat plates are transformed in the shape of parabola using Eq. (1) [5] as follows.

$$x^2 = 4 \times F \times y \tag{1}$$

$$S_a = 2 \times F \times \left[(\sec \frac{\varphi_r}{2} \times \tan \frac{\varphi_r}{2}) + \left(\ln(\sec \frac{\varphi_r}{2} + \tan \frac{\varphi_r}{2}) \right) \right] \tag{2}$$

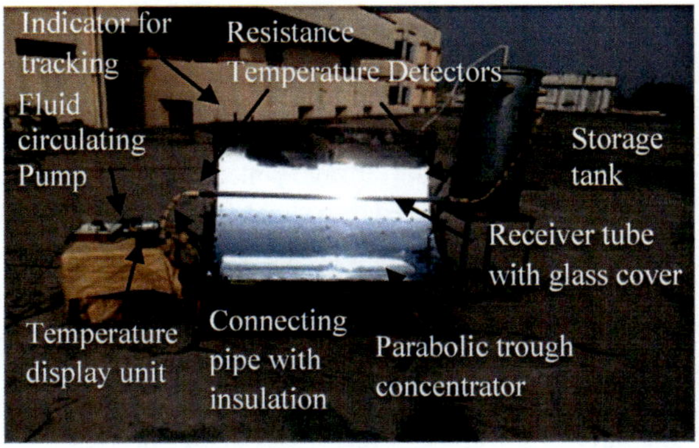

Fig. 1 Parabolic trough collector built-in with storage tank

The value of F is obtained as 0.273 m, when putting arc length $S_a = 1.255$ m and chosen rim angle in Eq. (2) [17]. Aperture half width value of parabola ($x =$ 0.546 m) is found using Eq. (1). For 90° rim angle, parabola depth value (y) is equal to the value F. Henceforth, the PTC aperture area value is obtained as 1.33 m^2. Collector supporting structure fabrication is done in two parts. In upper first part, as per required aperture area of PTC, marking of square pipe is done. Then, cutting and welding are carried out. The straight flat plates of MS are transformed into the parabolic shape which is consistent to the parabola drawn on paper piece manually. Here, three parabolic MS plates are used in which one is welded at middle and two on the inner left and right side of upper frame. These plates are identical and parallel to each other. So, it will create parabolic structure on which reflector sheet is mounted. Leg part is the second part that takes up the upper part of supporting structure and is kept on the ground.

The PTC reflector reflects and concentrates the incoming radiation on to the absorber tube that is positioned at the focus line of concentrator. The absorber assembly consists of tube of copper that is housed in a concentric borosilicate glass cover and is painted by PU-matt black paint. The coating of black paint on the absorber surface enhances its absorptivity for focusing sun energy and decreases the reflectance. The absorber absorbs the concentrated solar irradiance and afterward transfers to circulating heat transfer fluid which is flowing through absorber. In order to have airtight enclosed space between receiver and glass cover, seals of rubber are provided at the ends of glass tube. Two inner collars, two outer collars and two ball bearings between inner and outer collars are used to supporting both ends of receiver tube.

Ball bearing facilitates for easy rotation of collector during tracking mode without disturbing the connecting pipes.

Detail about the collar, ball bearing and supporting rod is provided in Fig. 2. The flexible pipes with insulation are utilized to join the receiver, circulating pump, copper pipe existing within the storage tank and valves. The HTF flowing through the absorber tube firstly transfers its heat to copper pipe of storage tank, and subsequently, it is delivered to the fluid stored into the storage container. Water which is consisting of hydrogen and oxygen atoms has been utilized as HTF in the present study. The PTC specification details are provided in Table 1.

2.1.2 Thermal Storage Tank and Circulating Pump

A 52-L capacity thermal storage tank comprises two tanks of cylindrical shape which are made up of galvanized iron sheet. The outer tank is kept around the inner tank. The gap between two tanks is filled by glass wool in order to provide healthy insulation around the inner tank in which water/liquid is filled. A copper pipe which acting as a heat exchanger is placed in at inner tank bottom. The main advantage of such heat exchanger in the storage tank is that the other working fluids like nano-fluids can also be used to know their performance with the PTC system or vice versa. The inlet end of this pipe is joined to receiver outlet through the flexible insulated plastic

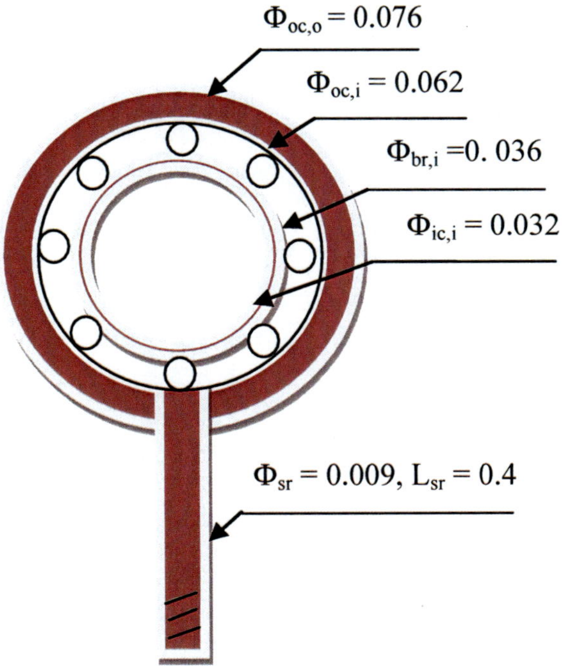

Fig. 2 Detail about the collar, bearing and supporting rod

Table 1 Parabolic trough collector specifications

Design parameters of PTC	Value
Collector length	1.22 m
Collector aperture width	1.09 m
Aperture area of collector	1.33 m^2
Focal length	0.273 m
Receiver external diameter	0.0318 m
Receiver internal diameter	0.0284 m
Outer diameter of glass cover	0.0556 m
Inner diameter of glass cover	0.0506 m
Concentration ratio	10.60
Reflector material	Anodized aluminum
Receiver tube material	Copper
Supporting structure material	Mild steel

pipe and pipe fittings. The outlet end fixed is connected to the absorber entrance via the circulating pump, pipe fittings and valve. So, the pump is circulating the HTF within the close circuit of absorber, copper pipe and plastic pipe during the whole PTC system working in order to deliver thermal energy of receiver to the water stored inside the tank. In open circuit, approximately 1 L of water as a HTF is sucked by

Table 2 Specifications of the storage tank

Construction parameters	Value
Outer and inner tank heights	0.655 and 0.615 m
Outer and inner tank diameters	0.41 and 0.33 m
Storage tank capacity	52 L
Inner and outer copper pipe diameters in tank	0.0045 and 0.0055 m
Copper pipe length in tank	12 m
Thickness of insulation	0.04 m
Insulation material	Glass wool
Storage tank material	Galvanized iron sheet (24 gauge)

the pump initially to fill pipes and receiver. The circuit is closed during working of PTC system.

The storage tank top side is closed by an insulated cover of GI sheet for preventing the stored water heat loss from topmost end. A small hole is made around the central of cover in which a hollow pipe of PVC having five small holes at various heights along its length is passed and put into the storage tank in vertical direction. At these holes, detectors are placed to detect the stored water temperature in tank at different heights. For monitoring continuously, detectors are connected to temperature indicator kit. Storage tank specification details are provided in Table 2.

2.2 *Measuring Instruments and Devices*

Resistance temperature detectors (RTDs) are used to show the temperature of HTF. These detectors are connected to the digital indicator which shows the temperature with resolution of 0.1 °C and error of ±0.3%. Ambient air temperature is recorded by a mercury-in-glass thermometer with ±1 °C resolution. Volume measuring beaker (least count 10 ml) and stopwatch (least count 1 s) are utilized for measurement of HTF flow rate initially. Afterward, circuit is closed. Mass flow rate 0.0325 kg/s remains constant during obtaining every set of data. Calibrated solar meter (TENMARS TM-207) with resolution 1 W/m^2 and accuracy ±5% is used to measure the solar irradiance by maintaining its plane and collector aperture plane in parallel to each other [18].

2.3 *PTC System Operation*

PTC system testing is done under two modes of working at TIT&S, Bhiwani (28°78′ N latitude and 76°13′ E longitude), a city located in India's northern region. In first

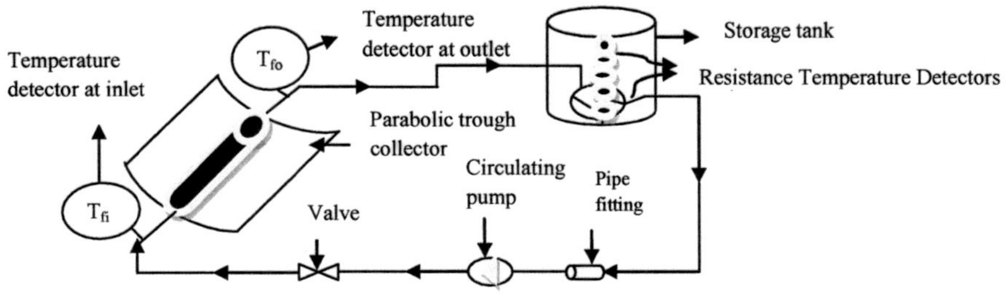

Fig. 3 Experimental setup schematic layout

mode, aperture plane is inclined with an angle equal to latitude angle of testing location and facing due south and collector axis is orientated in E–W direction. Second mode is the tracking mode in which PTC rotates horizontally about the N–S axis. One-axis tracking is used in direction east to west manually with interval of 10 min to track the sun so that maximum sunlight can be focused into the absorber. For tracking purpose, an indicator is attached which helps to tract the sun. The testing apparatus schematic layout is provided in Fig. 3.

2.4 Testing Procedure

Firstly, the surface of reflector and glass tube is cleaned in order to remove the dust particle, etc. Then, storage tank is filled by the freshwater. As per working mode at least 30 min before, PTC is positioned and directed toward the insolation to start the testing. The starting of pump is at 9:30 A.M. Water flow rate remains constant in each experiment and measured initially. The testing is carried out in November month by circulating HTF through the collector. Taking the reading is started from 10:00 A.M. onward. The readings are taken in a tabular form from 10:00 to 16:00 h at each hour. The pump is switched off at 4:00 P.M. because afterward heat losses of HTF during circulation in the circuit are significant as compared to collected energy by collector.

2.5 Parametric Performance

2.5.1 Useful Heat Gain (Q_u)

$$Q_u = \dot{m} \times C_{pf} \times (T_{fo} - T_{fi}) \tag{3}$$

2.5.2 Instantaneous Thermal Efficiency (η_{I})

$$\eta_I = \frac{Q_{\mathrm{u}}}{I \times A_{\mathrm{a}}} = \frac{\dot{m} \times C_{\mathrm{pf}} \times (T_{\mathrm{fo}} - T_{\mathrm{fi}})}{I \times A_{\mathrm{a}}} \tag{4}$$

2.5.3 Collected Energy (E_{co}) and Stored Energy (E_{st})

$$E_{\mathrm{co}} = \frac{\dot{m} \times C_{\mathrm{pf}} \times \left[(T_{\mathrm{fo}} - T_{\mathrm{fi}})_{k+1} + (T_{\mathrm{fo}} - T_{\mathrm{fi}})_{k}\right] \times 3600}{2} \tag{5}$$

where $(T_{\mathrm{fo}} - T_{\mathrm{fi}})_k$ and $(T_{\mathrm{fo}} - T_{\mathrm{fi}})_{k+1}$ show the temperature difference of HTF between outlet and inlet of receiver at any time k instantly and after a gap of 1 h from the time kth, respectively.

$$E_{\mathrm{st}} = m \times C_{\mathrm{p,st}} \times (T_{\mathrm{st},k+1} - T_{\mathrm{st},k}) \tag{6}$$

where $T_{\mathrm{st},k}$ and $T_{\mathrm{st},k+1}$ show the stored water mean temperature in storage tank at kth time and after 1 h time interval from the time kth, respectively.

2.5.4 Charging (η_{ch}) and Overall Efficiency (η_{ov})

$$\eta_{\mathrm{ch}} = \frac{E_{\mathrm{st}}}{E_{\mathrm{co}}}; \quad \eta_{\mathrm{ov}} = \frac{E_{\mathrm{st}}}{I \times A_{\mathrm{a}}} \tag{7}$$

2.5.5 Uncertainty Analysis

Root square method has been used to carry out the uncertainty analysis as the error is propagated during experimental measurements. Table 3 indicates the various measuring instrument uncertainties which are used in the experiments. The collector efficiency uncertainty w_η can be obtained on the basis of Eq. (4) by utilizing the equation as follows [19]:

$$w_\eta = \sqrt{w_{\mathrm{m}}^2 + w_{\mathrm{I}}^2 + w_{\Delta T}^2} \tag{8}$$

Table 3 Measuring instrument uncertainties

Parameters	Uncertainties %
Insolation	±5
Mass flow rate	±5
Temperature (resistance temperature detector)	±0.3

The uncertainties in A_a and C_{pf} are accounted as insignificant. The collector efficiency uncertainty is found as around 7.08%, and it is approximately 5% in the instantaneous stored energy and useful heat gain.

3 Results and Discussion

The variation of solar beam irradiance in relation to time measured during the south-facing and tracking modes of experimental work on November 23 and 24, 2015, respectively, at Bhiwani, India, is shown in Fig. 4. It is noticed that solar beam irradiance is recorded to be 270 and 264 W/m^2 at 10:00 h and it raises up to 436 and 291 W/m^2 till noon on south-facing and tracking modes, respectively. After the noontime, rate of solar beam irradiance decreases and it is 138 W/m^2 on south-facing mode and 180 W/m^2 during tracking mode of experiment at 16:00 h. There is gradual variation in beam irradiance found from 10:00 to 15:00 h during tracking mode beyond which significant variation is recorded.

The variation of HTF temperature at inlet and exit of PTC receiver and the ambient air temperature with time during south-facing and tracking modes is shown in Figs. 5 and 6. It is noticed that small variation in ambient air temperature is recorded during both modes of working. The range of variation of ambient temperature is found to be 24–30 °C in south-facing mode and 23–30 °C in tracking mode during the testing.

For both modes, the maximum rise in HTF temperature is found during noontime because of higher solar energy availability during this time and minimum rise is found during evening time, i.e., at 16:00 h. It is due to the lower beam irradiance obtained at evening time which results in less gain of heat by working fluid during its circulation through receiver. The entrance and exit temperatures of HTF at PTC inlet and exit increase up to 15:00 h and then start reducing gradually. At the end of

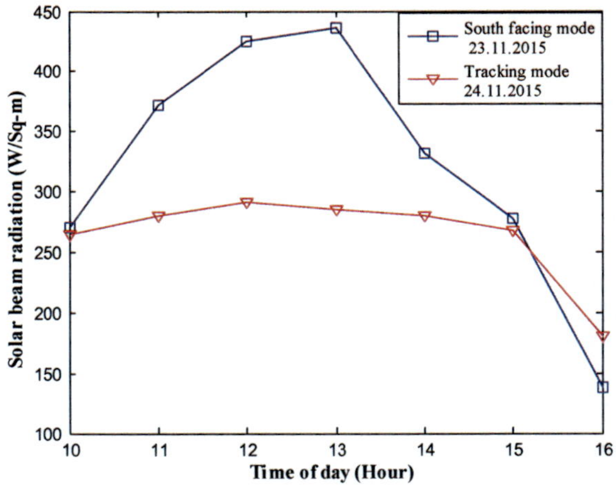

Fig. 4 Variation of solar beam radiation with respect to time

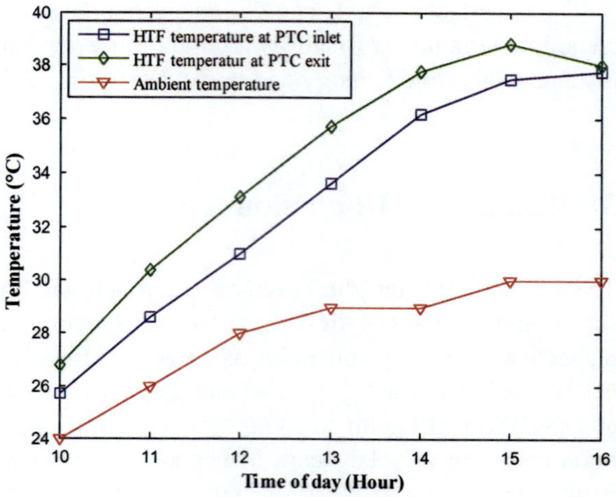

Fig. 5 Variation of HTF and ambient temperature with time during south-facing mode

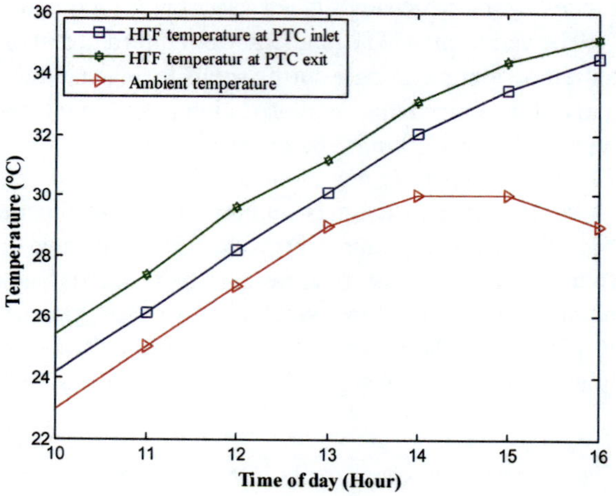

Fig. 6 Variation of HTF and ambient temperature with time during tracking mode

experiment, they come near to other due to reduced beam radiation and more heat losses from the receiver tube. Highest temperature of HTF at exit of the receiver is achieved as 38.8 °C during south-facing mode. In tracking mode, highest temperature of HTF at receiver exit is achieved as 35.1 °C.

Figures 7 and 8 present the variation in performance of PTC in terms of instantaneous thermal efficiency and useful heat gain with time during south-facing and tracking modes, respectively. It is observed from Fig. 7 that the instantaneous thermal efficiency and useful heat gain increase at higher rate from 44.84 to 53.27% and 150.2 to 245.7 W during 10:00 to 11:00 h interval of time, beyond which again rise is there till noon at slow pace, and then these begin decreasing. The highest useful heat gain and efficiency are 286.7 W and 54.4% at noon in case of south-facing working mode. The initial efficiency rise is because of high heat extraction by working fluid between

Fig. 7 Variation of collector thermal efficiency and useful heat gain with time during south-facing mode

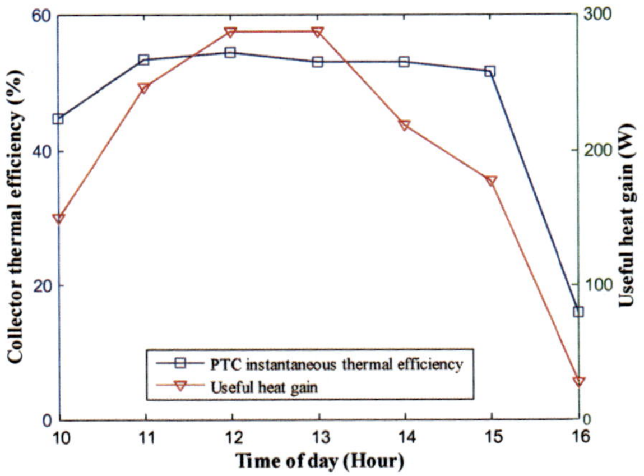

Fig. 8 Variation of collector useful heat gain and thermal efficiency with time during tracking mode

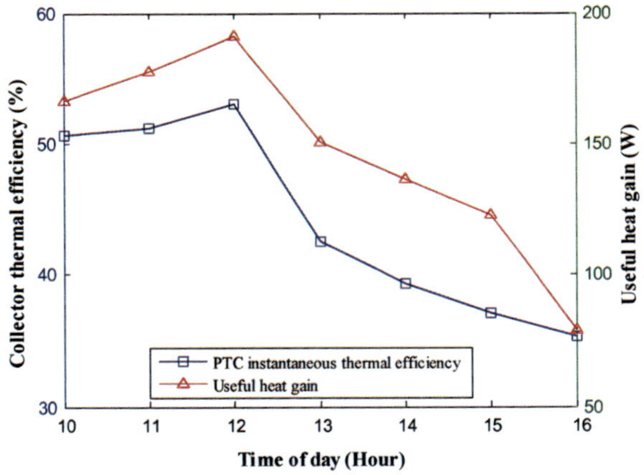

10:00 and 11:00 h. The high heat extraction is due to highest increase in rate of irradiance during this time period. Thus, the efficiency and useful heat gain of SPTC depend on the intensity of beam irradiance and this input operating parameter is more encouraging in the time interval between 10:00 and 11:00 h. After 16:00 h, there are no significant useful heat gain and efficiency obtained, so the system is switched off after this hour.

From Fig. 8, it is noticed that instantaneous thermal efficiency and useful heat gain increase sharply from 50.65 to 53.04% and 165.8 to 191.1 W during the time period of 10:00 to 12:00 h, beyond which these start decreasing. The peak efficiency and useful heat gain are found to be 53.04% and 191.1 W at noon with tracking mode. These obtained efficiencies and useful heat gains are comparable as reported by other investigator for different PTCs in previous works [6]. It shows the construction consistency of present PTC.

In the storage tank, stored water temperature variation with time during the charging process for south-facing and tracking modes is provided in Figs. 9 and 10, respectively. $X/H = 0$ shows the inner tank base of storage system. The five thermocouple positions from the inner tank bottom are $X/H = 0.10, 0.34, 0.58, 0.83$ and 0.99.

It is observed from the figures that there is sharp increase in stored water average temperature within the storage tank during the time interval from 10:00 to 14:00 h for both working modes. After 14:00 h, gradual rise is obtained. It is because that higher heat is transferred to storage water by the circulating HTF for time interval of 10:00–14:00 h and later on, smaller heat is delivered. First reason for high and smaller heat delivery is that higher rate of solar irradiance is achieved up to the noontime and then slower rate is gained till the end of experiments. Second reason is that less heat loss takes place from the HTF to the atmosphere up to the noontime

Fig. 9 Variation of average temperature within the storage tank with time during south-facing mode

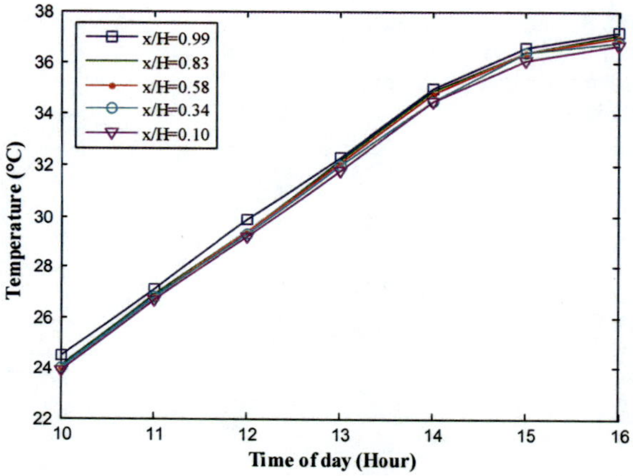

Fig. 10 Variation of average temperature within the storage tank with time during tracking mode

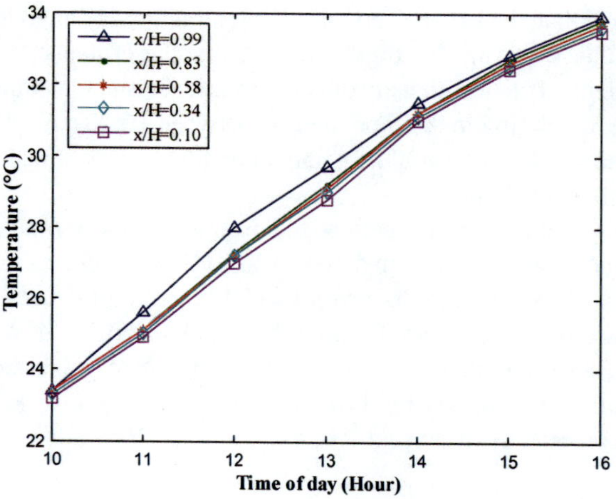

as compared to heat loss after the noontime. The average water temperature rise in the storage container is obtained to be around 13 and 10.4 °C with south-facing and tracking modes, respectively, from 10:00 to 16:00 h. It shows that temperature rise of water is achieved from PTC in the storage tank during both modes of operation.

Moreover, Figs. 11 and 12 show hourly based collected and stored energy of PTC system during the process of charging. The maximum hourly stored and collected energies are obtained as 615.9 and 992.25 kJ in case of south-facing mode during noontime, whereas these energies are 637.9 and 480.5 kJ with tracking mode. The hourly total storage and collected energies are found to be 2808.70 and 4512.44 kJ with south-facing mode, and 2262.62 and 3071.25 kJ with tracking mode during the system operation from 10:00 to 16:00 h.

Fig. 11 Variation of hourly collected and stored energy, with time during south-facing mode

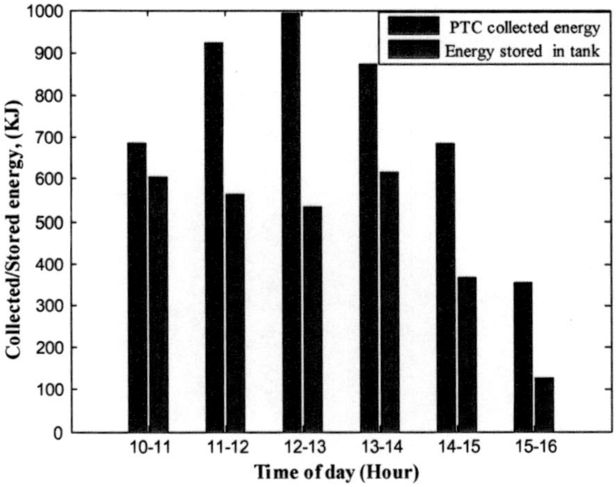

Fig. 12 Variation of hourly collected and stored energy, with time during tracking mode

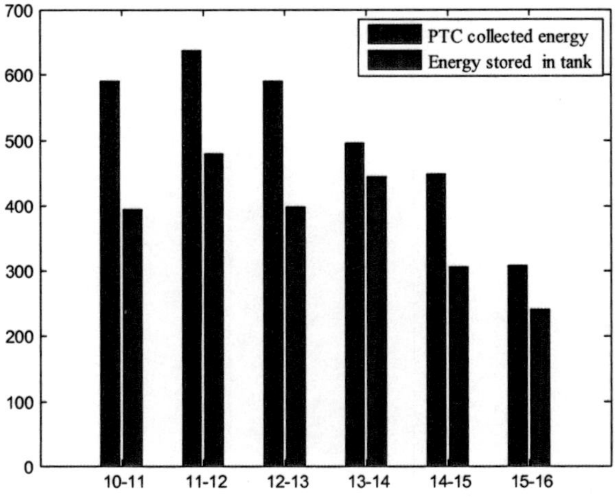

The system overall and charging efficiency variation in relation to time is shown in Figs. 13 and 14, respectively.

Figure 13 shows that charging and overall efficiencies start from 87.98 and 42.07% (at 10:00–11:00 h) which are highest in case of south-facing mode. Then, these decrease to 53.71 and 27.7% up to the time 13:00 h. During 13:00–14:00 h, there is slight increase in both efficiencies. After 14:00 h, again significant decrease in efficiency is found up to evening time (16:00 h). This increase and decrease in both efficiencies are due to variation of energy collected, solar beam radiation intensity and thermal losses from the PTC system during the concern time of operation. After 14:00 h, energy loss is more as compared to energy collected and irradiance is also reduced. So, the reducing trend is observed with charging and overall efficiencies after the 14:00 h.

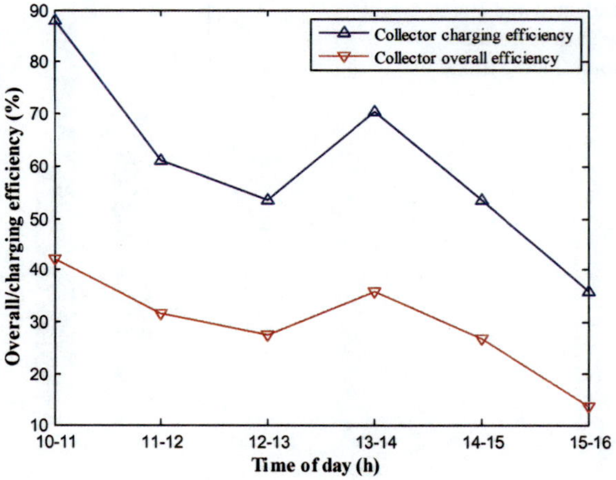

Fig. 13 Variation of system overall and charging efficiency, with time during south-facing mode

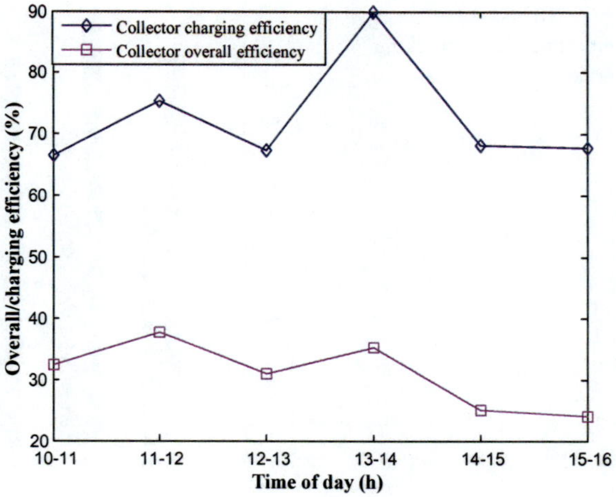

Fig. 14 Variation of system overall and charging efficiency, with time during tracking mode

Figure 14 shows that charging and overall efficiencies start from 66.56 and 32.38% (at 10:00 h–11:00 h) in case of tracking mode. These efficiencies vary in increasing and decreasing trend up to 14:00 h due to variation of energy collected, heat losses and solar beam irradiance from the system. After 14:00 h, again decreasing trend is obtained in these efficiencies up to evening time (16:00 h). The highest charging and overall efficiencies are found as 89.80 and 37.63% during the tracking mode of operation.

4 Conclusions

PTC system has been constructed with simplified structure, and its performance is evaluated during the south-facing and tracking modes of operations. The maximum useful heat gain and thermal efficiency achieved by the PTC are 286.65 W, 54.4% and 191.1 W, 53.04% during south-facing and tracking modes of operation, respectively, in the month of November (start of winter time). But for tracking mode, additional cost of tracking mechanism is also required. The highest PTC system charging efficiency is obtained to be 87.98% during south-facing mode of operation and 89.80% during tracking mode. The variations of solar beam radiation intensity and heat losses have considerable effect on the system thermal performance. From the above results, it is also noted that these collectors are proficient to heat the stored water with reasonable performance with south-facing mode also in which there is no need of tracking. Such collectors can ideally be suitable for utilization in isolated areas of developing nations.

References

1. Baharoon DA, Rahman HA, Omar WZW, Fadhl SO (2015) Historical development of concentrating solar power technologies to generate clean electricity efficiently—a review. Renew Sustain Energy Rev 41:996–1027
2. Villicana-Ortiz E, Gutierrez-Trashorras AJ, Paredes-Sanchez JP, Xiberta-Bernat J (2015) Solar energy potential in the coastal zone of the Gulf of Mexico. Renew Energy 81:534–542
3. Kalogirou S, Lloyd S (1992) Use of solar parabolic trough collectors for hot water production in Cyprus—a feasibility study. Renew Energy 2:117–124
4. Fernandez-Garcia A, Zarza E, Valenzuela L, Perez M (2010) Parabolic-trough solar collectors and their applications. Renew Sustain Energy Rev 14:1695–1721
5. Heiti RV, Thodos G (1983) An experimental parabolic cylindrical concentrator: its construction and thermal performance. Sol Energy 30(5):483–485
6. Hamad FAW (1988) Performance of a cylindrical parabolic solar concentrator. Energy Convers Manage 28(3):251–256
7. Dudley VE, Kolb GJ, Mahoney AR, Mancini TR, Mattews CW, Sloan M, Kearney D (1994) Test results: SEGS LS-2 solar collector. Report of Sandia National Laboratories (SAND94-1884)
8. Zarza E, Valenzulea L, Leon J, Hennecke K, Eck M, Weyers HD, Eickhoff M (2004) Direct steam generation in parabolic troughs: final results and conclusions of the DISS project. Energy 29:635–644

9. Brooks MJ (2005) Performance of a parabolic trough solar collector. Thesis, Master of Science, University of Stellenbosch
10. Reddy KS, Kumar KR, Ajay CS (2015) Experimental investigation of porous disc enhanced receiver for solar parabolic trough collector. Renew Energy 77:308–319
11. Padilla RV, Demirkaya G, Goswami DY, Stefanakos E, Rahman M (2011) Heat transfer analysis of parabolic trough solar receiver. Appl Energy 88(12):5097–5110
12. Li M, Wang LL (2006) Investigation of evacuated tube heated by solar trough concentrating system. Energy Convers Manage 47:3591–3601
13. Kumar D, Kumar S (2015) Year-round performance assessment of a solar parabolic trough collector under climatic condition of Bhiwani, India: a case study. Energy Convers Manage 106:224–234
14. Saini G, Singh H, Saini K, Yadav A (2015) Experimental investigation of the solar cooker during sunshine and off-sunshine hours using the thermal energy storage unit based on a parabolic trough collector. Int J Ambient Energy 1–12. http://dx.doi.org/10.1080/01430750.2015.1023836
15. Arinze EA, Schoenau GJ, Besant RW (1985) Thermal performance evaluation of active and passive water heat-storage schemes for solar energy applications. Energy 10:1215–1223
16. Dincer I, Dost S, Li X (1997) Performance analysis of sensible heat storage systems for thermal applications. Int J Energy Res 21:1157–1171
17. Arasu AV, Sornakumar T (2007) Design, manufacture and testing of fiberglass reinforced parabola trough for parabolic trough solar collectors. Sol Energy 81:1273–1279
18. Sukhatme SP (1996) Principal of thermal collection and storage, 2nd edn. TMGH Pub Com Ltd, New Delhi
19. Kumar DS (2000) Mechanical measurements and control, 3rd edn. Metropolian Book Co. Pvt. Ltd., New Delhi

Numerical Simulation of an Inertance Pulse Tube Refrigerator Using a Mixture of Refrigerant

Debashis Panda, M. Kumar, A. K. Satapathy, and Sunil Kr Sarangi

Abstract A numerical study is conducted to investigate the impact of mixture of refrigerant on the cooling performance of an inertance pulse tube refrigerator (IPTR). The influence of helium–hydrogen mixture on the cooling capacity is investigated numerically. Conservations of mass, momentum and energy equations are solved in the computational domain of an IPTR by finite volume method using FLUENT. Properties of helium–hydrogen mixture have been estimated by utilizing ideal gas law and are provided to the FLUENT as user-defined functions. It is seen that with an increase in the molar percentage of hydrogen than helium in a helium–hydrogen mixture, it enhances the cooling capacity, attains a maximum value and then decreases. It is examined that at a composition of about 50% hydrogen and 50% helium, the cooling capacity increases by about 11.51%, and further increase in hydrogen percentage than helium reduces the capacity.

Keywords IPTR · Cooling capacity · Temperature · Mixture · He-H_2

Nomenclature

E	Energy
h	Enthalpy
k	Thermal conductivity
p	Pressure
v	Velocity
φ	Porosity
ρ	Density
∇	Gradient

D. Panda (✉) · M. Kumar · A. K. Satapathy
National Institute of Technology, Rourkela 769008, Odisha, India
e-mail: debashis1003@gmail.com

S. K. Sarangi
C.V. Raman College of Engineering, Bhubaneswar 751019, Odisha, India

M. Ramgopal et al. (eds.), *Advances in Air Conditioning and Refrigeration*, Lecture Notes in Mechanical Engineering, https://doi.org/10.1007/978-981-15-6360-7_6

1 Introduction

Due to the compact and simple structural configuration of an inertance pulse tube refrigerator (IPTR), it is extensively used in cooling of military helicopters, night vision cameras, SQUIDS, space satellites, missile guidelines, etc. [1]. At the late twentieth century, Gardner and Swift [2] replaced the orifice valve of an orifice pulse tube refrigerator (OPTR), by a small diameter and large-length inertance tube and renamed it as an inertance pulse tube refrigerator (IPTR). A number of design and analysis have been done to improve the refrigeration power of an IPTR (such as exergy analysis [3], thermodynamic analysis [4], isothermal model [5]). Additionally, one-dimensional and multi-dimensional numerical analyses of IPTR have been carried out using codes like SAGE [6], FLUENT [7–9] to visualize the fluid flow and thermal characteristics occurring inside the tube. Chen et al. [10] used 1D numerical code to investigate the effect of mixture of refrigerant ($He + H_2$) on the cooling power of an IPTR. In this paper, the refrigeration performance of an IPTR has been investigated by using a mixture of helium and hydrogen in place of pure helium using the commercial CFD code ANSYS FLUENT®.

2 Numerical Analysis

Here, the details about the numerical modelling are presented in a concise manner. Numerical analysis is done with ANSYS FLUENT®, and geometrical modelling is done through ANSYS DM®. More details about the geometry creation, meshing and solution procedure are explained below.

2.1 Computational Domain

A 2D axisymmetric model has been created by using ANSYS DM® to conceive the three-dimensional thermo-hydrodynamic influences of helium–hydrogen mixture, on the rate of cold production inside an IPTR. A schematic diagram is shown in Fig. 1. Dimensions of structural parameters are considered to be same as those of references [7–9]. Since regenerator and heat exchangers contain wire meshes, there

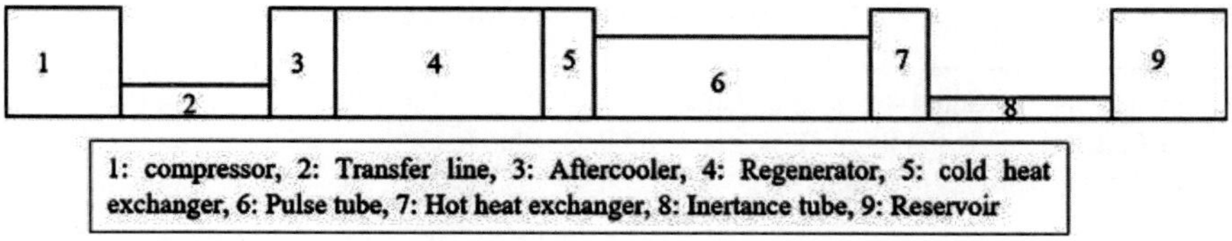

Fig. 1 Computational domain of IPTR

exists an extra pressure drop and also a bigger rate of heat transfer. To account the aforementioned phenomena occurring inside the regenerator and heat exchangers in the analysis stage, separate surface sketches are created for different components and are joined with each other during the geometry creation stage. The reason for doing so is to implement different cell zone conditions to individual surface sketches (for regenerator and heat exchangers) in the FLUENT® solver.

The mesh is generated using ANSYS MESH®, and attention is focused to generate a global non-uniform mesh with finer mesh at the junction of different surface sketches (i.e. two components). After cooler, cold and hot heat exchangers are modelled with unstructured mesh, whereas for the remaining parts, a rectangular structured meshing is adopted. Structured meshing is an essential requirement for the compressor to avoid the "negative cell volume detected" error. Inflation is used at the junction of the reservoir and inertance tube with a coarse mesh to save the computational time. However, finer meshing has been opted for the regenerator, cold and hot heat exchangers and pulse tube.

2.2 Numerical Methodology

The generated mesh is exported from the ANSYS MESH® and imported into the FLUENT® solver separately, due to the use of compiled UDF (for compressor mesh motion) and interpreted UDF (for calculation of temperature-dependent fluid properties). After importing the mesh file into the FLUENT® solver, firstly all the mesh interfaces created by default in solver are deleted and then manually interfaces are created in order to avoid "mesh checking fail" error. The compiled UDF which outlines the sinusoidal volume variation of the compressor is loaded to the FLUENT® solver, and this is used to create the dynamic mesh motion of the compressor wall [9]. The mixture fluid is assumed to behave like an ideal gas, whose specific heat at constant pressure and volume (C_p and C_v) are estimated by using ideal gas laws. Density is estimated from the pressure and volume by changing the molecular weight of the mixture. The temperature-dependent thermal conductivity and viscosity equations are provided into the solver as interpreted UDF. The $k - \epsilon$ turbulent model is opted to account for the turbulent flow, which occurs at the inertance tube. On the other hand, the flow regime inside the porous media (regenerator and heat exchangers), and pulse tubes are assumed to be as laminar. To consider the excessive pressure drop and heat transfer in the regenerator and heat exchangers, the numeric value of inertial resistance, viscous resistance and porosity is provided. Thermal non-equilibrium model (TNEM) is opted to study the temperature variations of the solid. The procedure of calculating all these parameters was already explained by Banjare et al. [11].

All the walls are modelled as no-slip boundary conditions. The boundary condition for wall temperatures is provided in reference [8, 9]. To model the dynamic meshing of compressor wall (piston), the loaded compiled UDF is hooked with the wall and modelled as a deforming body, whereas side-walls are identified as a rigid body. A pressure-based segregated solver with implicit formulation has been opted for the solution of the governing equations. The first-order formulation is opted for transient terms, whereas an upwind method has been opted for the momentum equation. The convergence criteria opted for continuity, turbulent k-ε equation are 10^{-3}, whereas 10^{-6} for u-momentum, v-momentum and energy equations. The governing differential equations are:

Continuity equation:

$$\frac{\partial}{\partial t}(\varphi \rho_f) + \nabla.[\varphi \rho_f \vec{v}] = 0 \tag{1}$$

Momentum equation:

$$\frac{\partial}{\partial t}(\varphi \rho_f \vec{v}) + \nabla.[\varphi \rho_f \vec{v}\vec{v}] = -\varphi \nabla p + \nabla.[\varphi \bar{\tau}] - \left(\frac{\mu}{\alpha}\vec{v} + \frac{c_f}{2}\rho_f \vec{v}|\vec{v}|\right) \tag{2}$$

where

$$\bar{\tau} = \mu\left[\left(\nabla\vec{v} + \nabla\vec{v^T}\right) - \frac{2}{3}\nabla.\vec{v}I\right] \tag{3}$$

Energy equation:

$$\begin{aligned} &\frac{\partial}{\partial t}(\varphi \rho_f E_f + (1-\varphi)\rho_s E_s) + \nabla.[\vec{v}(\rho_f E + p)] \\ &= \nabla[\{\varphi k_f + (1-\varphi)k_s\}\nabla T + (\varphi \bar{\tau}.\vec{v})] \end{aligned} \tag{4}$$

where

$$E = h - \frac{p}{\rho} + \frac{v^2}{2} \tag{5}$$

where ρ is density, E is energy, h is enthalpy, p is pressure, v is volume, T is temperature, $\vec{v}$ is velocity vector, and is the porosity. For the non-porous zones, the value of is taken as unity, and the last term on right-hand side of Eq. (2) is neglected. More details of these may be found out from reference [9]. The solution is initialized by applying a standard initialization method with default initial conditions. A total of 59 iterations per each time step are provided for better convergence. The model is solved in a PC of a 3GHZ processor and 4 GB of RAM, and it takes approximately 40 days to get a steady-state solution.

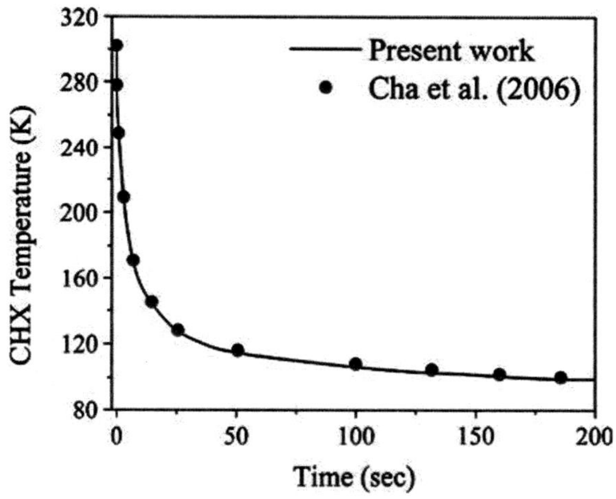

Fig. 2 Validation of numerical model with Cha et al. [7]

2.3 *Validation of the Model*

The model is first validated with the results of Cha et al. [7] to ensure the correctness of the model. The variations of cycle average temperature at CHX versus time for present model and Cha et al. [7] are depicted in Fig. 2.

2.4 *Grid Independence Test*

Grid independence test is carried out to determine the optimal number of grid points required to make a balance between accuracy of results and computational time. This is because, with increase in mesh size the accuracy of the results increases up to certain extent, whereas its computational time also increases. Here, it is portrayed that after 3975 number of grid points (as shown in Fig. 3), there does not exist any appreciable change in CHX wall temperature; thus, 3975 number grid point mesh has been opted for further analysis. The optimal grid size for different components (i.e. regenerator sub-domain, pulse tube sub-domain and reservoir sub-domain) is depicted in Fig. 4.

3 Results and Discussion

The influence of mixture of helium and hydrogen on the refrigeration temperature is studied. It is seen that a lowest refrigeration temperature of about 65.8 K has been achieved with a mixture of 50% helium + 50% hydrogen. Further increase in hydrogen percentage in the mixture increases the no-load refrigeration temperature.

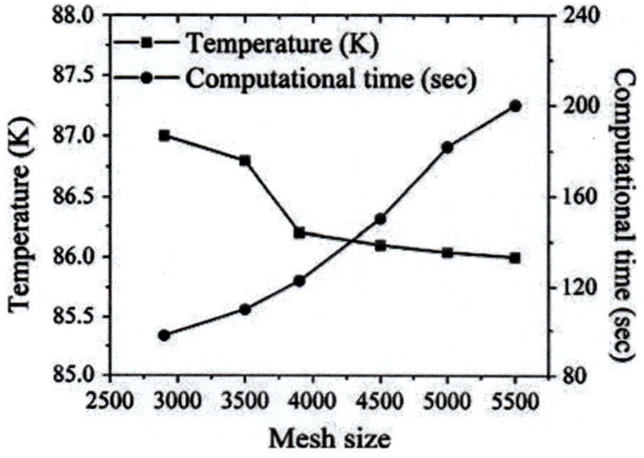

Fig. 3 Grid independence test results of IPTR

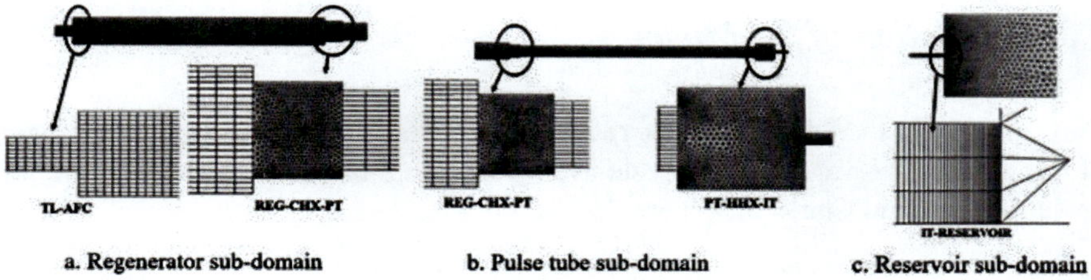

Fig. 4 Optimal grid size and connection among the node points in an IPTR

Axial and radial temperature distributions for a mixture of He and H_2 (50 + 50%) are presented in Fig. 5a. Temperature contour shown in Fig. 5a states that temperature

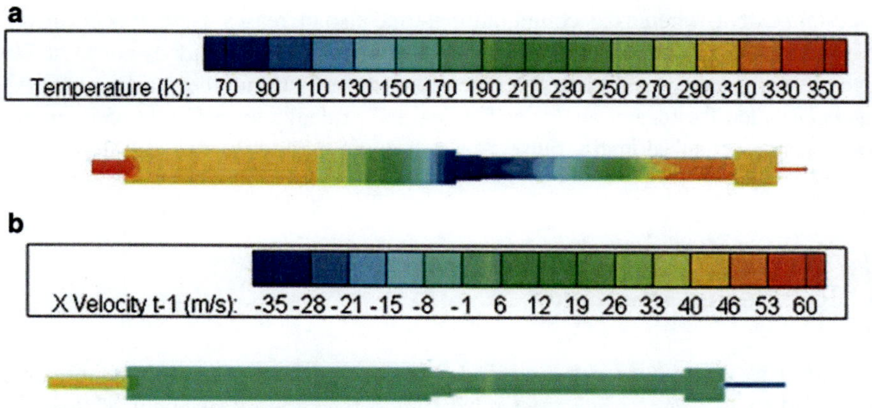

Fig. 5 **a** Temperature contour of IPTR (50% He + 50% H_2). **b**. Velocity contour of IPTR (50% He + 50% H_2)

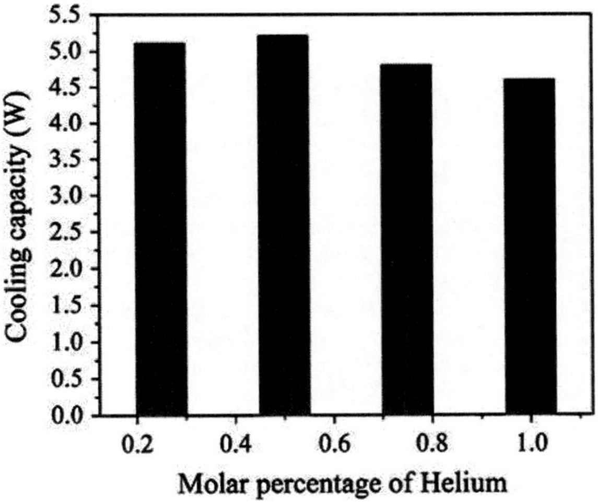

Fig. 6 Cooling capacity versus molar percentage of helium

fluctuation is almost straight along the regenerator, whereas it remains constant at the after cooler, cold and hot heat exchangers. However, there exists a radial temperature gradient inside the pulse tube. The results also explain that the temperature profile close to the cold heat exchanger and hot heat exchanger inside the pulse tube is nearly isothermal in nature, which obeys the isothermal model developed by Zhu et al. [5]. Velocity contour shows (in Fig. 5b) that velocity inside the tube lies in the range from −10 to 10 m/s, which states that the flow regime is laminar inside the pulse tube and the flow regime is plug flow. In the legend bar, which is shown for the velocity contour, the positive sign indicates that the fluid is flowing from the compressor to the reservoir and vice versa.

Effect on He and H_2 mixture on cooling capacity is presented in Fig. 6, and this figure shows that cooling capacity is maximum at about 50 + 50% mixture of hydrogen and helium. Further increase in hydrogen content (75% of hydrogen) in the mixture decreases the cooling capacity; these results are also analogous to that of Chen et al. [10].

4 Conclusion

Numerical simulation is performed using CFD code FLUENT® to study the influence of mixture of helium and hydrogen on refrigeration temperature and refrigeration capacity of IPTR. It is concluded that an increase in mole fraction of hydrogen on helium decreases the no-load refrigeration temperature up to certain percentage, and further increase in hydrogen composition increases refrigeration temperature thereafter. Similarly, increase in mole fraction of hydrogen than helium increases cooling capacity and becomes maximum at about 50% of hydrogen, and further increase in hydrogen content decreases the cooling capacity.

References

1. Radebaugh R (2009) Cryocoolers: the state of the art and recent developments. J Phys Condens Matter 21(16):164219
2. Gardner D, Swift G (1997) Use of inertance in orifice pulse tube refrigerators. Cryogenics 37(2):117–121
3. Razani A, Roberts T, Flake B (2007) A thermodynamic model based on exergy flow for analysis and optimization of pulse tube refrigerators. Cryogenics 47(3):166–173
4. De Waele A (2011) Basic operation of cryocoolers and related thermal machines. J Low Temp Phys 164(5–6):179
5. Zhu S, Chen Z (1994) Isothermal model of pulse tube refrigerator. Cryogenics 34(7):591–595
6. Wu X, Chen L, Liu X, Wang J, Zhou Y, Wang J (2019) An 80 mW/8 K high-frequency pulse tube refrigerator driven by only one linear compressor. Cryogenics 1017–1011
7. Cha J, Ghiaasiaan S, Desai P, Harvey J, Kirkconnell C (2006) Multi-dimensional flow effects in pulse tube refrigerators. Cryogenics 46(9):658–665
8. Rout SK, Choudhury BK, Sahoo RK, Sarangi SK (2014) Multi-objective parametric optimization of Inertance type pulse tube refrigerator using response surface methodology and non-dominated sorting genetic algorithm. Cryogenics 6271–6283
9. Panda D (2016) Mathematical modelling and design software for pulse tube cryocoolers. M.Tech (Res.) Thesis, NIT Rourkela
10. Chen G, Tang K, Huang Y, Gan Z, Bao R (2004) Refrigeration performance enhancement of pulse tube refrigerators with He–H2 mixtures and Er3NiHx regenerative material. Cryogenics 44(11):833–837
11. Banjare Y, Sahoo R, Sarangi S (2009) CFD simulation of a Gifford–McMahon type pulse tube refrigerator. Int J Therm Sci 48(12):2280–2287

Structural and Thermal Analysis of Cold-Head Cylinder of a GM Cryocooler

Debashis Panda, A. K. Satapathy, Sunil Kr Sarangi, and Ranjit K. Sahoo

Abstract To suffice the cryogenic cooling requirements for various academic and commercial applications, high refrigerating capacity GM cryocoolers have been used widely because of its compact size. Numerous innovative design approaches have been proposed and implemented to reduce the loss mechanisms to enhance its refrigeration capacity. In this paper, the structural and thermal effects of a cold-head cylinder are studied numerically. In place of a cylinder of constant diameter along the axial direction, three different types of stepped cylinder configurations have been used. Variations of von Mises stress, temperature, equivalent strain, maximum and minimum principal stress for different cylindrical configurations have been examined. It is seen that after reducing the cylinder thickness, total deformation along the radial direction of cylinder increases. Also, the von Mises stress increases, but its value is much lower than that of yield strength of SS-304. The results will be helpful for better design of cold-head cylinders for GM cryocooler from structural and thermal point of view to avoid structural failures.

Keywords GM cooler · Von mises stress · Radial deformation · Pumping loss

1 Introduction

Over the past 40 years, Gifford–McMahon (GM) coolers are most popularly used for cooling of momentous scientific equipment, because of its higher cooling capacity. Its major application includes liquefaction and storage of nitrogen, argon, helium, neon, etc. [1, 2] Additionally, many geometrical changes are also conducted on the existing GM coolers to enhance its refrigeration performance by reducing loss mechanisms [3]. Among them, Bao et al. [4] proposed that a certain quantity of refrigeration power loss occurs due to the axial conduction along the cylinder wall of cold head.

D. Panda (✉) · A. K. Satapathy · R. K. Sahoo
National Institute of Technology, Rourkela 769008, Odisha, India
e-mail: debashis1003@gmail.com

S. K. Sarangi
C.V. Raman College of Engineering, Bhubaneswar 751019, Odisha, India

M. Ramgopal et al. (eds.), *Advances in Air Conditioning and Refrigeration*, Lecture Notes in Mechanical Engineering, https://doi.org/10.1007/978-981-15-6360-7_7

To overcome this, they have proposed cylinder of stepped wall thickness along the axial direction in place of constant thickness cylinder. This reduction in the thickness reduces the conduction loss. In this paper, numerical simulation is performed to study the variations of temperature, von Mises stress, its equivalent strain, maximum and minimum principal stress along the cylinder walls for four different stepped cylinder configurations.

2 Numerical Analysis

The numerical analysis is conducted by using ANSYS workbench [5]. Static thermal and static structural workbench models of ANSYS have been opted to find out the variations of temperature, von Mises stress, radial deformation, maximum and minimum principal stresses of the cylinder.

2.1 Computational Domain

Figure 1 shows four different varieties of cylinder geometries for the cold-head cylinder of a single-stage GM cooler. Dimensions of benchmark model are taken from reference [6, 7]. Case 1 consists of a cylinder of uniform wall thickness along the axial direction. In case 2, the wall thickness is fixed at the warm end, and after a particular length, its thicknesses decreases to half of its original thickness, along the radial direction. The cylinder shown in case 3 consists of three different wall thickness, whereas in case 4, the wall thickness is constant at both warm and cold

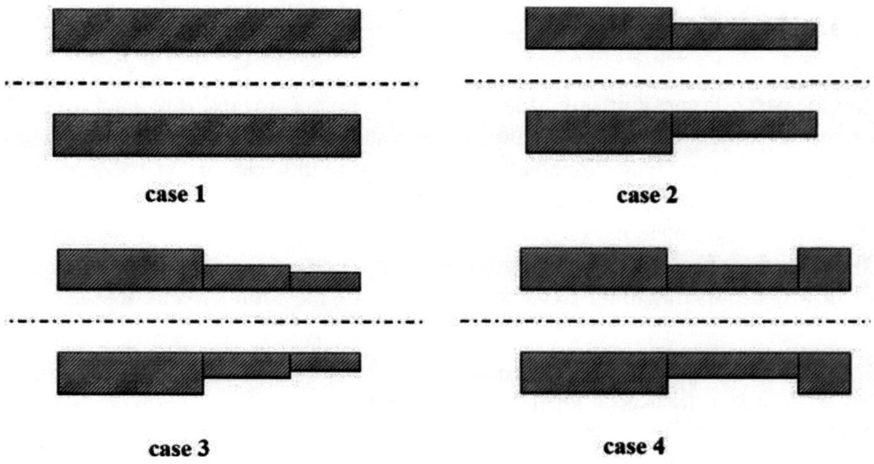

Fig. 1 Four different modified shapes of expander cylinder (dimensions are not to scale)

Table 1 Thickness and length of cold-head cylinders

Cases	Inner diameter (mm)	Thickness (mm)			Length (mm)		
		1st	2nd	3rd	1st	2nd	3rd
Case 1	82	3.25	–	–	192.5	–	–
Case 2	82	3.25	1.625	–	96.25	96.25	–
Case 3	82	3.25	2.75	2.25	65	65	62.5
Case 4	82	3.25	2.5	3.25	25	142.5	25

ends, while a different wall thickness is provided at the middle portion of cylinder. The cylinder wall thickness, its inner and outer diameters as well as length are presented in Table 1.

2.2 *Numerical Methodology*

The geometry and meshing are created by using the dimensions provided in Table 1 through ANSYS DM and ANSYS mesh, respectively (shown in Fig. 1). The governing equations for the structural problem are thus similar to that of given in reference [5]. During the thermal analysis, steady-state temperature values of 300 and 40 K are provided as boundary conditions at the warm and cold ends of the cylinder, respectively.

For the inner surface cylinder, a linear temperature variation is provided, whereas for the outer surface of the cylinder, an adiabatic boundary condition is adopted. Likewise, for the structural analysis, it is believed that both the top and bottom walls of the cylinder are fixed wall, and inside surface is subjected to a high pressure of 2.25 MPa. Temperature-dependent structural and thermal properties of SS-304 are provided due drastic variation of properties at the cryogenic temperature range. The solution is performed through a PC of a 3GHZ processor and 4 GB of RAM (Fig. 2).

3 Results and Discussion

3.1 *Results for the Case-1 Cylinder*

Figure 4 shows the changes in various thermo-mechanical parameters such as temperature, total deformation, von Mises stress, equivalent elastic strain, maximum and minimum principal stresses of a cylinder of uniform wall thickness (Fig. 3).

It is seen that almost a linear temperature variation exists along the axial length of the cold-head cylinder. Due to the adoption fixed-wall boundary conditions at both the ends, deformation is almost zero, whereas a maximum deformation of 0.00598 mm is

Fig. 2 Geometry of four different modified shapes of expander cylinder

seen along the radial direction of the cylinder. The maximum value of the von Mises stress is 56.56 MPa, which is much less than the yield stress value of 250 MPa. Thus, this cylinder could resist this high pressure without any structural failure. A maximum elastic deformation of 0.0001855 mm/mm is seen under those load conditions. The maximum and minimum values of principal stress are found to be 56.625 and 21.341 MPa under the above conditions.

3.2 Results for the Case 2 Cylinder

The results for case 2 cylinder is shown in Fig. 4. It is seen that the temperature variations along the cold-head cylinder follow a linear variation. But it is seen that the deformation in the cold part of cold-head cylinder (i.e., section with reduced wall thickness) is 0.012 mm, which is 51.66% greater than that of case 1 cylinder. It means by reducing the thickness of cylinder, its radial deformation increases. Since the value of von Mises stress is 70.072 MPa, which is less than that of its yield stress value, there will be no structural failure. But this value is more than that of previous value. Maximum value of the elastic strain is calculated as 0.00035. Maximum and minimum values of the principal stress are calculated as 92.148 MPa and 24.601 MPa, respectively.

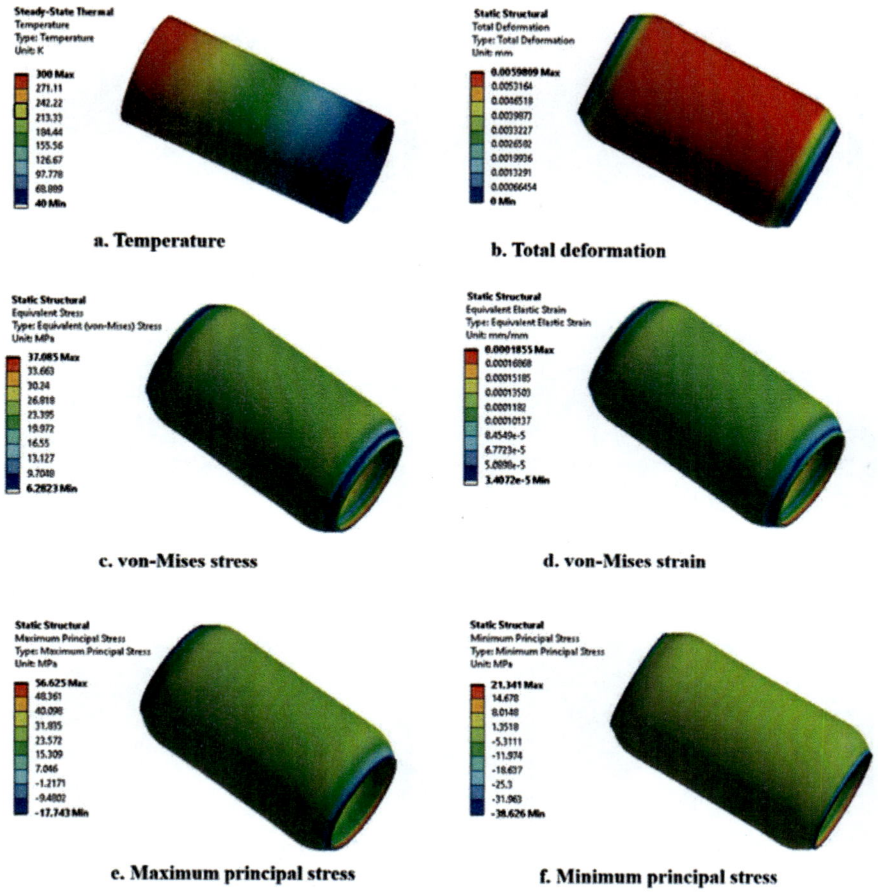

Fig. 3 Variation of structural and thermal parameters of case 1

3.3 Results for the Case 3 Cylinder

Figure 5 shows the results for temperature, von Mises stress, deformation, maximum and minimum principal stress variations. In this case, a linear temperature variations is also seen from the warm to cold end of the cylinder. Total radial deformation of 0.007339 mm is obtained at the cold end region of cylinder. Maximum and minimum principal stress values are calculated as 55.729 and 15.983 MPa. The value of von Mises stress is 39.439 MPa which is much less than the yield limit.

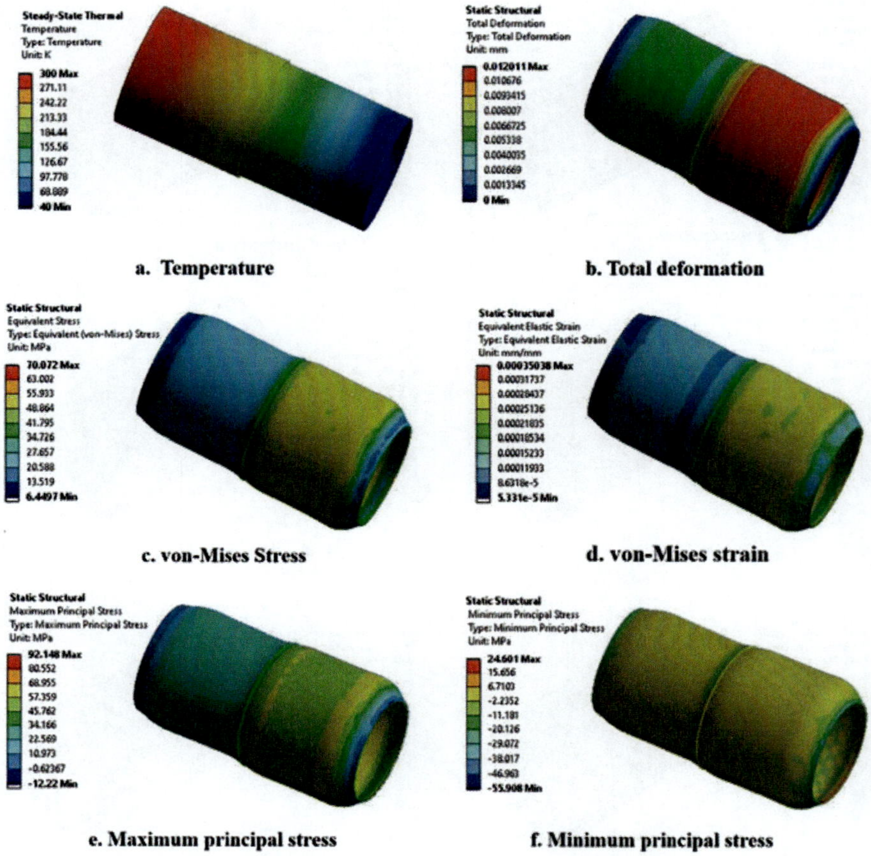

Fig. 4 Variation of structural and thermal parameters of case 2

3.4 Results for the Case 4 Cylinder

The results (such as temperature variation, radial deformation, von Mises stress, equivalent elastic strain, maximum and minimum principal stress) for the case 6 cylinder are shown in Fig. 6. It is clearly visualized that a maximum deformation of 0.006429 mm is obtained along the radial direction. Also, maximum value of von Mises stress is 39.642 MPa, which is quite lower than the yield stress limit.

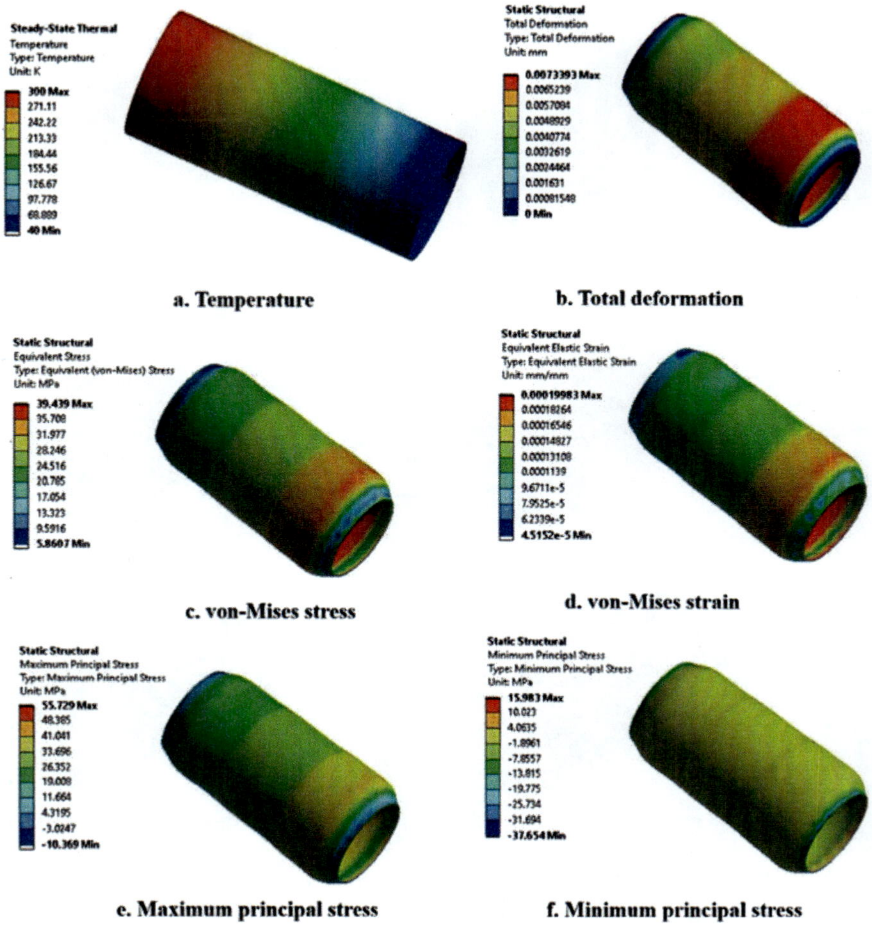

Fig. 5 Variation of structural and thermal parameters of case 3

4 Conclusion

Numerical simulation is performed using ANSYS structural and thermal work benches to study the effect of variations of various structural and thermal parameters on a cylinder of uniform thickness and cylinders of variable wall thickness. The variations of temperature, stress and strains along the axial length for various cylindrical configurations have been studied. It is seen that:

- Using a cylinder of variable wall thickness in place of uniform wall thickness reduces the heat loss.
- By decreasing the thickness von Mises stress, equivalent elastic strain and radial deflection increase.

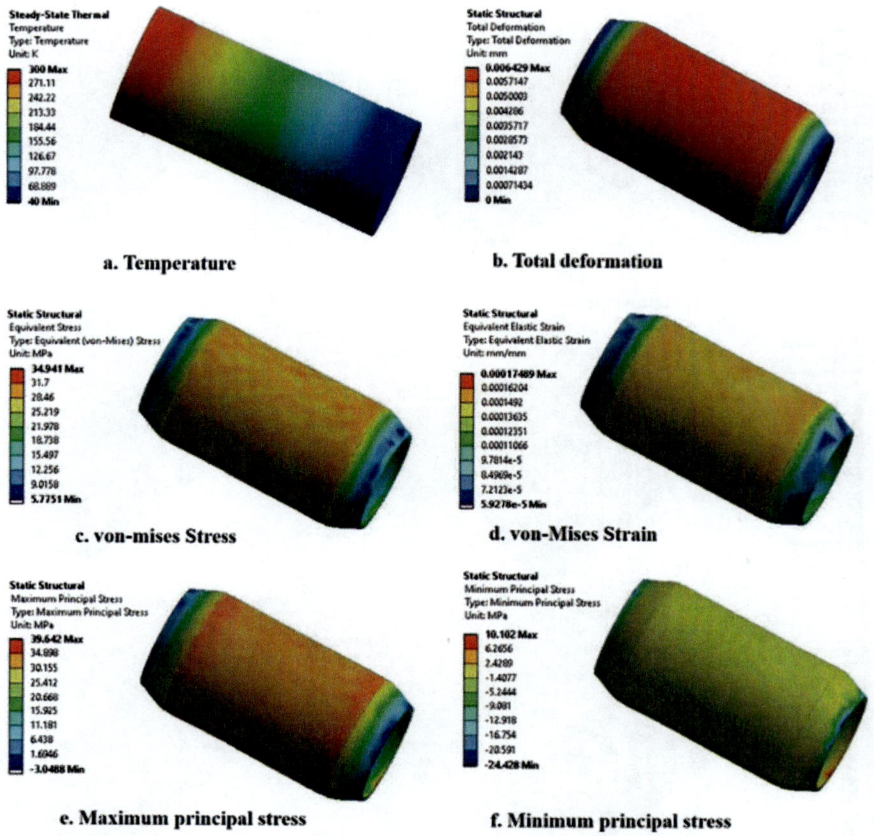

Fig. 6 Variation of structural and thermal parameters of case 4

References

1. Radebaugh R (2009) Cryocoolers: the state of the art and recent developments. J Phys: Condens Matter 21(16):164219
2. De Waele A (2011) Basic operation of cryocoolers and related thermal machines. J Low Temp Phys 164(5–6):179
3. Bao Q, Xu M, Tsuchiya A, Li R (2015) Recent development status of compact 2 K GM cryocoolers. IOP Conference Series: Materials Science and Engineering. IOP Publishing, p 012136
4. Xu M, Bao Q, Tsuchiya A, Li R (2015) Development of compact 2 K GM cryocoolers. Phys Proc 67491-6
5. Workbench A. V13. 0 User's manual. Ansys Inc
6. Xu M, Morie T (2012) Numerical simulation of 4 K GM cryocooler. Cryocoolers 17253-9
7. SHI cryogenics group, cryocooler product catalogue http://www.shicryogenics.com/products/4kcryocoolers/rdk-408d2-4k-cryocooler-series/ (Last accessed 19.12.2019)

CFD Analysis to Envisage the Fluid Flow Inside a Turboexpander Operating at Cryogenic Temperature

Manoj Kumar, Ranjit K. Sahoo, Debashis Panda, and Suraj Kr Behera

Abstract The one-dimensional design methodology of radial turbine blade profile has a substantial role in the advancement of an efficient liquefaction cycle (cryogenic fluids) because of growing demand in research and various industrial applications. The main part of the current study is to obtain an optimum design of radial turboexpander. The CFD analysis of a turboexpander is carried out to characterize the flow filed inside it. The initial blade profile is generated using ANSYS Blade-Gen which is further modified based on CFD analysis. The pressure, temperature, velocity, static enthalpy and entropy at various cross sections are reported.

Keywords Turboexpander · Cryogenics · Numerical analysis

1 Introduction

The generation of ultra-low temperature using expander unit is a significant milestone achieved in cryogenic history. Primarily, the cryogenic temperature or liquefaction of gases is produced through step-by-step development, starting from Siemens and Kirk's engine acting as an expander to throttle valve based on the Joule–Thomson effect. In 1896, K. Onnes suggests the theory for using a radial turbine in the liquefaction process of gas. Claude implements radial turbine on behalf of an expansion device in his experiment for air liquefaction-based cycles. It is a modified form of Linde–Hampson cycle, later on, renowned as a Claude cycle, which has higher system efficiency. The operational cost of the Claude cycle-based liquefaction cycles is lower due to relatively less power consumption as compared to that of the systems based on the Linde–Hampson cycle. Thereafter, the research is focused on the optimal design of cryogenic turboexpander system. In the present scenario, researchers are interested in the optimal design of cryogenic turboexpander as a replacement of Joule–Thomson valves to save the energy [1].

M. Kumar (✉) · R. K. Sahoo · D. Panda · S. K. Behera
National Institute of Technology, Rourkela 769008, Odisha, India
e-mail: manojbeg526@gmail.com

M. Ramgopal et al. (eds.), *Advances in Air Conditioning and Refrigeration*, Lecture Notes in Mechanical Engineering, https://doi.org/10.1007/978-981-15-6360-7_8

Table 1 Turboexpander design data

Inlet total pressure	8 bar
Total temperature	150 K
Turbine diameter	24.86 mm
Number of turbine blades	13
Turbine inlet blade height	1.36 mm
Pressure ratio	3.86
Turbine axial length	10.52 mm
Number of nozzles	17

Kumar et al. [2–6] conduct the computational fluid dynamic (CFD) and experimental analysis of helium and nitrogen turboexpander. Harinck et al. [7] implement the three-dimensional (3-D) CFD analysis of a nozzle, radial expansion turbine and diffuser used in organic Rankine cycle (ORC) cycle. Pasquale et al. [8] focus on to optimize the blade profile using various optimization methods. Chang et al. [9] suggest the two-stage helium refrigerator based on reversed Brayton cycle in which two radial turbines are used. Kumar et al. [10, 11] design and develop an experimental investigation and artificial intelligence techniques to predict the performance of a cryogenic turboexpander. The design procedure of nozzle is also explained which has substantial use in cryogenic applications [12]. The study is extended to conduct the mean line design and CFD analysis of helium turboexpander at different inlet conditions [13, 14].

In this study, the mean line design procedure of a radial expansion turbine, nozzle and diffuser is conducted. Thereafter, CFD investigation is performed to determine the flow behaviour of nitrogen fluid at cryogenic temperature. Firstly, the turbine blade profile is acquired from Blade-Gen using the operational conditions as stated in Table 1. The high-resolution computational mesh is generated in ICEM followed by the 3-D CFD analysis implemented to envisage the flow and thermal analysis using ANSYS CFX® computational tool. The variation of flow (pressure, Mach number, velocity) and thermal (temperature, static enthalpy and entropy contours) parameters is obtained at various cross sections along the axial distance of the turboexpander.

1.1 Computational Domain and Numerical Methodology

The Reynolds-averaged Navier–Stokes (RANS) equation is solved to analyse the fluid flow inside the turboexpander which is established for finite volume method. The governing equation terms are discretized using second-order backward Euler scheme. Figure 1 presents the three-dimensional computational model and its mesh.

Total pressure and total temperature have opted at the inlet boundary (nozzle). The mass flow rate is imposed at the diffuser outlet. The wall is assumed to be adiabatic and no slip. Table 2 presents the other relevant boundary conditions.

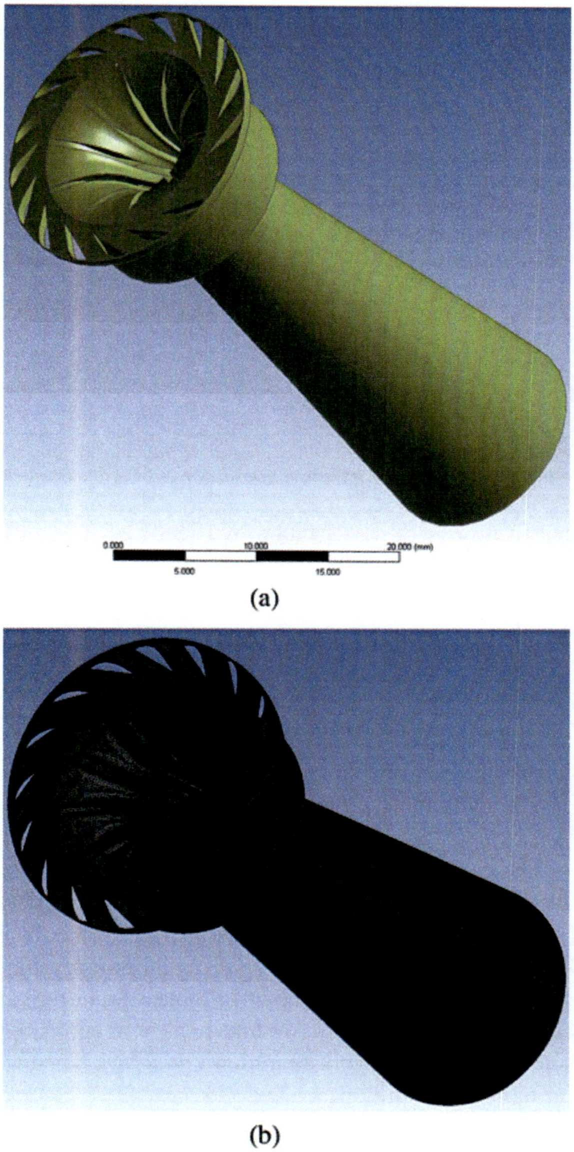

Fig. 1 **a** 3-D model and **b** computational mesh

Table 2 Boundary conditions

Inlet total pressure	8 bar
Inlet total temperature	150 K
Rotational speed	119, 196 rpm
Mass flow rate	0.05 kg/s
Turbulence kinetic energy	5%
Walls	Adiabatic, no-slip

The governing differential equations are:

Continuity equation:

$$\frac{\partial \rho}{\partial t} + \nabla.(\rho U) = 0 \tag{1}$$

Momentum equation:

$$\frac{\partial(\rho U)}{\partial t} + \nabla.(\rho U \otimes U) = -\nabla p + \nabla.\tau + S_{\mathrm{M}} \tag{2}$$

Energy equation:

$$\frac{\partial(\rho h_{\mathrm{tot}})}{\partial t} - \frac{\partial p}{\partial t} + \nabla.(\rho U h_{\mathrm{tot}}) = \nabla.(\lambda \nabla T) + \nabla.(U.\tau) + U.S_{\mathrm{M}} + S_{\mathrm{E}} \tag{3}$$

The $k - \omega$ SST transport equations are as follows (Eq. 4–5) [5]:

$$\frac{\partial}{\partial t}(\rho k) + \frac{\partial}{\partial x_i}(\rho k u_i) = \frac{\partial}{\partial x_j}\left(\Gamma_k \frac{\partial k}{\partial x_j}\right) + G_k - Y_k + S_k \tag{4}$$

$$\frac{\partial}{\partial t}(\rho \omega) + \frac{\partial}{\partial x_i}(\rho k \omega u_i) = \frac{\partial}{\partial x_j}\left(\Gamma_k \frac{\partial \omega}{\partial x_j}\right) + G_\omega - Y_\omega + S_\omega \tag{5}$$

The additional details of governing equations are presented in our previous studies [13, 14].

The convergence criteria are 10^{-6} for CFD analysis which takes approximately 140 h to obtain a converged solution on Dell workstation (5810).

The grid independence test is also performed for reliability of the numerical simulation. In this regard, three different grid resolutions are used to discretize the computational domain as shown in Table 3. Since the variation of isentropic efficiency between second and third row are very less as compared to time taken for a converged solution on a Dell workstation 64 GB RAM. Therefore, second row is considered for numerical solution.

Table 3 Grid independence test

No. of nodes (millions)	Isentropic efficiency	CPU time (h)
1.61	0.72	43
1.72	0.75	49
1.89	0.76	57

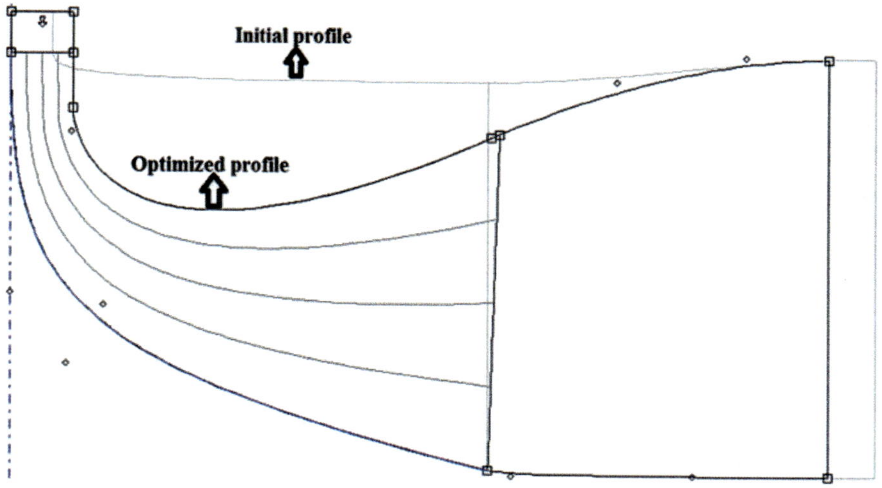

Fig. 2 Initial and optimized meridional profile

1.2 Results and Discussion

Figure 2 presents the final optimized meridional profile of radial turbine which is obtained by reducing the high-pressure regions. The fluid flow and their thermal behaviour are very critical at cryogenic temperature because of variation in density and viscosity of the fluid. The temperature drop is the primary aim of the cryogenic turboexpander. Figure 3 represents the pressure and temperature contour at different cross sections of the nozzle and turbine. It is noticed that the decrease in temperature is approximately 26 K for 8 bar and 150 K inlet pressure and temperature, respectively. The temperature drop is obtained because of static enthalpy drop which finally produces the cooling capacity of the turboexpander. Figure 4a presents the enthalpy contours of the turboexpander at different cross sections along the axial distance. It is noticed that after $z = 2$ mm, the enthalpy drop reduces in a more significant way. It happens due to higher pressure drop. Also, the entropy increases in these regions. It is found that the rate of entropy generation is higher inside the turbine due to rotation of the blades which finally provides the higher pressure drop in these regions which is clearly observed in Fig. 4b. These entropy generations are also obtained because the secondary losses and vortex formation in between the blade passages are obtained from CFD analysis of the initial profile obtained from Blade-Gen. Thereafter, vortex formation, high-pressure and separation regions are modified to get the optimized blade profile (Fig. 2) for which numerical simulations are performed and presented hereafter.

Figure 5 represents the Mach number and velocity distribution at different planes. The highest Mach number is obtained near the turbine inlet which is in subsonic zone (0.72) due to which the fluid achieves the maximum velocity (168 m/s).

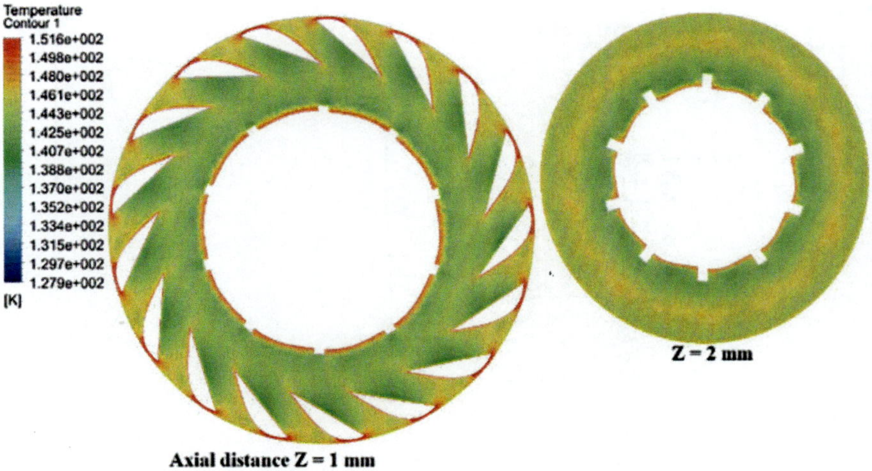

(a)

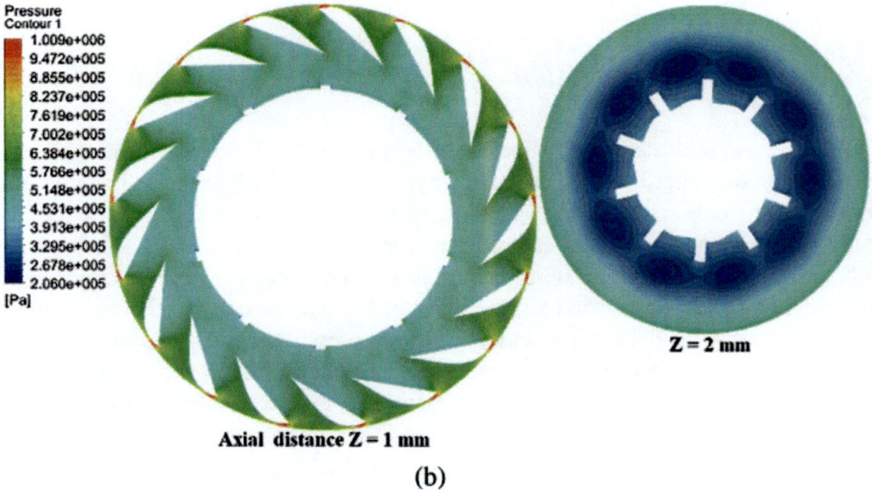

(b)

Fig. 3 Temperature and pressure distribution at various locations

2 Conclusions

This study presents the mean line design and CFD analysis of a turboexpander to envisage the fluid flow characteristics at different cross sections along the axial distance. The initial turbine blade profile is created using Blade-Gen which is further modified based on the numerical results. The variation of pressure, temperature, Mach number, etc., is determined at various axial distances of the turboexpander. It is noticed that the highest temperature drop of 26 K takes place at a pressure ratio of 3.86 which is a significant amount for a cryogenic process. The Mach number (0.72)

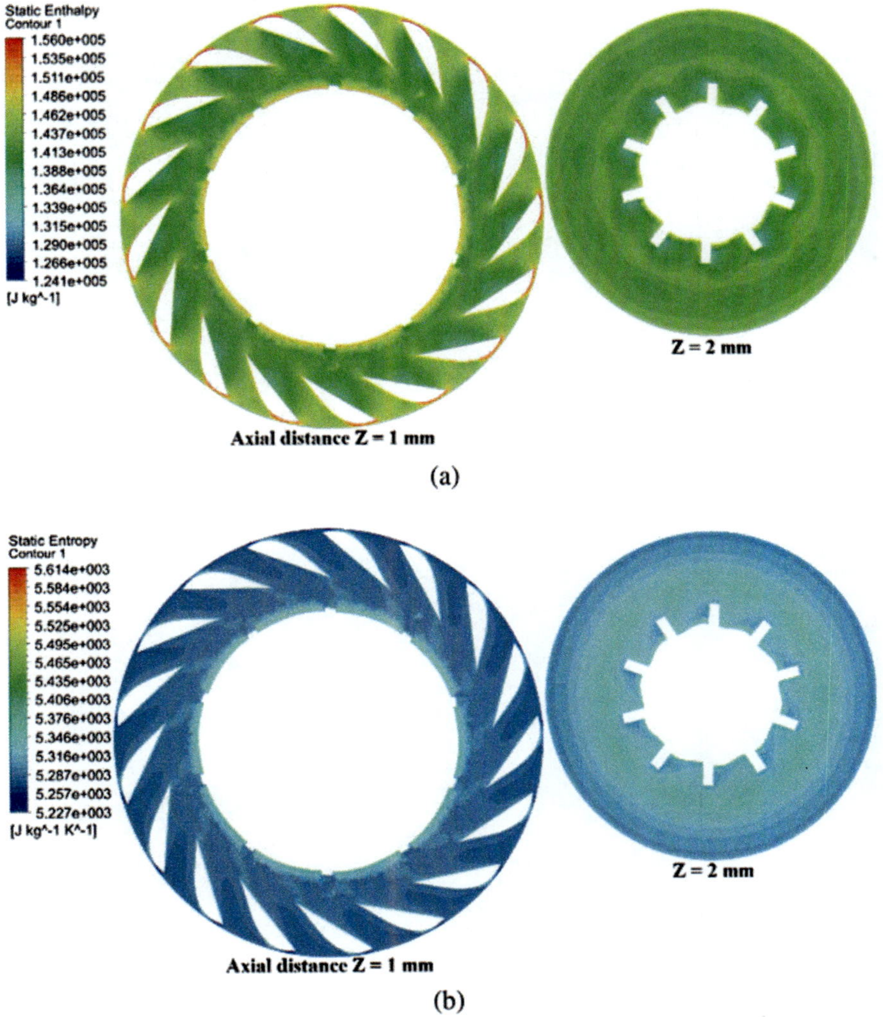

Fig. 4 Static enthalpy and static entropy distribution at various locations

contours show that the fluid flow is in the subsonic zone throughout the turboexpander which is an essential condition for adequate design of such type of systems.

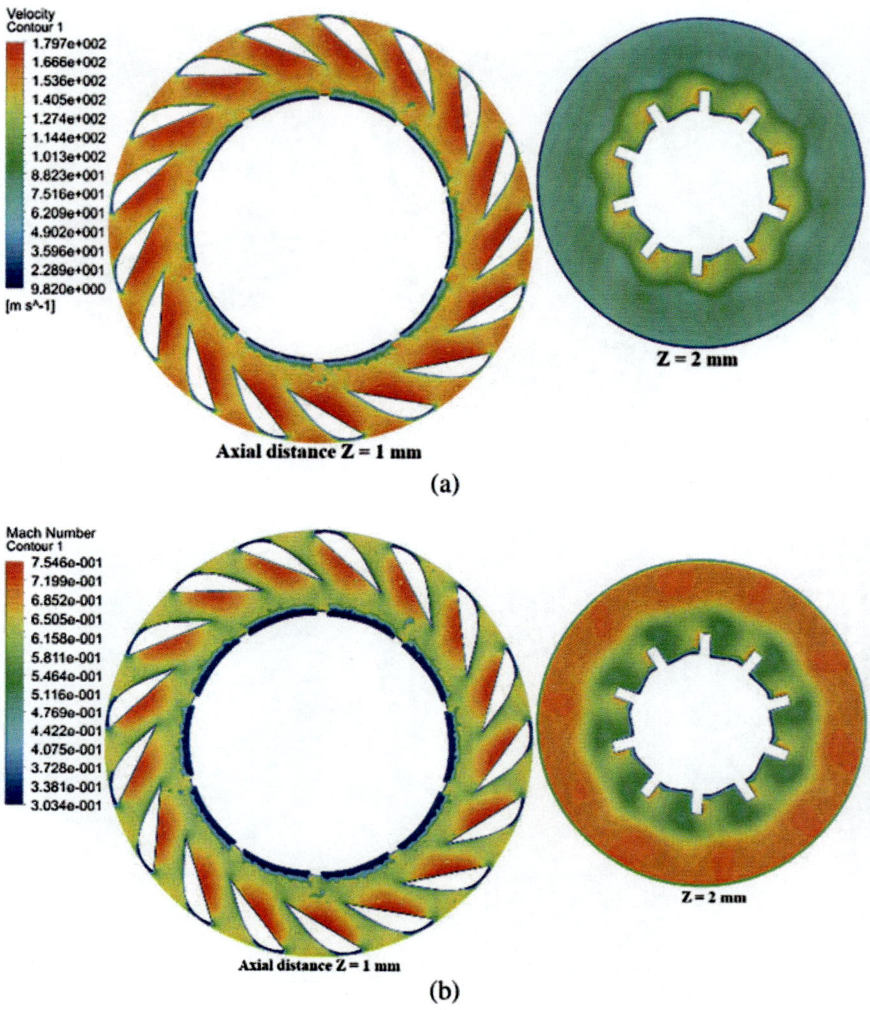

Fig. 5 Velocity and Mach number distribution at various locations

References

1. Chiu C-H, Kimmel HE (2001) Turbo-expander technology development for LNG plants. LNG-13 Conference Seoul, Korea, May 2001
2. Kumar M, Behera SK, Kumar A, Sahoo RK (2019) Numerical and experimental investigation to visualize the fluid flow and thermal characteristics of a cryogenic turboexpander. Energy 189:116267
3. Kumar M, Kumar P, Sahoo RK (2019) Design and numerical analysis to visualize the fluid flow pattern inside cryogenic radial turbine. Advances in fluid and thermal engineering. Springer, Singapore, pp 179–187
4. Kumar M, Sahoo RK (2017) Development and numerical analysis to visualise the flow pattern of cryogenic radial turbine for helium gas. ISHMT Digital Library. Begel House Inc.

5. Kumar M, Sahoo RK, Behera SK (2019) Design and numerical investigation to visualize the fluid flow and thermal characteristics of non-axisymmetric convergent nozzle. Eng Sci Technol Int J 22(1):294–312
6. Manoj K, Panchal RR, Sahoo RK (2018) Numerical investigation to visualize the fluid flow characteristics of cryogenic radial turbine. In: Proceedings of the 7th International and 45th National Conference on Fluid Mechanics and Fluid Power (FMFP) December 10–12, 2018, IIT Bombay, Mumbai, India, IIT Bombay, Mumbai, India
7. Harinck J et al (2013) Performance improvement of a radial organic Rankine cycle turbine by means of automated computational fluid dynamic design. Proc Inst Mech Eng Part A: J Power Energy 227(6): 637–645
8. Pasquale D, Ghidoni A, Rebay S (2013) Shape optimization of an organic rankine cycle radial turbine nozzle. J Eng Gas Turbines Power 135(4):042308
9. Chang H-M, Ryu KN, Baik JH (2018) Thermodynamic design of hydrogen liquefaction systems with helium or neon Brayton refrigerator. Cryogenics 91: 68–76
10. Kumar M, Panda D, Behera SK, Sahoo RK (2019) Experimental investigation and performance prediction of a cryogenic turboexpander using artificial intelligence techniques. Appl Therm Eng 114273
11. Kumar M, Panda D, Sahoo RK, Behera SK (2019) Performance prediction, numerical and experimental investigation to characterize the flow field and thermal behavior of a cryogenic turboexpander. Heat Mass Transf 1–22
12. Kumar M, Biswal R, Behera SK, Sahoo RK (2019) Design and numerical investigation to predict the flow pattern of non-axisymmetric convergent nozzle: a component of turboexpander. J Traf Transp Eng 7:264–281
13. Kumar M, Panda D, Sahoo RK, Behera SK (2019) Preliminary design, flow field, and thermal performance analysis of a helium turboexpander: a numerical approach. SN Appl Sci 1(11):1482
14. Kumar M, Panda D, Kumar A, Sahoo RK, Behera SK (2019) A methodology for the performance prediction: flow field and thermal analysis of a helium turboexpander. J Brazilian Soc Mech Sci Eng 41(11):484

Analysis of Thermal Efficiency of Solar Flat Plate Collector Using Twisted Tape

Arun K. Behura, Ashwini Kumar, V. C. Todkari, Gaurav Dwivedi, and Hemant K. Gupta

Abstract Solar energy is clean, renewable, abundant and free. The light and heat energy can be harnessed by using several technologies such as solar heating, photovoltaic cell and photosynthesis. Solar collectors are used for collecting solar thermal energy by heating and reheating a stream of water. This paper focuses on improving performance of solar flat plate collector using twisted tape. Flat plate collectors have a wide range of usage in industrial, public as well as residential sectors. Solar water heaters are environment-friendly, cheap to operate and need very little when compared with other solar applications. Many researchers are working to enhance the thermal efficiency of solar water heaters. The thermal efficiency enhancement can be done by modification of design of parts of solar flat plate collector such as absorber plate, STs, working fluid, thermal insulation, efficiency of fin and selective coating of an absorber plate. Use of twisted tape is an efficient method to increase the thermal efficiency by increasing the heat transfer by increasing the contact surface area between the water and metal tubes (inserted tape tubes) and also by promoting turbulence of flow.

Keywords Solar flat plate collector · Heat transfer coefficient · Twisted tapes · Friction factor · Pressure drop

A. K. Behura
School of Mechanical Engineering, VIT Vellore, Vellore, Tamil Nadu, India

A. Kumar · V. C. Todkari
Department of Mechanical Engineering, H.S.B.P.V.T's GOI, College of Engineering—Parikrama, Kashti, Ahmednagar, Maharashtra, India

G. Dwivedi
Energy Centre, MNIT, Bhopal, India

H. K. Gupta (✉)
Department of Mechanical Engineering, SRICT, Vataria, Bharuch, Gujarat, India
e-mail: hemant.gupta@srict.in

M. Ramgopal et al. (eds.), *Advances in Air Conditioning and Refrigeration*, Lecture Notes in Mechanical Engineering, https://doi.org/10.1007/978-981-15-6360-7_9

Nomenclature

TT	twisted tape
ST	storage tank
T	temperature
A	area
D	diameter of an absorber tube
N	number of tubes
M	mass of water
C_p	specific heat
S	solar flux

1 Introduction

Solar energy is free and clean energy, and it is available in abundant quantity in India. Solar water heater (SWH) based on flat plate collector is a useful technique to use this clean energy in an effective way. A method of enhancing the efficiency of the solar flat plate collector is insertion of twisted tape on the fluid flow path. The enhanced surface area and enhanced turbulence lead to increased heat transfer and thus higher efficiency of the water heating system. In plain tube arrangement, heat transfer is quite low. Insertion of TTs gives better results in heat transfer. Most of the heat transfer phenomena are associated with air or gas flow through pipes, which require more efficient heat transfer. Use of artificial roughness of different geometries has been extensively investigated to enhance the heat transfer rate in pipes and ducts. The recent works of Ashwini et al. [1], for three sides' artificial roughened solar air heater, have been reported for such enhancement in heat transfer coefficient. Many more investigations have been conducted to show the enhancement of heat transfer coefficient by using different methodologies, e.g. use of external vibration [2], use of booster mirror, using transverse wires as roughness element, wire screen and reviews on methodology for such enhancement of heat transfer coefficient. Performance can be increased using different techniques such as surface extension, modification, coating, additives and insertion. Insertion of twisted tapes provides better results in heat transfer [3]. According to Raja and Suresh [4], twisted tape has uniform twist ratio along the length and at $y/w = 3$ effectiveness with twisted tape is found to be more than others. Kumar and Prasad [5] focused on heat transfer enhancement by using twisted tapes in the twist ratio between 3 and 12 which results in increment in 10–50% rise in heat transfer. Jaisankar et al. [6] investigated that enhancement of heat transfer rate when Reynolds number was set in between 3000 and 23,000 and TT of different ratios TWIST RATIO = Y/W was conducted in which the ratio is 3 and pressure loss was found to be less than other cases. During the 3 experimentation in which left-right, straight helical screw and plain tube was conducted and in which left-right helical screw turns out to maximum heat transfer rate resulted in better results by Jaisankar et al. [7]. During the experimentation in which twofold whole spaced Twisted Tape and twofold regular spaced Twisted Tape was conducted in which whole spaced tape resulted in more heat transfer investigated by Sivashanmugam and Nagarajan [8]. Study and research of non-conventional energy have become most significant after the Kyoto protocol by Shukla et al. [9]. Water heating through solar clean energy is very good technology to convert solar clean energy directly into thermal useful energy. It has been recognized and is being commercialized among the industries as well as residential purposes. Performance of solar water heater will be increased by methods like geometrical alteration on the absorber plate, use of nano-fluids and selective coats. Twisted tape can be used in heat exchange as well as solar water heater to improve convection.

2 Experimental Set-up and Methodology

To see the effect of vibration on heat transfer coefficient in forced convection, an experimental investigation was made. Figure 1 represents complete set-up of flat plate collector on which experiments are conducted. All the tubes are filled with water for measuring the water flow rate and temperature readings using digital thermometer in regular interval and determine the thermal efficiency. Insert twisted tape inside

Fig. 1 Complete set-up of flat plate collector

Fig. 2 Flat plate collector

Fig. 3 Aluminium TTs

the tubes, and repeat the experiments for determining the thermal efficiency and compared well.

Figure 2 shows the flat plate collector available for experimentation. Figure 3 represents the twisted tapes made up of aluminium of 1-mm thickness. Figures 4 and 5 represent pyranometer and pyranometer reader which measures the solar radiation.

Fig. 4 Pyranometer

Fig. 5 Pyranometer reading

3 Results and Discussion

Tables 1 and 2 show the readings of flat plate collector without inserting twisted tapes and with inserting twisted tapes, respectively. Figure 6 shows the comparison between the efficiencies of before and after the insertion of twisted tapes. Figure 7 represents the variation of solar flux with respect to time during the day.

Determine the efficiency of FPC without inserting twisted tapes

- Area (A) = 2 m * 1 m = 2 m^2
- Mass of water (M) is in L/h
- T = degree Celsius
- Specific heat enthalpy (C_p) = 1005
- Solar flux (S) in W/m^2
- Pyranometer reading in mv
- Efficiency in % = $M * C_p * (T2 - T1)/S * A$
  Efficiency = 16%

Determine the efficiency of FPC with inserting TTs.

- Area (A) = 2 m * 1 m = 2 m^2
- Mass of water (M) is in L/h
- T = degree Celsius
- Specific heat enthalpy (C_p) = 1005
- Solar flux (S) in W/m^2
- Pyranometer reading in mv
- Efficiency in % = $M * C_p * (T2 - T1)/S * A$
- Efficiency = 22%.

4 Conclusions

On the basis of above results and discussion, the following measure conclusions have been found:

- In the plain tube condition, heat transfer is low when compared to twisted tape insert. We can insert a twisted tape, so that the contact surface area increases and convection occurs faster.
- As we know heat flows faster in metal–metal and then metal–water, conduction is faster than convection.
- Insertion of twisted tapes gives better results in heat transfer increasing overall efficiency by 5%.

Table 1 Readings without inserting the twisted tapes

Sl. No.	Time	Mass of water	T1 (inlet)	T2 (outlet)	T3 (glass)	T4 (ambient)	Solar flux	Efficiency	T (absorber)	Pyranometer reading
1	12:35	54	32.4	58.2	50	36	831.52	14.5	49.6	11.7
2	12:40	54	32.5	51.2	53	36.3	847.45	14.8	50.6	11.8
3	12:45	54	32.8	52.2	59.1	38.8	870.56	16.56	50.7	11.2
4	13:00	76.5	33.5	48.2	57.4	39.98	868.5	16.63	47.4	10.8
5	13:05	76.5	33.3	48	58.6	40	867.8	16.93	48.4	10.9
6	13:10	76.5	33.5	49.5	60.6	39.8	866.56	16.95	48.9	11.1

Table 2 Readings with inserting the twisted tapes

Sl. No.	Time	Mass of water	T1 (inlet)	T2 (outlet)	T3 (glass)	T4 (ambient)	Solar flux	Efficiency	T (absorber)	Pyranometer reading
1	12:35	54	33.4	59.5	50.2	34	825.42	22.6	59.8	9.6
2	12:40	54	33.5	54.3	52.6	34.2	830.33	22.73	60.7	9.5
3	12:45	54	33.6	53.6	58.7	35.6	835.61	22.8	60.8	9.7
4	13:00	76.5	33.8	49.6	53.7	34.9	839.72	22.85	57.7	9.4
5	13:05	76.5	33.9	51	55.6	36	840.5	22.88	58.8	9.7
6	13:10	76.5	33.8	51.8	58.3	36.7	845.53	22.89	59.8	9.8

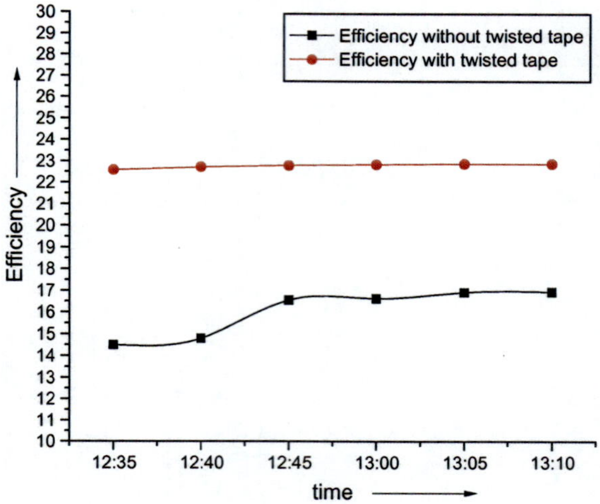

Fig. 6 Efficiency comparison of flat plate collector with and without inserting twisted tape

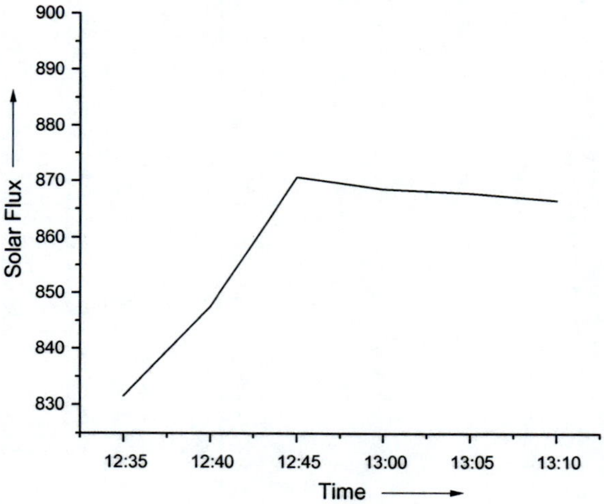

Fig. 7 Variation of solar flux with respect to time

References

1. Kumar A (2018) Optimal thermohydraulic performance in three sides artificially roughened solar air heaters, PhD Thesis, NIT Jamshedpur
2. Tusi Y (1953) The effect of vibration on heat transfer coefficient, PhD Thesis, Ohio State University
3. Hiware RT, Kongre S (2016) Performance improvementtechniquesin solar flat plate collector: a review. IJARIIE 2:3969–3974
4. Raja M, Suresh Israve R (2016) Thermal augmentation in flat plate collector solar water heater using modified twisted tape. Int J Appl Eng Res 11:695–700
5. Kumar A, Prasad BN (2000) Investigation of twisted tape inserted solar water heaters heat transfer, friction factor and thermal performance results. Renew Energy 19:379–398

6. Jaisankar S, Radhakrishnan TK, Sheeba KN (2009) Experimental studies on heat transfer and friction factor characteristics of forced circulation solar water heater system fitted with helical twisted tapes. Solar Energy 83:1943–1952
7. Jaisankar S, Radhakrishnan TK, Sheeb KN (2009) Experimental studies on heat transfer and friction factor characteristics of thermos-syphon solar water heater system fitted with spacer at the trailing edgeof twisted tapes. Appl Therm Eng Int J Eng Technol Manage Appl Sci 29:1224–1231
8. Sivashanmugam PK Nagarajan (2007) Studies on heat transfer and friction factor characteristics of laminar flow through a circular tube fitted with right and left helical screw-tape inserts. Exp Thermal Fluid Sci 32:192–197
9. Ruchi Shukla K, Sumathy Phillip Erickson, Gong Jiawei (2013) Recent advances in the solar water heating systems: a review. Renew Sustain Energy Rev 19:173–190
10. Kumar A, Kumar A (2018) Performance characteristics of solar air heaters by using CFD approach. Scholar's Press, European Union, ISBN: ISBN 978-620-2-31166-3

Performance Enhancement of Domestic Refrigeration System Using R-134a Refrigerant Blended with Graphene as Nano Additives

Amar Kumar Das and Ritesh Mohanty

Abstract Refrigeration system involved with evaporative heat transfer intends to intensify the cooling effect in domestic applications. In order to reduce the energy consumption and improve the cooling rate, various attempts have been undertaken. In this experiment, the exploitation of a refrigerating system using R-134a as refrigerant has been studied. The study also explains the thermal conductivity, dynamic viscosity and rate of heat transfer of graphene used as nano additives in evaporator tube of a vapour compression refrigeration (VCR) system. The thermal conductivity of base refrigerant rises with rise of temperature to the optimal concentration value of 0.6–0.9% by wt. in nano additives. In addition, significant improvement in the performance of refrigeration system due to the addition of nano additives in domestic refrigerant has been addressed. However, the apprehension of global warming due to potential use of R-134a as refrigerant is still a matter of concern. Hence, the paper encourages the use of nano refrigerant that leads to improvise the effectiveness, durability and energy efficiency of refrigeration without any system modification.

Keywords Refrigeration system · Nanofluid · VCR · NanoGraphene oxide

1 Introduction

Modern world undergoing a phase of liberization, urbanization and industrializations is confronted with challenges for energy security and sustainability. In order to regulate the energy loss, stringent energy policies have been implemented by governments to put a ban on energy wastes that leads to reduce energy consumption [1]. To meet such energy requirements, improvisation in performance and energy efficiency

A. K. Das (✉)
Department of Mechanical Engineering, Gandhi Institute for Technology, Bhubaneswar, India
e-mail: amar.das@gift.edu.in

R. Mohanty
Department of Mechanical Engineering, DRIEMS Polytechnic, Tangi, Cuttack, India
e-mail: ritesh.rmohanty@gmail.com

M. Ramgopal et al. (eds.), *Advances in Air Conditioning and Refrigeration*, Lecture Notes in Mechanical Engineering, https://doi.org/10.1007/978-981-15-6360-7_10

of a refrigeration system becomes a matter of concern. Extensive inspection has been supervised to scrutinize the thermal conductivity of the nanofluids. However, there is least scope of literature to explain thermal conductivity of nano refrigerant and its improvement methods [2]. Thermophysical properties of base fluid can be increased by accumulating low concentration of nanoparticles in it. This may be due to higher dispersion quality of nanoparticles that leads to augment the thermal conductivity of the refrigerant. Nanoparticles have been used as supplements in refrigerants and being accepted as a promising method of improvising the performance of the vapor compression refrigeration system (VCRS) without modifying the system components. It has been mentioned that nano refrigerant having higher thermal conductivity due to better dispersion behavior and stability can be used significantly in the refrigeration system [3].

Studies on potential properties like thermal conductivity and viscosity of nano refrigerants in comparison with conventional refrigerant used in refrigeration system have been illustrated. Various researches on nano additives encourage its use as nano refrigerants due to its promising tribological performance of compressor in a refrigeration system. According to Selvam and his colleague [4], thermal conductivity of nano refrigerant has been improved using silver nanofluid in ethylene glycol in water as base fluid. However, the thermal conductivity is found increasing with the increase in the concentration of nanoparticles and temperature. Akilu et al. [5] investigated the use of TiO_2–CuO nano composite in ethylene glycol as base fluid. The trends of increasing thermal conductivity by increase in nanoparticles volume concentration and temperature have been followed. It has been observed that an enrichment of 16.7 and 80% was gained at 2.0% volume concentration compared to base ethylene glycol at 40.4 °C. Akhavan et al. [6] discussed on the implementation of different mass fraction of CuO nanoparticles as nano refrigerant compared to R600a as base fluid. They have reported that the heat transfer of the nano refrigerant has been improved around 83% than the pure refrigerant. Nabil et al. [7] also studied heat transfer performance of TiO_2 nanoparticles in water-ethylene glycol (EG) mixture. The results have been reported on different operating temperature conditions. Prime factors like viscosity and thermal conductivity of the refrigerant were maintained at temperature range of 30–80 °C. There was a substantial rise of thermal conductivity of 15.4% at 1.5 vol.% nano concentration at 60 °C as reported. Wang et al. [8] also investigated how to improve the coefficient of performance of refrigeration system using a mixture of fullerenes (C70) and $NiFe_2O_4$ dispersed in compressor lubricant oil after solid grinding. The coefficient of performance of the refrigeration system was also found increasing by 1.23%. Bhattacharya et al. [9] investigated on the usage of Al_2O_3 as nano refrigerant and declared that a better energy saving could be achieved as compared to pure R-134a. In addition, extensive researches are desirable on thermophysical properties of hybrid nanoparticles in hydrocarbon mixtures and R-134a with GON refrigerants as additives in vapor compression refrigeration systems.

Hence, this study aims to explore the means and ways of how to improve the thermal conductivity and viscosity of the nanographene/R-134a as refrigerant. In addition, the coefficient of performance of the refrigeration cycle using the nano refrigerant is found promising. This may be due to higher thermophysical properties of the nano refrigerant that leads to enhance the rate of heat transfer in an evaporator tube.

2 Materials and Approaches

2.1 Refrigerant Properties

A refrigerant is a substance having desirable thermodynamic properties used in the refrigeration cycle. In this experiment, the thermophysical properties of nano refrigerant are investigated by considering graphene as better nano additives in conventional refrigerant, R-134a. Table 1 shows the thermophysical properties of the graphene nanoparticles (GNO) used as additives in base refrigerant. The combined study of the thermal properties (thermal conductivity and specific heat) and the rheological properties (viscosity) of various refrigerant has been tested in the laboratory. The thermal conductivity of all nanofluids was measured using a KD2 Pro conduct meter (Decagon Devices Inc.) with the help of the transient hot wire technique [10].

The specific heat for each nanofluid was measured in a differential scanning calorimeter (DSC). The viscosity and rheological behavior of nanofluids were obtained by conducting tests under steady-state conditions using a HAAKE RheoStress rotational remoter.

Table 1 Thermophysical properties of nano refrigerants

Nano refrigerant	Density	Viscosity	Thermal conductivity
	(kg/cm^3)	(Pa s)	(W/m K)
Copper oxide (CuO)	6320	–	32.9
R-134a	1202.6	0.00019336	0.0139
Aluminum oxide (Al_2O_3)	3880	–	40
R245fa	1339	0.402×10^{-3}	0.081
Graphene nano-oxide (GNO)	2400	–	2000

2.2 *Preparation Method of Nano Refrigerant*

2.2.1 One-Step Method [11]

According to Eastman et al., one-step method is used to condense the vapor of copper nano refrigerant with ethylene glycol as base fluid. This method will help to reduce the undesirable agglomeration in a fluid. This may be possible due to controlling the processes like drying, storage, transportation and dispersion of nanoparticles. This will lead to enhance the stability of the fluid. Such method helps in preparing nanofluid by uniformly dispersing nanoparticles into the base fluid. It results in a stable nanofluid with suspended particle without aggregation.

2.2.2 Two-Step Method [12]

Unlike the previous method, the preparation of nanofluids using two-step method is widely accepted. The first step of this method includes grinding or milling of nanoparticles into a dry powder form, either physically or chemically. After preparing in required nano size, it is allowed to disperse into the base fluid. Such, nanofluids are prepared following the dispersion techniques like magnetic stirring, ultrasonic agitation, homogenizing, ball milling and high shear mixing. Two-step method has been implemented in industry for large-scale production due to its economic viability and stability. In addition, surfactants are being used in this method in order to reduce the tendency of aggregation because of nanoparticles having a high surface area and surface activity. However, the undesirable effect of surfactant at high-temperature application still remained as a matter of concern.

2.3 *Thermal Conductivity of Nanofluid*

A nanofluid is being prepared by making a colloidal solution of nanoparticles in a base fluid. A typical nanofluid consists of nano powder of metals, oxides, carbides and carbon nanotubes. In order to determine the conductivity of a heterogeneous mixture, Maxwell has proposed a model of thermal conductivity. Thermal conductivity model has been proposed basing upon the continuity and discontinuity of a phase of base fluid. So, the effective thermal conductivity of nanofluid has been expressed by Maxwell as follows:

$$\frac{K_{\text{eff}}}{K_{\text{f}}} = \frac{K_{\text{S}} + 2K_{\text{f}} - 2\phi(K_{\text{f}} - K_{\text{p}})}{K_{\text{S}} + 2K_{\text{f}} + \phi(K_{\text{f}} - K_{\text{p}})} \tag{1}$$

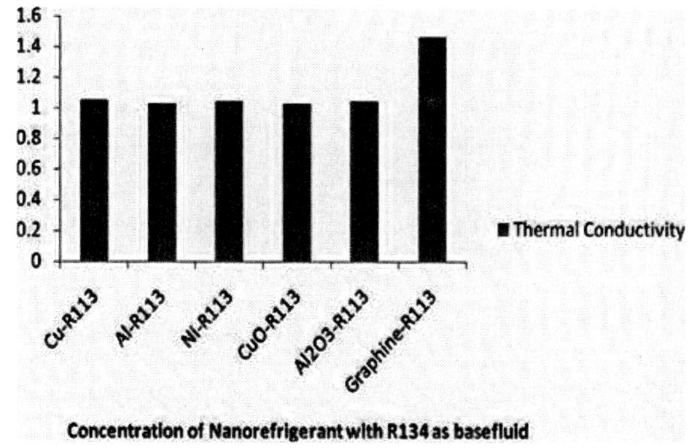

Fig. 1 Effect of temperature on thermal conductivity

where k_{eff} is the effective thermal conductivity of the mixture, and k_s and k_f are thermal conductivity of particle and base fluid, respectively. ϕ is the volume fraction of particles in the base fluid (Fig. 1).

2.4 Zeta Potential Analysis Techniques

The significance of zeta potential analysis is to determine the stability of a nanofluid. In this technique, the electric potential difference between the dispersion medium and the stationary layer of fluid attached to the dispersed particle can be measured. So, colloids with higher zeta potential (negative or positive) are electrically stabilized, while colloids with lower zeta potentials tend to coagulate in the fluid. In general, a value of 25 mV (positive or negative) can be taken as the arbitrary value that separates low-charged surfaces from highly charged surfaces. The colloids with zeta potential from 40 to 60 mV are regarded as stable, and those with more than 60 mV have excellent stability.

2.5 Experimental Analysis

The refrigeration system comprises compressor, condenser, expander and evaporator. The compressor used in this system is meant to compress the refrigerant to a high pressure that leads to reach high temperature than ambient. At such high temperature, heat is rejected to the surrounding through the condenser that leads to condense the refrigerant. Then the condensate is undergone an expansion or throttling process by passing through the expander resulting in a decrease in temperature as well as pressure. Then the refrigerant having low pressure and temperature is allowed to pass through the evaporator. The refrigerant absorbs the heat and gets converted into vapor and again compressed by compressor. Thus, the cycle is repeated. In this

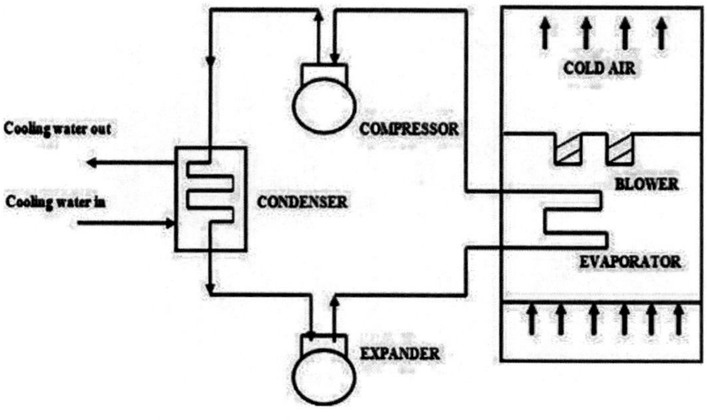

Fig. 2 Schematic diagram of refrigeration system

experiment, graphene nano-oxide with R134a has been used as the refrigerant due to its high conductivity and stability. The pressure and temperature of the refrigerant have been recorded in entire cycle (Fig. 2).

2.5.1 Coefficient of Performance of the Vapor Compression Refrigeration Cycle

It is regarded as the performance of a refrigeration system. It may be expressed as the refrigerating effect produced due to the work input. So, COP may be mathematically expressed as

$$\mathrm{COP} = Q/W \tag{2}$$

where Q = refrigerating effect (evaporator) and W = amount of work input. (compressor)

$$\begin{aligned}\mathrm{COP} &= \frac{C_\mathrm{p}(T_1 - T_4)}{\frac{n}{n-1} X C_\mathrm{v}(\gamma - 1)[(T_2 - T_1)] - [(T_3 - T_4)]} \\ &= \frac{\gamma(T_1 - T_4)}{\frac{n}{n-2}(\gamma - 1)[(T_2 - T_1)] - [(T_3 - T_4)]} \\ &= \frac{(T_1 - T_4)}{\frac{n}{n-2}(\gamma - 1)/\gamma[(T_2 - T_1)] - [(T_3 - T_4)]}\end{aligned} \tag{3}$$

where $T1, T2, T3, T4$ are the temperatures of air at different positions in refrigeration γ is taken as Cp/Cv (Fig. 3).

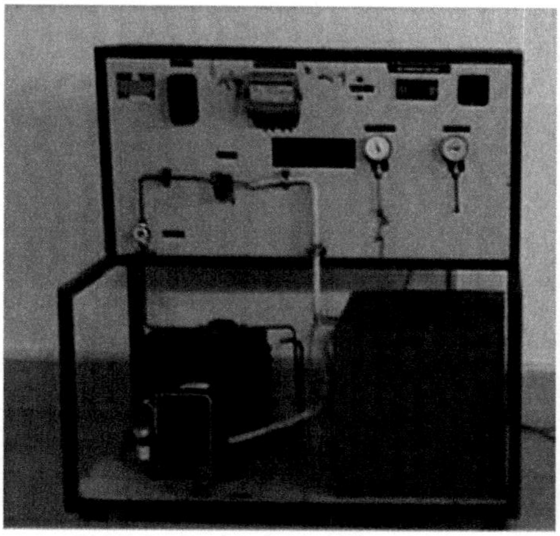

Fig. 3 Experimental setup of VCR system

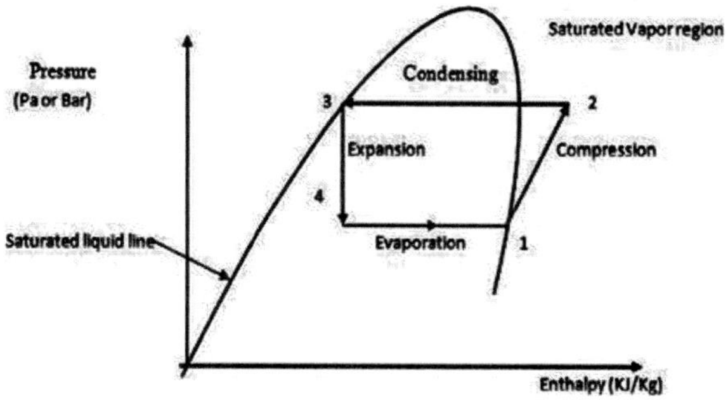

Fig. 4 P–h diagram of refrigeration cycle

The experiment has been conducted in a vapor compression refrigeration (VCR) test rig using R134a as refrigerant with 0.6–0.9% by wt. of nano graphene oxide (Fig. 4; Table 2).

Pressure and enthalpy of a refrigerating system are very important factors to explain the performance of a vapor compression cycle. The compression and expansion processes of a refrigerant are undergoing at different pressure line. The phase change of refrigerant takes place with the pressure line.

3 Result and Discussion

3.1 Effect on Coefficient of Performance (COP)

The performance of a refrigerating cycle can be well explained by its coefficient of performance (COP) value. COP may be regarded as the effective refrigeration by consuming input work. COP has been evaluated using the observation data generated

Table 2 Technical specification of test rig

Equipment	Specifications
Capacity of evaporator	165 L
Compressor	Kirloskar Hermitically sealed
Refrigerant	R-134a, Graphene nano refrigerant
Condenser type	Natural convention, air cooling
Expansion device	Copper capillary tube
Flow rate measurement	Glass tube rotameter
Energy meter	Voltas, single phase, 220 V AC, 50 Hz
Evaporator	Fin and copper tube type cooling coil
Temperature indicator	RTD thermocouple
Pressure indicator	Suction and discharge pressure gauge
Test environmental temperature	27 °C
Expansion device	Capillary tube

from the experimental setup. Refrigeration effect is calculated by the help of an energy meter which is assembled to heater incorporated with the setup. Evaporator captured the heat from hot water of heater and released the same quantity of heat by refrigerant under steady-state condition.

Figure 5 shows the variation of COP with the concentration change of graphene nano-oxide (GNO) in the base fluid as R134a.COP of pure R134a is evaluated as 2.69. On the other hand, the trend of increase in COP by adding graphene as nano refrigerant in base fluid has been observed. The COP is found increasing by 45 and

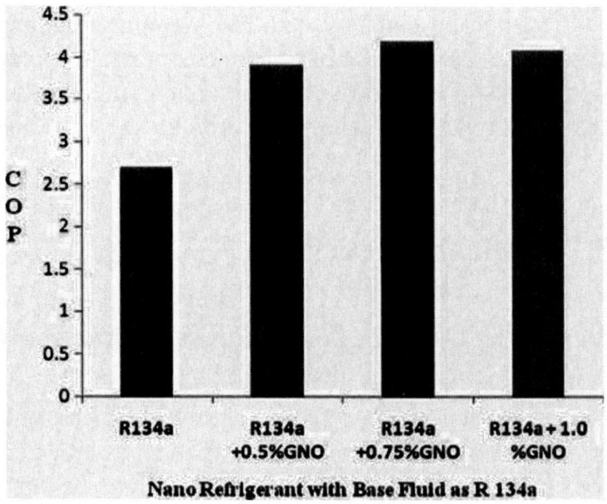

Fig. 5 Effect of concentration of GNO on COP

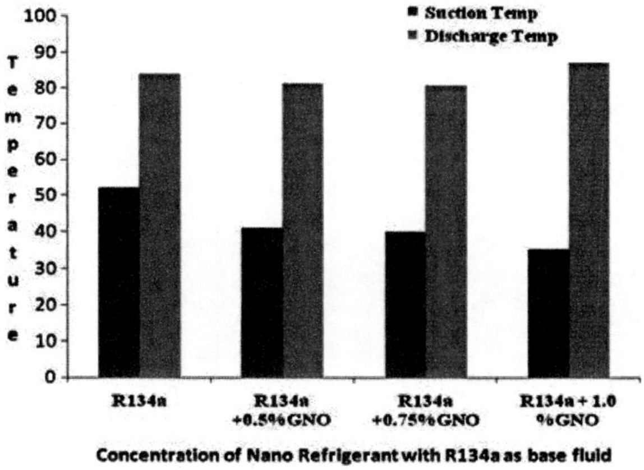

Fig. 6 Effect of concentration of GNO on suction and discharge temperature

56% with base fluid adding graphene 0.5 and 0.75%, respectively. However, the same trend is followed up to 0.9% of the addition but by adding 1% by wt. with base fluid, the COP is found decreasing by around 4.4% from the previous.

3.2 Effect on Suction and Discharge Temperatures of Compressor

Figure 6 shows the effect of adding nano refrigerant on the suction and discharge temperature of the compressor. The discharge temperature for pure R134a is 84 °C. It is found decreasing by 3.1, 3.8% with adding the graphene with base R134a by 0.5 and 0.75%, respectively. However, the discharge temperature is found increasing by 4.2% by adding 1% by wt. of nano graphene. The suction temperature for pure R134a is 52 °C. It is decreased by 21.03 and 23.4% by adding the graphene with base R134a by 0.5 and 0.75%, respectively. On adding 1% by wt., the suction temperature is decreased by 31.84%. Higher discharge temperature results in high condensing temperature and low evaporating temperatures.

3.3 Effect on Suction and Discharge Pressure

Figure 7 shows the variation of condenser and evaporator pressure by adding graphene nano additives in base refrigerant R134a. For this experiment, the suction and discharge pressures have been recorded and observed that suction pressure exhibits a decrement and discharge pressure an increment in pressure on addition of graphene nano-oxide (GNO) refrigerant in R134a as base fluid. The suction pressure shows a decrement of 31.24, 33.25 and 22.3% on adding GNO by 0.5, 0.75 and

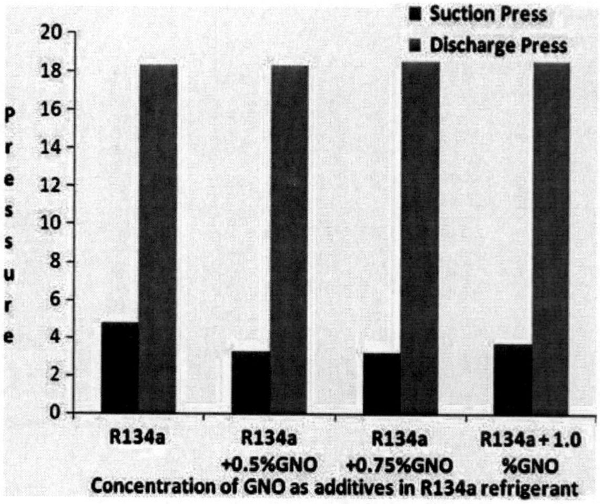

Fig. 7 Effect of concentration of GNO on suction and discharge pressure

1% by wt., respectively. In addition, the discharge pressure shows an increment of 0.08, 1.20 and 1.23% on adding GNO by 0.5, 0.75 and 1% by wt., respectively. The increase in discharge pressure (condenser) may be due to excess frictional resistances offered by the fluid in the tube and decrease in suction pressure (evaporator) may be due to resistance of flow of refrigerant. Lower condenser pressure is always expected for higher refrigeration effect and lower power consumption.

3.4 *Effect on Compression Ratio*

Figure 8 shows higher compression ratios results of higher condensing pressures and lower evaporator pressures. So, in order to achieve lower condensing pressures or

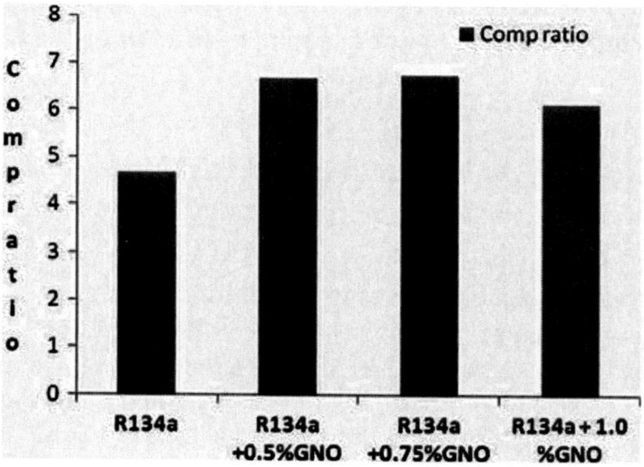

Fig. 8 Effect of concentration of GNO on compression ratio

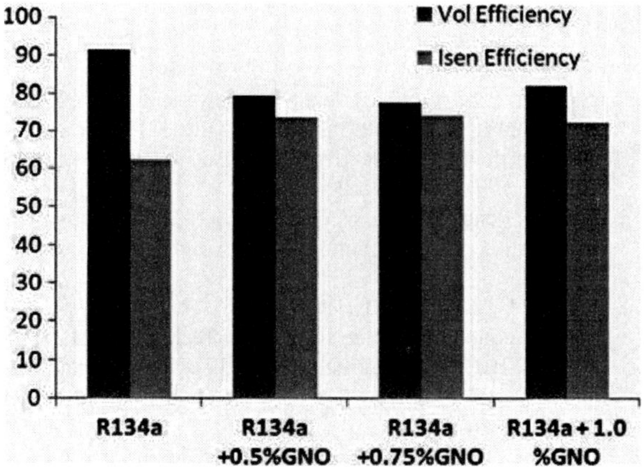

Fig. 9 Effect of concentration of GNO on volumetric and isentropic efficiency

higher evaporator pressures, lower compression ratios of the refrigerant are desirable. The compression ratio for pure 134a is 4.68. The compression ratio is found to be increased by 42.12, 44.23 and 31.83% on adding 0.50, 0.75 and 1% by wt., respectively.

3.5 Effect on Isentropic Efficiency and Volumetric Efficiency

Figure 9 shows the variation of concentration on volumetric and isentropic efficiency of the cycle. The volumetric efficiency for pure R134a is 91.27. The volumetric efficiency is decreased by 13.21, 15.28 and 10.24% on adding GNO by 0.5, 0.75 and 1% by wt., respectively. Higher volumetric efficiency leads to intake more amount of refrigerant into the cylinder, thus resulting in higher refrigeration. In addition, the isentropic efficiency is increased by 18.21, 23.22 and 21.20% on adding GNO by 0.5, 0.75 and 1% by wt., respectively.

4 Conclusion

After the completion of different experimental analysis results, it was adumbrated that addition of graphene nano-oxide in the base refrigerant R134a makes a satisfactory improvement in the comprehensive performance of the VCRS than that of pure base refrigerant. However, increase in the % of nanoparticles in the base refrigerant would result a decrease in system performance. The significance of various parameters affecting the performance of the refrigeration system was well studied and addressed. Hence, the result showed that by taking 0.5–09% of graphene nanoparticles (GNO) with R134a, the efficiency of the VCR system is significantly improved as compared to the pure base refrigerant.

References

1. A. M. K. P. T. Ã. (2010) Science review of internal combustion engines 36:4657–4667
2. Mahbubul IM, Saidur R, Amalina MA (2013) Influence of Particle concentration and temperature on thermal conductivity and viscosity of Al_2O_3/R141b nano refrigerant. Int Commun Heat Mass Transfer 43:100–104
3. Ding G, Peng H, Jiang W, Gao Y (2009) The migration characteristics of nanoparticles in the pool boiling process of nanorefrigerant and nanorefrigerant-oil mixture. Int J Refrig 32(1):114–123
4. Selvam C, Lal DM, Harish S (2016) Thermo physical properties of ethylene glycol-water mixture containing silver. J Mech Sci Technol 1(3):1271–1279
5. Akilu S, Baheta AT, Sharma KV (2017) Experimental measurements of thermal conductivity and viscosity of ethylene glycol-based hybrid nanofluid with TiO_2–CuO/C inclusions. J Mol Liq 246:396–405
6. Ghorbani B, Akhavan-behabadi MA, Ebrahimi S, Vijayaraghavan K (2017) Experimental investigation of condensation heat transfer of R600a/POE/CuO nano-refrigerant in flattened tubes. Int Commun Heat Mass Transf 88:236–244
7. Nabil MF, Azmi WH, Abdul Hamid K, Mamat R, Hagos FY (2017) An experimental study on the thermal conductivity and dynamic viscosity of TiO_2–SiO_2 nanofluids in water: Ethylene glycol mixture. Int Commun Heat Mass Transf 86:181–189
8. Wang R, Zhang Y, LiaoY (2017) Performance of rolling piston type rotary compressor using fullerenes (C70) and $NiFe_2O_4$ nanocomposites as lubricants additives
9. Bhattacharyya S, Das P, Haldar A, Rakshit A (2018) Performance ana lysis of a geothermal air conditioner using nanofluid. In: Anand G, Pandey J, Rana S (eds) Nanotechnology for energy and water. ICNEW 2017. Springer proceedings in energy. Springer, Cham
10. Eastman JA, Choi SUS, Li S, Yu W, Thompson LJ (2001) Anomalously increased effective thermal conductivities of ethylene glycol-based nanofluids containing copper nanoparticles. Appl Phys Lett 78(6):718–720
11. Subramani N, Prakash MJ (2011) Experimental studies on a vapour compression system using nano refrigerants. Int J Eng Sci Technol 3(9):95–102
12. Mukherjee S, Paria S (2013) Preparation and stability of nanofluids—a review. IOSR J Mech Civil Eng (IOSR-JMCE) 9(2):63–69 (2013)

Capillary Tube Flow Characterization of a Transcritical CO_2 Cycle Using Separated Two-Phase Flow Model

Abhijit Date and Neeraj Agrawal

Abstract Flow behaviour of a capillary tube in the transcritical CO_2 cycle is predicted based on homogeneous and separated two-phase flow models. Five different empirical correlations of void fraction available in open literature are used to predict the flow behaviour by separated two-phase model. Fauske [1] void fraction correlation matches reasonable well with homogeneous flow model with the discrepancy in capillary tube prediction as around 2%. It is observed that the homogeneous flow model is sufficient enough and recommended to predict the capillary tube flow in a transcritical CO_2 system used for capillary tube modelling due to its simplicity and relatively lesser computations.

Keywords Transcritical · Carbon dioxide · Homogeneous · Two phase

Nomenclature

L	Length (m)
x	Dryness fraction (–)
v	Velocity (ms^{-1})

A. Date (✉)
Mechanical Engineering Department, Finolex Academy of Management and Technology, Ratnagiri 415639, India
e-mail: abhijitdate1@gmail.com

N. Agrawal
Mechanical Engineering Department, Dr. Babasaheb Ambedkar Technological University, Lonere 402103, India
e-mail: neeraj.titan@gmail.com

M. Ramgopal et al. (eds.), *Advances in Air Conditioning and Refrigeration*, Lecture Notes in Mechanical Engineering, https://doi.org/10.1007/978-981-15-6360-7_11

Subscripts

sup	Supercritical
subliq	Subcooled liquid
tp	Two phase
g	Gas
l	Liq

1 Introduction

Capillary tube facilitates the throttling process in the vapour compression systems, used in relatively small capacity refrigeration systems. The flow dynamics are complex in nature where refrigerant gets accelerated and phase change takes place, liquid to vapour, simultaneously, while pressure drops due to friction and acceleration of the fluid. The momentum pressure drops dominate in two-phase region due to the presence of vapour. Proper selection of capillary tube with respect to its combination of diameter and length directly affects the system performance to avoid starvation and flooding of the evaporator under off design conditions. Characterization of flow helps to design the capillary tube. Various models and techniques have been developed and reported in open literature to characterize the capillary tube flow with conventional subcritical refrigerants. Extensive studies are available in open literature on characterization of the capillary tube with transcritical CO_2 cycle using homogeneous two-phase flow model. Use of separated two-phase flow model to characterize the capillary tube is less explored.

Swart [2] described the behaviour of a capillary tube in his pioneer work and suggested to use these principles to the actual refrigeration systems. In the pioneering work by Marcy [3], graphical integration technique is developed to compute length or mass flow rate for specified capillary tube. Rating charts for R-12 and R-22 are developed by Hopkins [4]. Whitesel [5, 6] studied choked flow in capillary tube in his pioneer work. Two computer-based codes are brought out by Erth [7] based on the work carried out by Hopkins and Whitesel. Kim [8] developed an iterative procedure for the determination of two-phase length of the capillary tube and examined the critical flow condition at the exit of the capillary tube. Wong and Ooi [9] carried out a comparative analysis of homogeneous and separated flow model and predicted that separated flow model with Lin friction factor and Miropolskiy's slip ratio predicts better than homogeneous flow model. Wongwises et al. [10] developed slip ratio correlations using separated flow model and concluded that the selection of appropriate correlation for frictional pressure gradient and slip ratio can be best suited to predict the two-phase flow behaviour of refrigerant. Choi et al. [11] developed correlation to predict mass flow rate through adiabatic capillary tube. Zhang and Ding [12] proposed two models M-solution to compute the mass flow rate and

L-solution capillary tube length. They have also analysed the choked and non-choked flow conditions.

Agrawal and Bhattacharyya [13] investigated flow characteristics of an adiabatic capillary tube in a transcritical CO_2 heat pump system employing the homogeneous model. Agrawal and Bhattacharyya [14] modelled capillary tube flow of transcritical CO_2 heat pump using separated two-phase flow model. It is reported that discrepancy between the separated flow model and homogeneous flow model is about 8–11%. Wang et al. [15] carried out characterization of straight and helically coiled capillary tube based on separated two-phase flow model including metastable flow with CO_2 refrigerant.

In the present work, capillary tube flow is modelled using homogeneous and separated two-phase flow models, and the results are compared based on the different void fraction empirical correlations available in open literature.

2 Modelling of Capillary Tube

The flow of refrigerant through capillary tube is divided into three different regions, namely supercritical region 1–2, transcritical region 2–3 and the subcritical region 3–4 as shown in Fig. 1. As represented on p–h diagram in Fig. 2, state '2' lies on critical temperature line, resulting subcooled liquid in the region 2–3. Unlike subcritical refrigerant capillary tube flow, the temperature decreases as the flow progresses in subcooled region in transcritical flow.

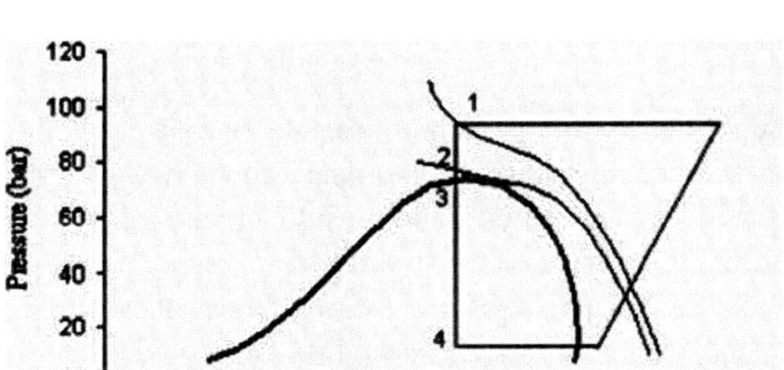

Fig. 1 Flow regions of adiabatic capillary tube [14]

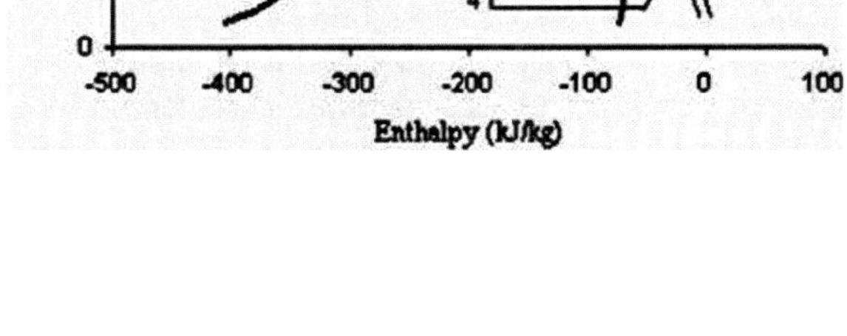

Fig. 2 P–h diagram of flow regions of capillary tube for CO_2 transcritical cycle [14]

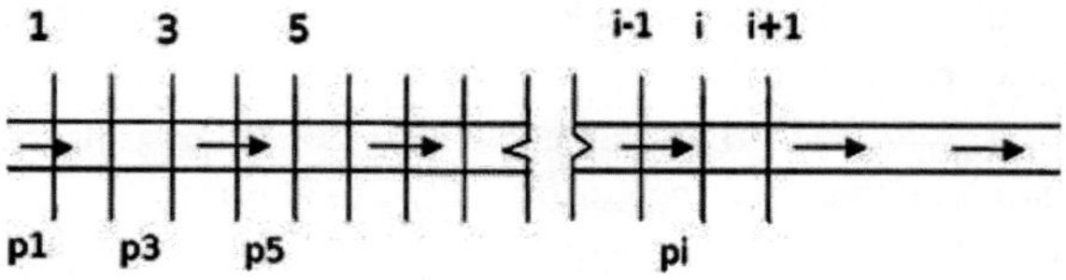

Fig. 3 Discretization for the capillary tube along length

Due to the inception of vaporization of refrigerant in the capillary tube, liquid flashes into vapour and enters into two-phase flow region. As flow progresses in the capillary tube, further pressure drop takes place due to acceleration of the vapour. The two-phase pressure drop is a combined result of the tube wall friction and the fluid acceleration, as shown below

$$\frac{\mathrm{d}p}{\mathrm{d}L} = \left(\frac{\mathrm{d}p}{\mathrm{d}L}F\right) + \left(\frac{\mathrm{d}p}{\mathrm{d}L}a\right)$$

The expansion process takes place from point 1–4 is represented in Fig. 2 on pressure–enthalpy (p–h) cycle plot. Modelling of each section is carried out independently applying conservation of mass, energy and momentum.

Combined length of capillary tube is expressed as

$$L = L_{\text{sup}} + L_{\text{subliq}} + L_{\text{tp}} \quad (1)$$

The capillary tube is discretized into a number of longitudinal elements along the length to capture sharp changes in CO_2 property as shown in Fig. 3. Capillary tube is modelled using homogeneous two-phase model with no slip condition and separated two-phase flow model with the slip conditions and corresponding void fraction taken into account.

Five void fraction correlations appeared in the open literature are used in the present study. These void fraction (α) and slip ratio (S) correlations are shown in Table 1.

3 Result and Discussion

Separated two-phase and homogeneous two-phase flow models are implemented for flow characterization of capillary tube. Five void fraction correlations available in open literature are used, and results are compared. Gas cooler pressure and temperature are taken as 100 bar and 313 K, respectively, while evaporator temperature is taken as 288 K. Mass flow rate of refrigerant is selected as 0.01 kg/s. A capillary tube of the specifications, $d_c = 1.0$ mm and $\varepsilon = 0.0015$ mm, is selected for the analysis.

A simulation code (FORTRAN) is developed which iteratively solves conservation equations and gives segmental increment in length for each iteration. Pressure is taken as the marching parameter, and capillary tube length is calculated for each

Table 1 Void fraction (α) and slip ratio (S) correlations

Investigator	Void fraction (α)	Slip ratio (S)
Fauske [1]	$\alpha = \frac{1}{1+\left(\frac{1-x}{x}\right)\left(\frac{v_l}{v_g}\right)^{0.5}}$	$S = \left(\frac{v_g}{v_l}\right)^{0.5}$
Zivi [16]	$\alpha = \frac{1}{1+\left(\frac{1-x}{x}\right)\left(\frac{v_l}{v_g}\right)^{0.67}}$	$S = \left(\frac{v_g}{v_l}\right)^{0.33}$
Thom [17]	$\alpha = \frac{1}{1+\left(\frac{1-x}{x}\right)\left(\frac{v_l}{v_g}\right)^{0.89}\left(\frac{\mu_l}{\mu_g}\right)^{0.18}}$	$S = \left(\frac{v_g}{v_l}\right)^{0.11}\left(\frac{\mu_l}{\mu_g}\right)^{0.18}$
Baroczy [18]	$\alpha = \frac{1}{1+\left(\frac{1-x}{x}\right)^{0.74}\left(\frac{v_l}{v_g}\right)^{0.65}\left(\frac{\mu_l}{\mu_g}\right)^{0.13}}$	$S = \left(\frac{x}{1-x}\right)^{0.26}\left(\frac{v_g}{v_l}\right)^{0.35}\left(\frac{\mu_l}{\mu_g}\right)^{0.13}$
Chisholm [19]	$\alpha = \frac{1}{1+\sqrt{1-x\left(1-\frac{v_l}{v_g}\right)}\left(\frac{1-x}{x}\right)\left(\frac{v_l}{v_g}\right)}$	$S = \sqrt{1-x\left(1-\frac{v_l}{v_g}\right)}$

segment using iteratively [20–23]. Properties code CO2PROP is used to calculate the thermodynamic as well as transport properties of CO_2.

Figure 4 shows the pressure variation along the length of capillary tube in two-phase flow region with different void fraction correlations. The flow characteristics, i.e. pressure variation along the capillary length with homogeneous model almost match with the separated flow model using Fauske [1] correlation. Discrepancy between models is about 2% for the chosen conditions. Deviation in pressure variation is seen towards the end of the capillary tube due to the existence of large proportion of vapours.

Variation of void fraction along the length of the capillary tube is exhibited in Fig. 5. It is seen that as pressure decreases, void fraction increases, initially it increases rapidly and then increases gradually. This is due to void fraction dependency on the quality of refrigerant. This may be due to transcritical operation of the cycle which results in refrigerant expansion closer to critical point where the quality of the refrigerant increases rapidly with drop in pressure. Among all the chosen correlations, Fasuke correlation exhibits the lowest value of void fraction. Highest value is encountered with Chisholm correlation. Low value of the void fraction indicates

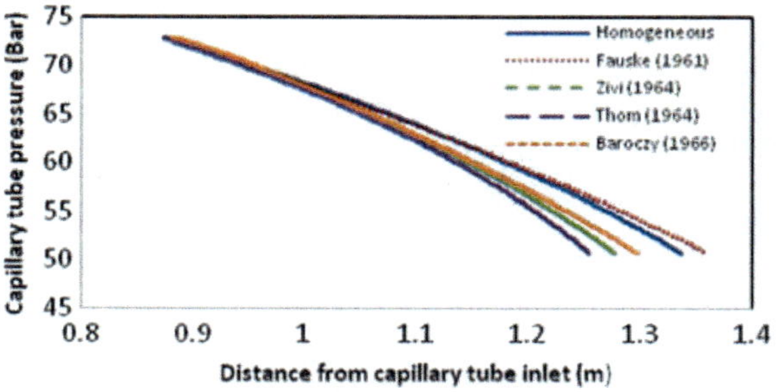

Fig. 4 Pressure variation along the length of capillary tube

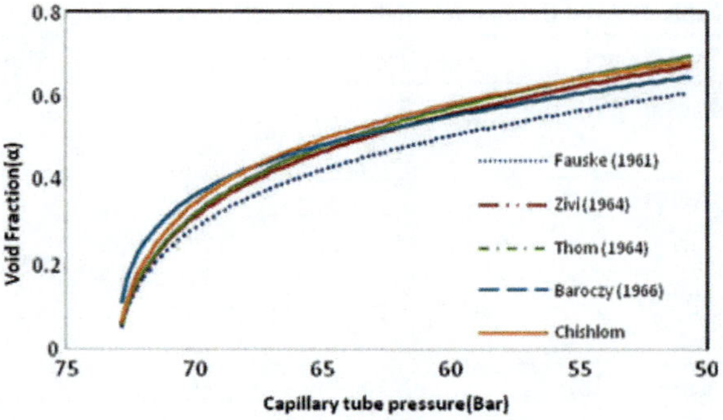

Fig. 5 Variation of void fraction along length of capillary pressure

gas-and liquid-phase flow in homogeneity which results in flow prediction almost similar to homogeneous two-phase model (Fig. 4).

Slip ratio quantifies the difference in velocities of the liquid and vapour phases when both phases are considered separately. Figure 6 shows the variation in slip ratio along the length of the capillary tube, which increases almost linearly. Fauske model exhibits relatively higher value of slip ratio due to large difference in the gas- and liquid-phase velocities.

Figure 7 exhibits the variation in gas velocities along the length of the capillary tube while expansion of the refrigerant. The variation follows the trend similar to slip ratio which implies that the slip ratio is the strong function of gas velocity. Though the slip ratio is higher in case of Fauske model, relatively voids are less and flow behaves like homogeneous.

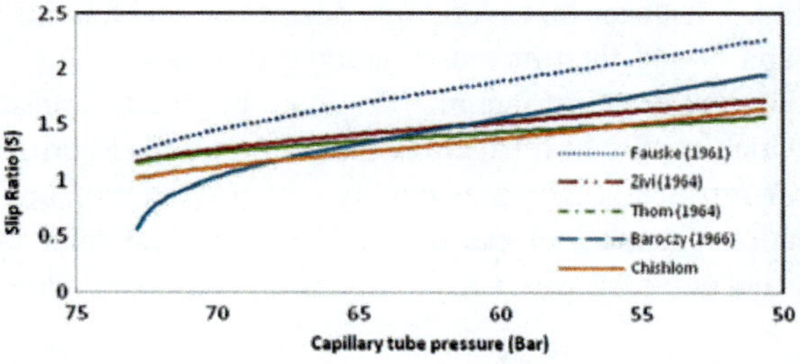

Fig. 6 Variation of slip ratio along the capillary tube length

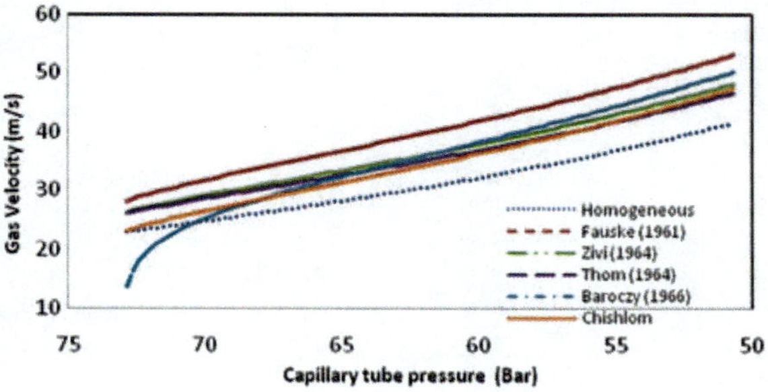

Fig. 7 Variation in gas velocity with capillary tube pressure

Table 2 Variation in parameters with void fraction correlations

Correlation	Length (L) m	Slip ratio (S)	Void fraction (α)	Liquid velocity (v_l) ms^{-1}	Gas velocity (v_g) ms^{-1}	Quality (x)	% Deviation length
Fauske [1]	1.3601	2.269	0.6086	23.46	53.23	0.4067	1.69
Zivi [16]	1.2771	1.717	0.6727	28.06	48.18	0.4069	4.51
Thom [17]	1.2542	1.558	0.6937	29.98	46.72	0.4069	6.22
Baroczy [18]	1.2973	1.945	0.6447	25.84	50.27	0.4069	3.00
Chisholm [19]	1.2357	1.639	0.6829	28.96	47.46	0.4069	7.60
Homogeneous	1.3374	–	–	41.54	41.54	0.4060	–

The typical values are summarized based on the chosen correlations are shown in Table 2.

4 Conclusions

Capillary tube flow is modelled in a transcritical CO_2 cycle using the homogeneous flow model and separated flow model. Flow behaviour and length of the capillary tube are predicted using various void fraction correlations available in open literature and compared with the homogeneous two-phase flow model. It is observed that separated two-phase model with Fasuke [1] void fraction correlation matches reasonably well with homogeneous flow model with the discrepancy in length prediction is around 2%. It implies that homogeneous two-phase model can predict reasonably well the flow behaviour of the capillary tube in a transcritical CO_2 cycle.

References

1. Fauske H (1961) Critical two-phase, steam-water flows. In: Proceedings of the 1961 heat transfer & fluid mechanics institute. Stanford University Press, Stanford, California, pp 79–89
2. Swart RH (1946) Capillary tube heat exchangers. Refrig Eng 221–224:248–249
3. Marcy GP (1949) Pressure drop with change of phase in a capillary tube. Refrig Eng 53–57
4. Hopkins NE (1950) Rating the restrictor tube. Refrig Eng 1087–1095
5. Whitesel HA (1957) Capillary two-phase flow. Refrig Eng 42–44:98–99
6. Whitesel HA (1957) Capillary two-phase flow, Part II. Refrig Eng 35–40
7. Erth RA (1970) Two-phase flow in refrigeration capillary tubes: analysis and prediction, Ph.D. Thesis. Purdue University
8. Kim RH (1987) Computer aided design of a capillary tube for the expansion valve of the refrigeration machine. ASHRAE Trans NT-87–08-5:1362–1369
9. Wong TN, Ooi KT (1996) Adiabatic capillary tube expansion devices: a comparison of the homogeneous flow and the separated flow models. Appl Thermal Eng 16(7):625–634
10. Wongwises S, Chan P, Luesuwanatat N, Purattanarak T (2000) Two phase-separated flow model of refrigerants flowing through capillary tubes. Int Commun Heat Mass Transf 27(3):343–356

11. Choi J, Kim Y, Kim HY (2003) Generalized correlation for refrigerant mass flow rate through adiabatic capillary tubes. Int J Refrig 26:881–888
12. Zhang CL, Ding GL (2004) Approximate analytic solutions of adiabatic capillary tube. Int J Refrig 27:17–24
13. Agrawal N, Bhattacharyya S (2011) Experimental investigations on adiabatic capillary tube in a transcritical CO_2 system for simultaneous water cooling and heating. Int J Refrig 34(2):476–483
14. Agrawal N, Bhattacharyya S (2008) Homogeneous versus separated flow models: capillary tube flow in a transcritical CO_2 heat pump. Int J Therm Sci 47:1555–1562
15. Wang J, Cao F, Wang Z, Zhao Y, Li L (2012) Numerical simulation of coiled adiabatic capillary tubes in CO_2 transcritical systems with separated flow model including metastable flow. Int J Refrig 35(8):2051–2358
16. Zivi SM (1964) Estimation of steady state steam void fraction by means of the principle of minimum entropy production. Trans ASME J Heat Transf 86:247–252
17. Thom JRS (1964) Prediction of pressure drop during forced circulation boiling of water. Int J Heat Mass Transf 7:709–724
18. Baroczy CJ (1966) A systemic correlation for two phase pressure drop. Chem Eng Progr Symp Ser 62(64):232–249
19. Chisholm D (1973) Pressure gradients due to friction during the flow of evaporating two-phase mixtures in smooth tubes and channels. Int J Heat Mass Transf 16(2):347–358
20. Christian JLH, Silva DL, Melo C, Gonçalves JM, Weber GC (2009) Algebric solution of transcritical carbon dioxide flow through adiabatic capillary tubes. Int J Refrig 32(5):973–977
21. Churchill SW (1977) Friction equation spans all fluid flow regimes. Chem Eng 84:91–92
22. Lin S, Kwok CCK, Li RY, Chen ZH, Chen ZY (1991) Local friction pressure drop during vaporization of R-12 through capillary tubes. Int J Multiphase Flow 17(1):95–102
23. McAdams WH, Woods WK, Bryan RL (1942) Vaporization inside horizontal tubes-II-Benzene-oil mixtures. Trans ASME 64:193–200

Experimental Study of the Effect of Al_2O_3 Nanoparticles on the Profitability of a Single-Slope Solar Still: Application in Southeast of Algeria

Mohammed El Hadi Attia, Ahmed Kadhim Hussein, Sachindra Kumar Rout, Jihen Soli, Elimame Elaloui, Zied Driss, Mebrouk Ghougali, Lioua Kolsi, and Ramesh Chand

Abstract Transforming the salt brackish water into fresh water is a real global problem. To solve this problem, a simple and economical solution in the form of the solar distillation was used. Traditional methods of the solar distillation did not succeed, because the yield of the solar still is low. So a solution is suggested in the present work to increase the profitability of the fresh water. One of the best ways is to include materials for the storage of the thermal energy, i.e., the temperature elevation of a material allows for the storage of energy and that is exactly the purpose of the present work. An aluminum oxide nanoparticles were prepared, with dimensions which are in the range of 6 nm, and it was applied successfully for the first time in a single-slope solar still under outdoors of El Oued city (Southeast of Algeria) climatic conditions. Three solar basins exposed to the sun under the same weather

M. E. Hadi Attia
Laboratory of LABTHOP, University of El Oued, El Oued 39000, Algeria

A. K. Hussein
College of Engineering, University of Babylon, Babylon City, Hilla, Iraq

S. K. Rout (✉)
C. V. Raman Global University, Bhubaneswar, India
e-mail: sachindra106@gmail.com

J. Soli · E. Elaloui
Unit of Materials Environment and Energy, Gafsa University Tunisia, Gafsa, Tunisia

Z. Driss
Laboratory of Electro-Mechanic Systems, ENIS, University of Sfax, Sfax, Tunisia

M. Ghougali
Laboratory of LEVRES, University of El Oued, El Oued 39000, Algeria

L. Kolsi
College of Engineering, Haïl University, Haïl City, Saudi Arabia

Ecole Nationale d'Ingénieurs, Monastir 5000, Tunisia

R. Chand
Government P. G. College, Dhaliara 177103, Himachal Pradesh, India

M. Ramgopal et al. (eds.), *Advances in Air Conditioning and Refrigeration*, Lecture Notes in Mechanical Engineering, https://doi.org/10.1007/978-981-15-6360-7_12

conditions and each of them containing two liters of the salt water were utilized in the experiments. The first solar still is used as a reference for the purpose of the comparison and contains the water only. The second still contains in its basin (1 g/L Al_2O_3), while the third one contains in its basin (2 g/L Al_2O_3). The results of this simple and inexpensive technique were improved that the productivity of the (second and third) solar stills was increased, respectively, in high percentages (i.e., 127 and 174%). It can be concluded that the suggested technique can be used efficiently to solve the shortage of the fresh water problem in El Oued city (Southeast of Algeria) by mixing Al_2O_3 nanoparticles with the salt water inside the solar still.

Keywords Solar energy · Al_2O_3 nanoparticles · Solar distillation · Energy storage · Fresh water

1 Introduction

Water can be considered as the main reason or a corner stone of any life over the Earth's planet. It is not easy nowadays to get healthy and fresh water especially in the poor countries and in rural and desert regions, where the water pipe network is not available especially in some places in Africa [1]. It is very important to mention that the water shortage in these regions can be considered as the main reason of the distribution of many dangerous diseases like cholera, diarrhea and guinea worm disease [2]. Although the Earth is covered by about 97.5% of the water, unfortunately it is almost salty and cannot be used for the humankind. From the other side, the fresh water sources begin to diminish gradually due to many reasons such as the pollution of rivers and natural water sources by the industrial components, the huge increase in the population and the dramatic variations in the climate especially the high increase in the environment temperature. Therefore, the only available solution to solve the scarcity of the water is to convert the salt water which is available in the sea and lakes into a fresh one by using the distillation techniques. Examples of these techniques include reverse or forward osmosis, vapor compression distillation, multistage flash distillation, membrane distillation, multi-effect distillation, electrodialysis, freeze desalination and desalination by pervaporation [3, 4]. Many of these techniques are needed a large amount of the energy, and therefore, it is considered expensive. Also, they emit toxic gases to the environment due to using the fossil fuel which make them unfriendly to it. Recently, interesting efforts were devoted by many researchers to the solar distillation which used the free and clean solar energy to produce the fresh potable water from the saline one. The device used to manage this process is called the solar distiller or a solar still. However,, there are many kinds of the solar still but the most cheap and easy construction one is called the single-basin single-slope solar still. This type can be utilized to purify the water containing a salt concentration up to ten-thousand parts per million. The productivity of the solar still depends significantly on the heat transfer process and the working fluid temperature which is usually the water [5]. Unfortunately, the thermo-physical properties of the water are poor especially the thermal conductivity of it. One of the most efficient solutions to solve this problem is add a metallic or nonmetallic nanoparticles to the water to improve their thermo-physical properties. The mixture of them is called the nanofluid. It is a term which was used for the first time by Choi [6] in 1995 to represent the combination of the water or any base fluid with the nanoparticles with a size of 1–100 nm dispersed in it [7–9]. Nanofluids can be classified as a new addition to the fluid families. Many researchers were investigated the effect of using the nanofluid to improve the productivity of the single-basin single-slope solar still. Gnanadason et al. [10] investigated experimentally the effect of adding the carbon nanotube to the water to improve the productivity of the single-basin solar still. They concluded that the productivity of the still was enhanced significantly by using the nanofluid compared with the pure water. Also, it was found that the use of the nanofluid in the basin surface of the still increased water temperature by increasing the heat transfer rate. Panitapu et al. [11] carried out an experimental study to enhance both the efficiency and the productivity of a single-basin single-slope solar still by adding TiO_2 nanoparticles to the water. The results indicated a significant enhancement in both the glass covers, vapor and water temperatures by using the nanofluid. Gupta et al. [12] made an experimental comparison between two single-slope solar stills filled, respectively, with the pure water and CuO–water nanofluid at a water depth of 5 and 10 cm. They concluded that the productivity of the still filled with the nanofluid was increased by about thirty percent at a water depth of 10 cm compared with the still filled by the pure water. Sain and Kumawat [13] investigated experimentally both effects of using alumina nanoparticles and the black paint of the base of the single-slope solar still on its performance. They concluded that the still productivity was improved by about thirty-eight percent by using these modifications. It was found also that the distilled water quantity given by the solar still was less than the quantity of the raw water which indicated a good quality of the distilled water. Gnanadason et al. [14] used the multiwalled carbon nanotubes in the basin surface to improve the productivity of the modified vacuum single-basin solar still. It was found that the distilled water production rate of the still depended on the depth of water, low pressure inside the still, still design, absorbing materials and the salt concentration. They concluded that the rate of evaporation was increased by using nanofluids. Sharshir et al. [15] studied the effects of flake graphite nanoparticles, phase change material and the film cooling on the performance of the single-slope solar still. They concluded that the productivity of the still was increased by about 50.28% by adding flake graphite nanoparticles to the water in the still basin. Moreover, it was found that the nanofluid increased the saturated vapor pressure of the water. Gnanadason et al. [16] presented experimental and theoretical analysis of the effects of addition of carbon nanoparticles to the water to improve both the productivity and the efficiency of the single-slope solar still. They concluded that both of them were enhanced significantly by using carbon nanoparticles compared with the pure water. Shankar et al. [17] made an experimental attempt to improve the performance of the single-slope solar still by using the aluminum oxide nanoparticles and painting the sidewalls of it. They deduced that both the nanoparticles and the white color increased the fresh water productivity of the still. Sharshir et al. [18] enhanced experimentally the single-slope solar still performance by using nanofluids and the glass cover cooling technique. Two types of nanoparticles (i.e., copper oxide and graphite) were used in their experiments. They concluded that the daily efficiency of the still was improved significantly by using both nanofluids and the glass cover cooling compared with the still filled with the water only. Elango et al. [19] presented an experimental study to increase the productivity of the single-basin single-slope solar still by using four different kinds of nanoparticles. The cost analysis of their study indicated that the payback period of the still filled with nanofluids was 2.85 years which was lesser than that of the still filled with the pure water. Rashidi et al. [20] suggested a volume of fluid model to investigate numerically the potential of Al_2O_3–water nanofluid to enhance the productivity of a single-slope solar still. Both the condensation and the evaporation phenomena were simulated by this model. Also, an entropy generation analysis was utilized to evaluate the system from the viewpoint of the second law of thermodynamics. It was found that both the entropy generation and the still productivity were increased by increasing the solid volume fraction of nanoparticles. Very recently, Nazari et al. [21, 22] presented an experimental and theoretical studies to improve the performance and the productivity of a single-slope solar still by employing Cu_2O nanoparticles and an external thermoelectric glass cover cooling channel. The effects of the solar radiation, nanoparticle concentration, volume flow rate of air in the cooling channel, ambient temperature on the daily thermal energy efficiency, distillated water productivity, exergy efficiency, glass cover, water basin and absorber plate temperatures. It was found that the productivity of the fresh water was increased by using both the thermoelectric cooling channel and adding Cu_2O nanoparticles to the basin brackish water.

The main objective of the present work is to investigate experimentally the effect of Al_2O_3 nanoparticles on the performance of a modified solar still in varying degrees under outdoors of El Oued city (Southeast of Algeria) climatic conditions. The used Al_2O_3 nanoparticles have a dimension in the range of 6 nm. Three solar basins each of them containing two liters of the salt water were utilized in the experiments. The first solar still is used as a reference for the purpose of the comparison and contains the water only. The second still contains in its basin (1 g/L Al_2O_3), while the third one contains in its basin (2 g/L Al_2O_3). The experiments were carried out at 23/6/2019 from 8 a.m. to 6 p.m. in the renewable energy laboratory of the University of El Oued in Algeria.

2 Experimental Methodology

2.1 Solar Still Description

The solar still apparatus is shown in Fig. 1 which consists of a wooden box of (50 cm) long, (50 cm) wide, (14 cm) height in the front, (6 cm) height in the back and (2 cm) wall thickness, while the angle of inclination is taken as 9°. To ensure a good thermal

Fig. 1 Photo of the solar still used in the experiments

insulation and to increase the temperature of the distillation chamber, it was painted with a black silicon. A transparent glass cover of 3 mm thick solar radiation was used in the still. A PCV tube was placed at the bottom of the distillation device. The tube collects the distilled water and returns it to the collection tank. It is simply like a greenhouse-based device.

2.2 *Principe of Operation*

Figure 2 presents a schematic representation of the present solar still. Temperature recording is done by using the thermocouple. Values of temperatures and the quantity of the water produced are measured every hour. Solar distillation principles were carried out according to the following steps:

- Firstly, the sun's rays pass through the glass cover to reach the black absorber at the bottom of the distiller's chamber.
- After that, the temperature of the absorber begins to increase and the heat transfer will be performed between the black absorber and the polluted water.

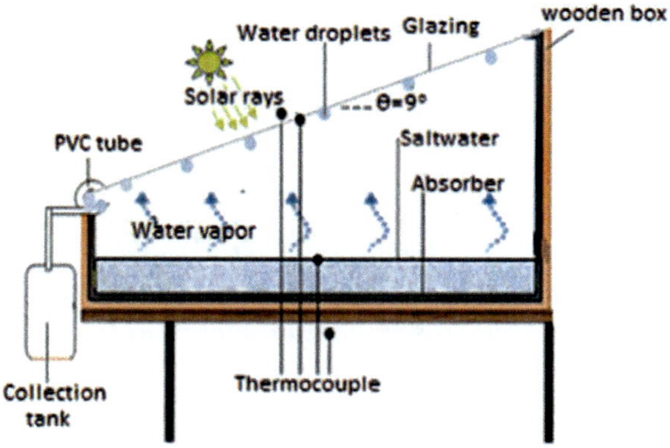

Fig. 2 Schematic representation of the experimental setup

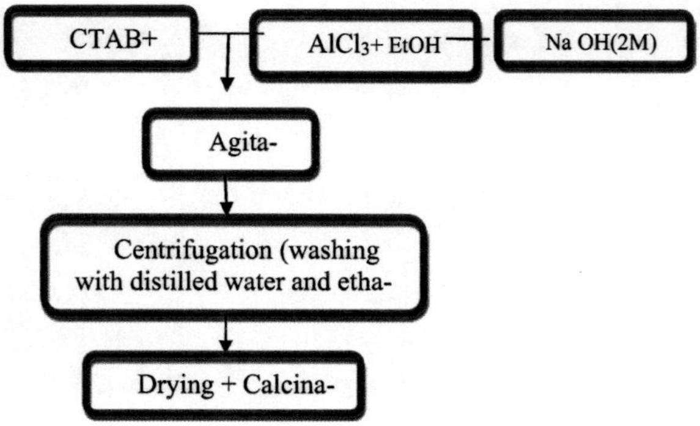

Fig. 3 Flow chart of the modified sol-gel method

- Then, the water starts to heat up and then it evaporates. The hot steam will arrive at the level of the cold glass.
- The condensation of this vapor will give droplets of water. Since the glass cover is inclined at an angle of 9°, these droplets will slip under the effect of gravity to accumulate in the PCV tube. The tube transmits the accumulated amount of fresh water to the accumulation tank.

3 Preparation of Al_2O_3 Nanoparticles

Generally, the sol-gel process [23] is a very simple way for the synthesis of nanomaterials. In fact, in recent years many metal oxides were synthesized such as SiO_2 [24], Al_2O_3 [25, 26], ZnO [27] and TiO_2 [28]. For the case of alumina, the method of the modified sol-gel cooperative self-assembly mechanism (CTM) is used. This method consists of polymerizing an inorganic precursor around the surfactant micelle. Following this method (see Fig. 3), the following protocol is proposed:

4 Characterization of Al_2O_3

4.1 Infrared Characterization

The characterization of alumina by infrared has been presented in Fig. 4, where the characteristic peaks can be observed at 635, 1741 and 1450 cm^{-1} which are attributed, respectively, to the Al–O, Al–O–Al and Al–OH–Al bonds [29, 30]. In addition, the peak at 1630 cm^{-1} corresponds to the hydroxyl molecule (HOH) [31]. However, the peak at 2490 cm^{-1} is also due to atmospheric CO_2 adsorption [32]. The presence of broadband at 3500 cm^{-1} is associated with the O–H group [33].

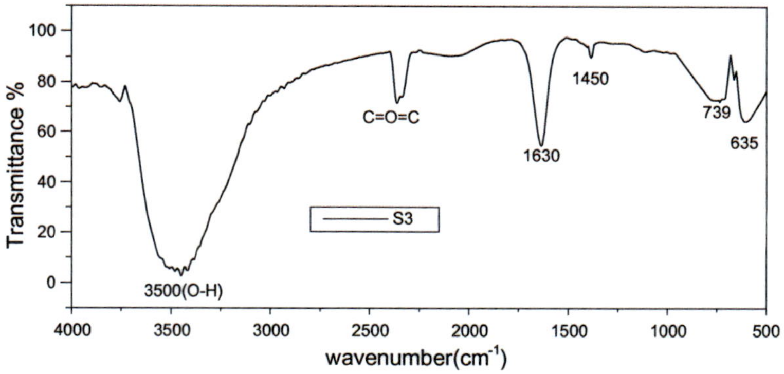

Fig. 4 IR spectrum of alumina

4.2 *Morphological Characterization by Scanning Electron Microscopy (SEM)*

Figure 5 illustrates the SEM image of the sample (S3) of Al_2O_3, which has spherical but aggregated particles. In order to give an idea about the chemical composition of alumina nanoparticles, an EDX analysis was carried out on this sample (S3) and presented in Fig. 5 and in Table 1. Thus, with this EDX graph, it has clearly seen that most of the spectra are corresponding to aluminum (Al) and oxygen (O).

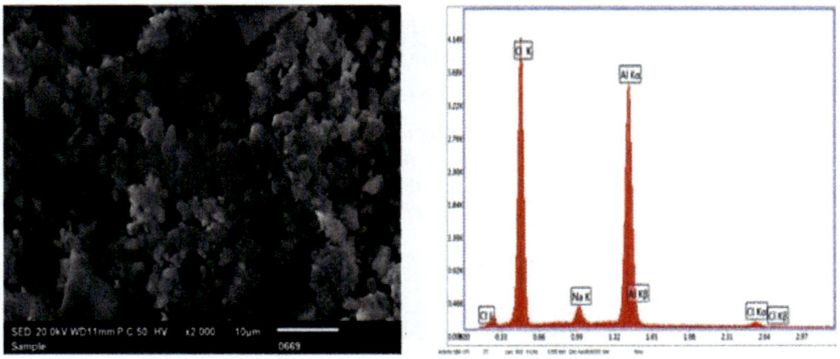

Fig. 5 SEM image of S3 sample of alumina and EDX

Table 1 Chemical composition of alumina

Element	Percentage (%)
O	37.96
Al	55.02
Na	3.27
Cl	3.75

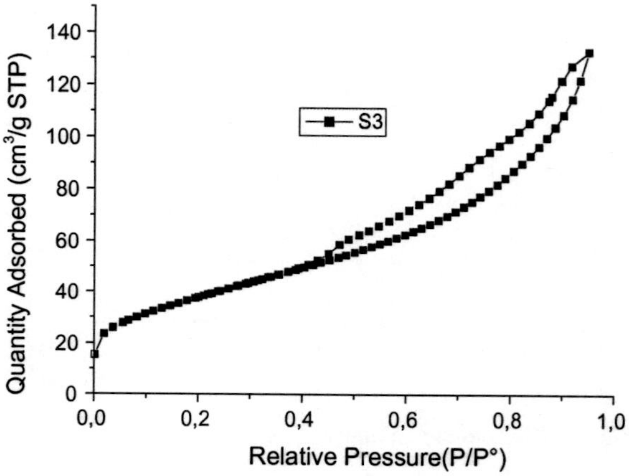

Fig. 6 Alumina adsorption-desorption isotherm

Table 2 Textual settings of Al_2O_3

Sample	S_{BET} (m^2/g)	Diameter of pore (nm)	Particle size (nm)
S3	136.9	6.1	43.7

4.3 Characterization by Adsorption/Desorption

The Al_2O_3 adsorption/desorption study shows that this alumina sample has a type IV isotherm (Fig. 6) which generally represents a characteristic of mesoporous materials with an H3 hysteresis cycle. This type confirms the morphology that has previously been described. On the other hand, this sample has a pore diameter equals to 6.1 nm (Table 2) (which is between 2 and 50 nm) and with a large specific surface area (about 136 m^2/g). If one added to this a particle size of 43.7 nm, it can be concluded and according to the UPAC classification that alumina is classified as mesoporous nanomaterials.

5 Method and Experiment

5.1 The Geographic Coordinates and Climatic Conditions

The climate of El Oued (Southeast of Algeria) is a Saharan climate characterized by a low rate of precipitation and high temperatures, significant evaporation and an excessive solar radiation. As a reference in the absence of climatic stations covering the whole Wilaya, the geographic coordinates of this station are altitude 64 m, longitude 06°47′ E and latitude 33°30′ N. The main factor of the solar distillation is the solar rays. To determine accurate experiments, the solar radiation, the ambient

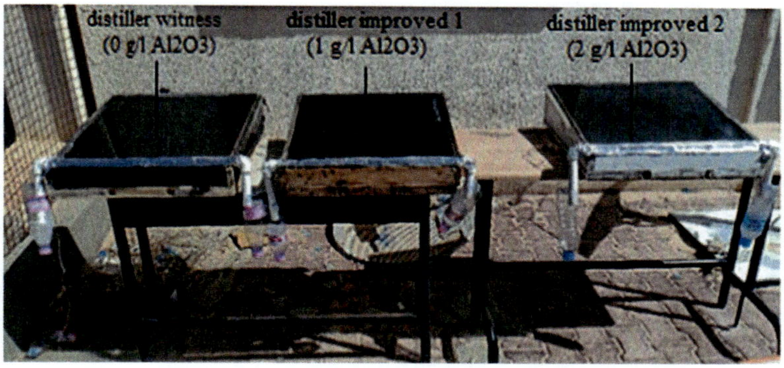

Fig. 7 Representation of the experimental rig

Table 3 Meteorological conditions

Date	23/6/2019
Sunrise	5:23 a.m.
Sunset	7:45 p.m.
Ambient temperature	31–46 °C
Humidity	15%
Wind	9 km/h

temperature, the temperatures of both sides of the glazing and the temperature of the water will be measured. The experiments of the three solar distillation basins are illustrated in Fig. 7. The first solar still is used as a reference for the purpose of the comparison and contains the water only (0 g/L Al_2O_3). The second still contains in its basin (1 g/L Al_2O_3), while the third one contains in its basin (2 g/L Al_2O_3). The experiments were carried out at 23/6/2019 from 8 a.m. to 6 p.m. in the renewable energy laboratory of the University of El Oued in Algeria with the climatic conditions indicated in Table 3.

5.2 *Meteorological Conditions of the Experiments*

The renewable energy laboratory of the University of El Oued also measured the pH and the electrical conductivity of the water used before and after the experiments. Table 4 summarizes the results obtained.

Table 4 Results of the analysis of the water used

Salt water (used water)	Distilled water (produced water)
pH = 8.82	pH = 6.84
$\sigma = 5955$ µs/cm	$\sigma = 28$ µs/cm

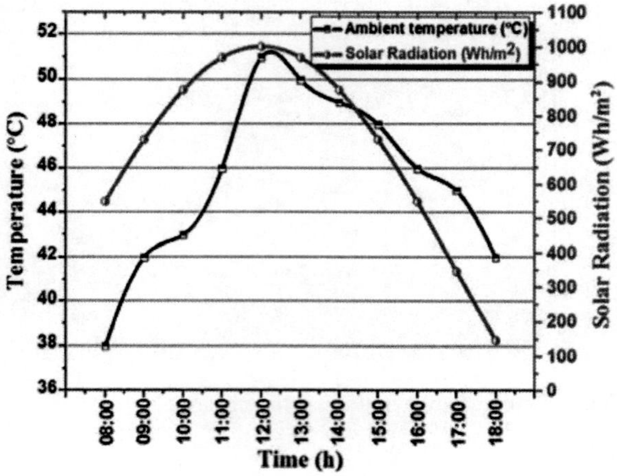

Fig. 8 Hourly temperature variation and the solar radiation

6 Results and Discussion

6.1 The Variation of the Solar Radiation and the Atmospheric Temperature

The variation of the solar radiation and the atmospheric temperature is shown in Fig. 8. It is clear that the profile of the incident solar radiation during the days of the testing has the same behavior. The solar radiation increases in the morning hours reaching its maximum values (i.e., 1002 Wh/m^2) around mid-day and then decreases in the afternoon. In addition, it can be observed in Fig. 8 that the maximum temperature is obtained during the period from 11 a.m. to 4 p.m. Moreover, it is noticed also that the temperatures at all points increase in the morning hours to reach a maximum value of 51 °C around mid-day before they start to reduce late in the afternoon.

6.2 The Variation of the Basin Salt Water Temperatures

Comparisons between the basin salt water temperatures for each of the three solar still basins are illustrated in Fig. 9. The first solar still is used as a reference for the purpose of the comparison and contains the water only (0 g/L Al_2O_3). The second still contains in its basin (1 g/L Al_2O_3), while the third one contains in its basin (2 g/L Al_2O_3). The temperature and the solar radiation profiles have the same tendency as illustrated from Fig. 9. It is obtained from this figure that the water and glass temperatures of the aluminum oxide nanoparticles (2 g/L) still are more than that of the conventional still (0 g/L Al_2O_3) by 0.5–4 °C and 0–2 °C, respectively. While the water and glass temperatures of the aluminum oxide nanoparticles (1 g/L) still are more than that of

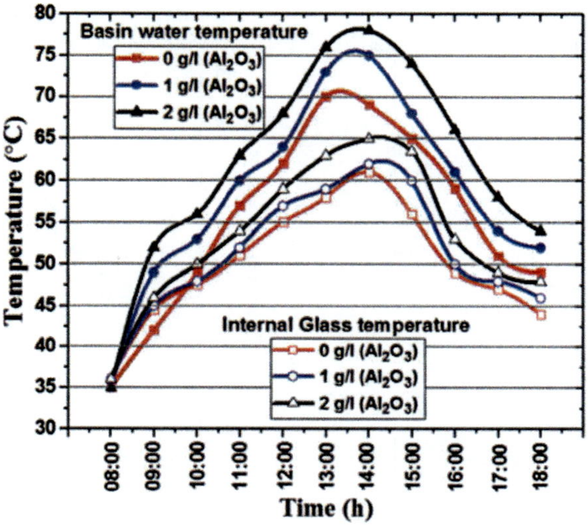

Fig. 9 Hourly variation of the basin water temperature

the conventional still (0 g/L Al_2O_3) by 0.5–6 °C and 0–3 °C, respectively. Hence, both the evaporation and the production rates are better in the nanofluid stills than that of the conventional still. Figure 9 shows also that the water temperature of the solar still tanks increases as the amount of nanostructures increases. The only difference that can be seen from this figure, is that the basin water temperature of the nanofluid stills are more than that of the water temperature of the conventional still. Therefore, the percentage of nanoparticles in the rate of rise of the salt-water temperature within the ponds containing the nanoparticles is higher compared to the basin, which contains the salty water only. It can be concluded that, the nanofluid enhances the transport and the heat transfer properties together with the water evaporation characteristics. Also, the addition of nanofluids in the basin surface increases the water temperature by increasing the heat transfer and the evaporation rates.

6.3 The Variation of the Internal and the External Glass Temperatures

The variation of the internal and the external glass temperatures of stills is shown in Fig. 10. It is clear that the profile of these incident temperatures during days of the testing has the same behavior. It can be observed from this figure that both temperatures at all points increase in the morning hours reaching their maximum values around mid-day and then decrease in the afternoon. Also, it can be noticed from Fig. 10 that the maximum internal and external glass temperatures are obtained during the period from 11 a.m. to 3 p.m. Furthermore, it can be seen that the internal and the external glass temperatures of the solar still tanks increase as the amount of

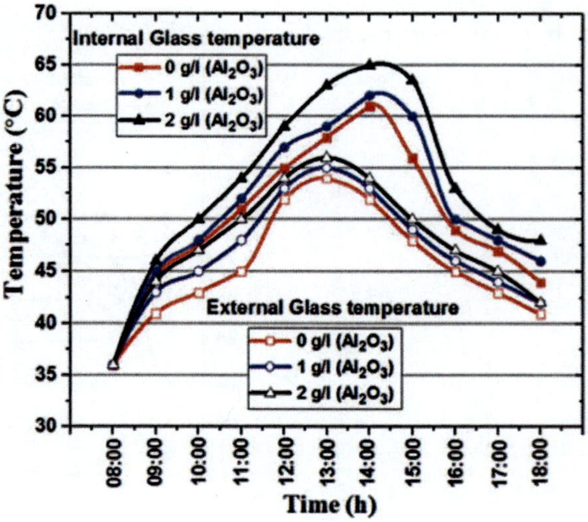

Fig. 10 Variation of the internal and external glass temperatures

nanoparticles increases. The glass temperature in the nanofluid still is less than that of the conventional still by about 1.5–3.0 °C.

6.4 The Variation of the Hourly Fresh Water Productivity

The variation of the hourly fresh water productivity is shown in Fig. 11. Comparisons between this parameter for each of the three solar still basins are illustrated

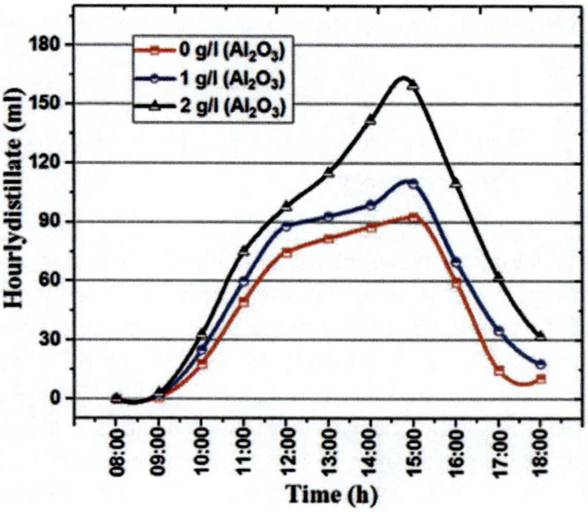

Fig. 11 Variation of the hourly fresh water productivity

in this figure. In general, it can be noticed from Fig. 11 that the hourly fresh water productivity increases during the daytime as the solar radiation intensity increases till it reaches the maximum value around mid-day. After that, in the afternoon period, the decrease in the temperature gradually reduces the productivity rate. Also from this figure, it can be observed that the water productivity is increased from zero value in the morning until it reached the maximum values in the afternoon. In addition, the maximum productivity occurs at the maximum temperature of the saline water (from 11 a.m. to 4 p.m). Note also that the difference is evident in the amount of the fresh water produced between the three still basins. This is mainly due to the existence of the aluminum oxide nanoparticles that caused a high heating rate inside the solar stills. Because of this difference, the ability of the condensation, the production rate in the nanofluid stills is more than that of the conventional still which is filled with the water only. In fact, the addition of nanoparticles in the basin surface increases the water temperature by increasing the heat transfer rate, and as a result, the evaporation rate increases also. It was found that the productivity was increased by using the aluminum oxide nanoparticles compared to the conventional still. In fact, when the aluminum oxide nanoparticles (1 g/L) and (2 g/L) are used, the productivity of the nanofluid still is increased, respectively, by about 127 and 174% compared with the conventional one.

7 Conclusions

In this work, the effect of Al_2O_3 nanoparticles on the performance of the modified solar still was investigated experimentally in varying degrees under outdoors of El Oued city (Southeast of Algeria) climatic conditions.

- The preparation of aluminum oxide nanoparticles was done by the authors in their laboratory, with dimensions in the range of 6 nm.
- It was shown that the salt water (pH = 8.82 and electrical conductivity = 5955 μs/m) has become a pure water (pH = 6.84 and electrical conductivity = 28 μs/m) after the solar distillation process.
- The technique used in the present work was proved to be environmental, ecological, economical and very simple.
- The climate of El Oued city (Southeast of Algeria) is very favorable for this technique.
- It was found that using aluminum oxide nanoparticles improves the evaporation and condensation rates by avoiding the effect of creating an amount of the water vapor content of the air above the saline.
- The higher performance of nanoparticles in the solar still is achieved for period from 11 a.m. to 3 p.m. (about 174% higher than the productivity of the conventional still).
- The productivity of the single-slope solar still can be increased by about 127 and 174% by mixing nanoparticles with the salt water inside the solar still.

References

1. Hussein AK (2015) Applications of nanotechnology in renewable energies—a comprehensive overview and understanding. Renew Sustain Energy Rev 42:460–476
2. Hussein AK, Walunj A, Kolsi L (2016) Applications of nanotechnology to enhance the performance of the direct absorption solar collectors. J Therm Eng 2(1):529–540
3. Li D, Li Z, Zheng Y, Liu C, Hussein AK, Liu X (2016) Thermal performance of a PCM-filled double-glazing unit with different thermophysical parameters of PCM. Solar Energy 133:207–220
4. Hussein AK (2016) Applications of nanotechnology to improve the performance of solar collectors—recent advances and overview. Renew Sustain Energy Rev 62:767–792
5. Hussein AK, Li D, Kolsi L, Kata S, Sahoo BA (2017) Review of nano fluid role to improve the performance of the heat pipe solar collectors. Energy Procedia 109:417–424
6. Choi U (1995) Enhancing thermal conductivity of fluids with nanoparticles. In: Siginer DA, Wang HP (eds) Developments and applications of non-Newtonian flows. FED 231:99–105
7. Chand R, Rana G, Hussein AK (2015) On the onset of thermal instability in a low Prandtl number nanofluid layer in a porous medium. J Appl Fluid Mechan 8(No.2):265–272
8. Chand R, Rana G, Hussein AK (2015) Effect of suspended particles on the onset of thermal convection in a nanofluid layer for more realistic boundary conditions. Int J Fluid Mech Res 42(No. 5):375–390
9. Hussein AK, Ahmed S, Mohammed H, Khan W (2013) Mixed convection of water-based nanofluids in a rectangular inclined lid-driven cavity partially heated from its left side wall. J Comput Theor Nanosci 10(9):2222–2233
10. Gnanadason M, Kumar P, Rajakumar S, Yousuf M (2011) Effect of nanofluids in a vacuum single basin solar still. Int J Adv Eng Res Stud 1:171–177
11. Panitapu B, Koneru V, Sagi S, Parik A (2014) Solar distillation using nano-material. Int J Sci Eng Technol 3:583–587
12. Gupta B, Shankar P, Sharma R, Baredar P (2016) Performance enhancement using nano particles in modified passive solar still. Proc Technol 25:1209–1216
13. Sain M, Kumawat G (2015) Performance enhancement of single slope solar still using nano-particles mixed black paint. Adv Nanosci Technol Int J 1:55–65
14. Gnanadason M, Kumar P, Jemilda G, Jasper S (2012) Effect of nanofluids in a modified vacuum single basin solar still. Int J Sci Eng Res 3:1–7
15. Sharshir S, Peng G, Wu L, Essa F, Kabeel A, Yang N (2017) The effects of flake graphite nanoparticles, phase change material, and film cooling on the solar still performance. Appl Energy 191:358–366
16. Gnanadason M, Kumar P, Wilson V, Hariharan G, Vinayagamoorthi N (2013) Design and performance analysis of an innovative single basin solar nanoStill. Smart Grid Renew Energy 4:88–98
17. Shankar P, Sharma R, Gupta B, Parmar H (2015) Effect of colour and Al_2O_3 nano particles on the efficiency of the solar still. SSRG Int J Therm Eng 1:1–6
18. Sharshir S, Peng G, Wu L, Yang N, Essa F, Elsheikh A, Mohamed S, Kabeel A (2017) Enhancing the solar still performance using nanofluids and glass cover cooling: experimental study. Appl Therm Eng 113:684–693
19. Elango T, Kannan A, Murugavel K (2015) Performance study on single basin single slope solar still with different water nanofluids. Desalination 360:45–51
20. Rashidi S, Akar S, Bovand M, Ellahi R (2018) Volume of fluid model to simulate the nanofluid flow and entropy generation in a single slope solar still. Renewable Energy 115:400–410
21. Nazari S, Safarzadeh H, Bahiraei M (2019) Performance improvement of a single slope solar still by employing thermoelectric cooling channel and copper oxide nanofluid: an experimental study. J Clean Prod 208:1041–1052
22. Nazari S, Safarzadeh H, Bahiraei M (2019) Experimental and analytical investigations of productivity, energy and exergy efficiency of a single slope solar still enhanced with thermoelectric channel and nanofluid. Renew Energy 135:729–744

23. Elaloui E, Pajonk G (2000) Synthèse Sol-Gel: Application à la Silice. J Soc Chim Tunisie 4:607
24. Pajonk G, Rao A, Parvathy N, Elaloui E (1996) Microstructural characterization of Silice aerogels using scanning electron microscopy. J Mater Sci 31:5683
25. Ponthieu E, Grimblot J, Elaloui E, Pajonk G (1993) Synthesis and characterization of pure and ytterium-containing alumina Aerogel. J Mater Chem 3:287
26. Elaloui E, Pierre A, Pajonk G (1997) Influence of Sol-gel processing method on the structure and the porous texture of alumina. J Catal 166:340
27. Elaloui E, Begag R, Pommier B, Pajonk G (2000) Thermostable yttrium doped inorganic oxide catalyst supports for high temperature reactions. Stud Surf Sci Catal 143:331
28. El- Mir L, Amlouk A, Elaloui E, Saadoun M, Pierre A (2008) Preparation and optical characterization of transparent, microporous TiO_2 xerogel monoliths. Mater Sci Eng B 146:69
29. Tabesh S, Davar F, Loghman-estarki M (2017) Preparation of γ-Al_2O_3 nanoparticles using modified sol-gel method and its use for the adsorption of lead and cadmium ions. J Alloys Compd 730:441–449
30. Zhang L, Wu Y, Zhang L, Wang Y, Li M (2016) Synthesis and characterization of mesoporous alumina with high specific area via co precipitation method. Vaccum 133:1–6
31. Peng Y, Nie J, Zhang W, Ma J, Bao C, Cao Y (2016) Effect of the addition of Al_2O_3 nanoparticles on the magnetic properties of Fe soft magnetic composites. J Magn Magn Mater 399:88–93
32. Farahmandjou M, Magahi S (2019) Sol-gel synthesis of ce-doped α-Al_2O_3: study of crystal and optoelectronic properties. J. Optic Commun 441:1–7
33. Tangsir A, Hafshejani L, Lähde A, Maljanen M, Hooshmand A, Naseri A, Moazed H, Jokiniemi J (2015) Water defluoridation using Al_2O_3 nanoparticles synthesized by flame spray pyrolysis (FSP) method. Chem Eng J 288:198–206

Heat Transfer in Triple-Concentric-Pipe Heat Exchanger: With/Without Corrugations

S. Beura, V. P. Mishra, S. N. Das, U. K. Mohanty, M. Mohapatra, and D. N. Thatoi

Abstract The present work involves the investigations related to heat transfer between three fluids which are at three different temperatures. For investigation process, an innovative triple-concentric-pipe heat exchanger is adopted. The heat exchanger is designed in the laboratory. Normal tap water, water at an elevated temperature and water below the room temperature are chosen to be the three fluids of interest. N-H-C and C-H-N flow configurations are considered for the experimentation. Different flow configurations (with/without corrugation on the outer surface of the middle pipe) are also considered in the investigation. For the different flow arrangements, the temperature distribution pertaining to the different fluids is measured experimentally. This is presented separately for different flow arrangements. Effective heat transfer is calculated for each of the arrangements and compared. It is evident that heat exchanger effectiveness is better in case of corrugated surface and also the N-H-C arrangement of heat exchanger is seen to be more effective in comparison with the C-H-N arrangement.

Keywords Triple concentric pipe heat exchanger · Corrugated surface · N-H-C · C-H-N · Heat exchanger effectiveness

1 Introduction

Energy conservation ensuring its optimum utilization is one of the prime concerns of the present times. This is sure to augment the advancement of the society in all respects making energy available for purposes starting from daily chores of household to the running of huge complicated machinery pertaining to continued functioning of the society and its various branches. Thermal energy constitutes a major part of energy necessary for the day today life of the human race and needs to be utilized to the optimum level with the minimum of losses. This can be ensured if

S. Beura (✉) · V. P. Mishra · S. N. Das · U. K. Mohanty · M. Mohapatra · D. N. Thatoi
Department of Mechanical Engineering, ITER, SOA (Deemed to be) University, Bhubaneswar, Orissa, India
e-mail: subhrajit.beura@gmail.com

M. Ramgopal et al. (eds.), *Advances in Air Conditioning and Refrigeration*, Lecture Notes in Mechanical Engineering, https://doi.org/10.1007/978-981-15-6360-7_13

one of the aspects concerning heat exchange can be made more effective and optimized. Keeping above in mind researchers are motivated to carry out fundamental research concerning transfer of heat in and between various systems. Improvements in the designing parameters leading to an improved performance of various heat exchangers related with space heating, air conditioning, waste-heat recovery, etc., have been their prime concern [1, 2]. Various food processing methods [3, 4] such as pasteurization, draying, etc., also depend a lot on the process of heat exchange for their efficient execution. Different types of heat exchangers, such as parallel or counter flow type, tube shape or plate type and direct or indirect type heat exchanger are in operation these days. It is, however, important to note that specification and application of a product determine the type of heat exchanger to be used. The heat transfer, in general, depends on the inherent characteristics of the fluid like its heat capacity, viscosity, etc. It also depends on the materials as well as the flow pattern adopted by fluid concerned. Many researchers have followed different methods for maximizing the useful work of a heat exchanger. Cost reduction is one of the prime motivating factors along with the reduction of the size of the heat exchanger [5–7] engaged a swirl element at the entry of the inner pipe of a concentric double-pipe heat exchanger. The researchers reported that such an arrangement improved the efficiency of the heat exchanger employed. In another study [8], the heat exchange tube was covered with metallic foam. The experimenters report that such a cover with metallic foam enhanced the successful functioning of the heat exchanger. A mathematical modeling that included the development of the governing differential equation together with possible solutions under simplified conditions for a triple-concentric-pipe heat exchanger was proposed by Unal [9]. Using these governing equations, theoretical analysis of triple-concentric-pipe heat exchanger concerning its design and performance estimations could be undertaken [10]. Based on their results, Ganji et al. [11] opined that the heat exchanger effectiveness can be enhanced by increasing the length and number of fins in the heat exchanger. Keeping the above in mind, the present work is proposed for an investigation pertaining to the study and analysis between three flowing fluids which is an important process in thermal engineering. The work includes the measurement and reflects the temperature distribution between the three fluids in the heat exchanger along its length. Also, the performance of triple-concentric-pipe heat exchanger with/without corrugations (on the external surface of the middle pipe) is estimated and compared.

2 Setup for Experimentation

The setup for experimentation consists of a triple-concentric-pipe heat exchanger with a brass inner pipe, copper middle pipe and mild steel outer pipe with 28.75, 61, 98.4 mm diameter, respectively. The length of the pipes was maintained at 1.9 m with a wall thickness of 1.5 mm. Water above the room temperature, water below the room temperature and water at room temperature could be fed into the heat exchanger from separate tanks. Separate copper pipes with/without corrugations were employed as the middle pipes. Flow meters were provided in the water duct for measuring the flow rate of the pumped fluid which could be altered by adjusting the gate valves provided separately for three tanks. A 3000 W electric heater immersed in hot water tank to heat water while ice cubes were used in cold water tank to cool water. The normal water tank was maintained at room temperature. T-type copper-Constantan thermocouples were employed at the inlet, outlet and different position in the heat exchanger for measuring the temperature which was recorded by temperature indicators. To start with, the system was maintained at a steady state. The setup for experimentation is presented in Fig. 1 photographically, while a line diagram of arrangement is provided in Fig. 4. Figures 2 and 3 show the middle copper pipe with and without corrugations, respectively.

1-Cold water tank	6- Pump for normal water tank
2-Hot water tank	7-Gatevalve (cold water flow)
3- Normal water tank	8- Gate valve (hot water flow)
4- Pump for cold water tank	9- Gate valve (normal water flow
5- Pump for hot water tank	10- Test rig
	11- Temperature Indicator

Fig. 1 Experimental setup of triple-concentric-helical pipe heat exchange. Apparatus used in setup

Fig. 2 Plane middle pipe

Fig. 3 Corrugated middle pipe

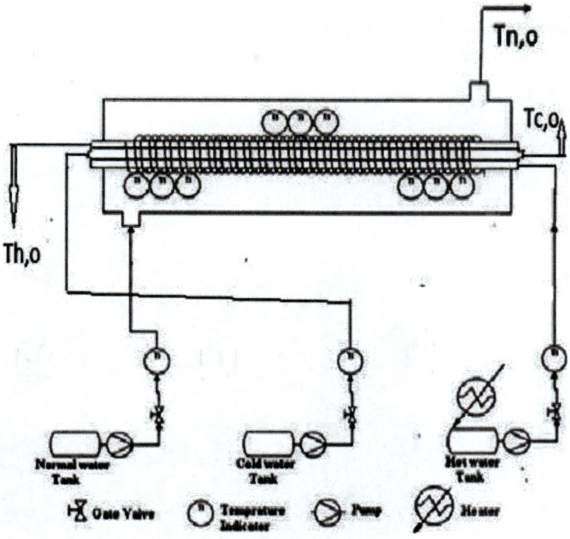

Fig. 4 Schematic line diagram for experimental setup of triple-concentric-pipe heat exchanger with corrugated surface (C-H-N arrangement)

2.1 Experimental Findings and Discussions

Temperature measurements were carried out with different arrangements (with/without the corrugation in the middle pipe) under different operating conditions (half and full throttle). Two separate configurations pertaining to the flow of hot/cold/normal water in the heat exchanger, namely N-H-C and C-H-N, were adopted for the conduct of the experiment. In the N-H-C configuration, water at room temperature passed through the pipe surrounded by the outer two pipe allowing hot water and cold water to flow through the annulus on the inner and the outer regions, respectively. In the C-H-N arrangement, the path of flow of cold water and normal water was interchanged with no alterations in the path of flow of the hot water that remained the same as in the N-H-C configuration.

$$\text{Effectiveness of heat exchanger } \in = \frac{\dot{Q}}{\dot{Q}\text{max}}.$$

where $\dot{Q}$ and $\dot{Q}\text{max}$ are the calculated heat exchanger rate of heat transfer and maximum possible heat transfer rate [10].

$$\dot{Q} = C_h(T_{h,i} - T_{h,o})$$
$$Q_{max} = C_{min}(T_{h,i} - T_{c,i})$$

where

$$C_{min} = C_h \text{ if } C_h \leq C_n + C_c \tag{1}$$

$$C_{min} = C_n + C_c \text{ if } C_h \geq C_n + C_c \tag{2}$$

where C_c, C_h and C_n are the heat capacity rate on the cold fluid side, hot fluid side and normal fluid side, respectively, and C_{min} is the minimum capacity rate. $T_{h,i}$ and $T_{h,o}$ are the inlet and outlet hot water temperatures, respectively, $T_{c,i}$ is the cold water temperature.

2.2 *Temperature Variations (C-H-N Configuration)*

In non-corrugated configuration, the flow rates of three fluids are different (V_h = 14 l/min, V_n = 26 l/min, V_c = 49 l/min). The hot water temperature drops from 70 to 43 °C, in normal water, the rise in temperature is from 30 to 35 °C, and in cold water, the rise in temperature is from 20 to 24 °C as observed. On account of the availability of less heat transfer surface, despite appreciable temperature differences, rise in the temperature of cold water is not appreciable (Fig. 5).

In corrugated configuration, the temperature of hot water drops from 70 to 37 °C, in water at room temperature, the temperature rise is from 29 to 34 °C, and in cold water, the rise in temperature is from 20 to 26 °C as observed. The heat transfer is more effective in comparison with non-corrugated due to the presence of corrugated

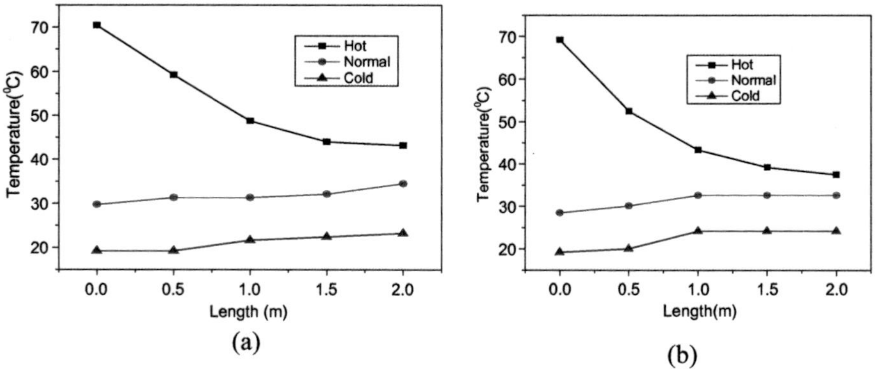

Fig. 5 Temperature distributions along the length of heat exchanger under (C-H-N) configuration (V_h = 14 l/min, V_n = 26 l/min, V_c = 49 l/min). **a** Non-corrugated and **b** Corrugated

surface which provided for an increase in the heat transfer surface. Also, more turbulence created which improved the heat transfer coefficient and resulted in better heat transfer between hot and normal water.

2.3 Temperature Variation (N-H-C Configuration)

For the non-corrugated middle pipe, the temperature variation results from experiments are presented for the heat exchanger under N-H-C configuration. The flow rates for all three fluids are $V_h = 14$ l/min, $V_c = 26$ l/min, $V_n = 49$ l/min. The hot water temperature drops from 70 to 39°C, in normal water, the temperature rise is from 32 to 35 °C, and in cold water, the rise in temperature is from 20 to 30 °C as a consequence of heat exchanges between the fluids. The temperature rise of cold water is compared to that in water at room temperature due to larger heat transfer area between the inner and outer annulus of hot and cold water, respectively (Fig. 6).

In corrugated arrangement, the temperature variation results from the experiments are presented for the heat exchanger under N-H-C configuration, and the flow rates for all three fluids are $V_h = 14$ l/min, $V_c = 26$ l/min, $V_n = 49$ l/min. The hot water temperature decreases from 70 to 34 °C, in normal water, the rise in temperature is from 32 to 34 °C, and in cold water, the rise in temperature is from 20 to 34 °C. The large diameter of the annulus contributing to the availability of large surface area between the hot and cold water as compared to that hot and the normal water further assists the heat transfer process, increasing the heat transfer rate to the cold water from hot water. As the consequence, the outlet temperature of the hot water is less than that of the normal water. Owing to the higher temperature difference along the entire length of the heat exchanger and the greater area for transfer of heat between hot and cold water, there is a continuous transfer of heat from hot to cold water finally

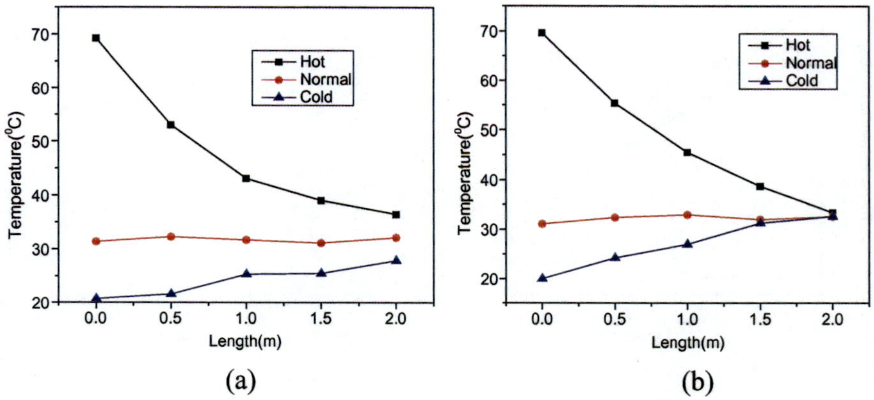

Fig. 6 Temperature distribution of three fluids along the length of heat exchanger under (N-H-C) ($V_h = 14$ l/min, $V_n = 26$ l/min, $V_c = 49$ l/min). **a** Non-corrugated and **b** Corrugated configuration

resulting in a fall of temperate of hot water which assumes values lower than that of normal water. In this N-H-C arrangement, the heat exchanger shows better heat transfer in comparison with C-H-N arrangement.

2.4 Temperature Variation (N-H-C Configuration)

A different temperature distribution was obtained pertaining to hot, cold and normal water when under full throttle conditions, a flow rate of 14, 26 and 49 l/min was maintained for the hot normal and cold water, respectively. Here, drop in temperature for hot water was from 70 to 34 °C, in normal water, the temperature rise was from 32 to 34 °C and that in cold water was from 20 to 34 °C as a consequence of heat exchanges between water under different thermal conditions (Fig. 7).

However, under half throttle, the hot water temperature dropped from 70 to 42 °C, the temperature rise in normal water was from 32 to 34 °C, and same for cold water was from 20 to 30 °C. It can be seen that compared to earlier cases, as a result of reduction in volume flow rates, the rise of temperature of the normal and cold water also reduced. In all these cases, the volume flow rate of hot water is the driving force for heat exchanges. Since the reduced flow rate results in a reduction of flow velocity of the water at the higher temperature, the coefficient of heat transfer on account of convection on the hot water side brings a reduction in the rate of heat transfer from hot water to the system.

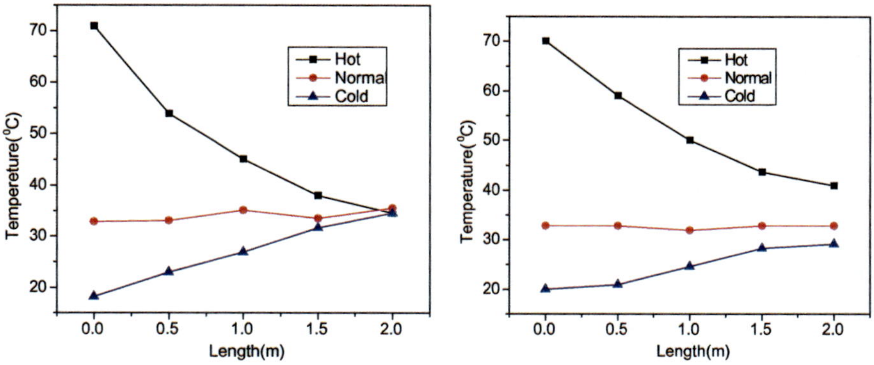

(a) (b)

Fig. 7 Temperature distribution along the length of heat exchanger under (N-H-C) (V_h = 14 l/min, V_n = 26 l/min, Vc = 49 l/min). **a** Full throttle and **b** Half throttle

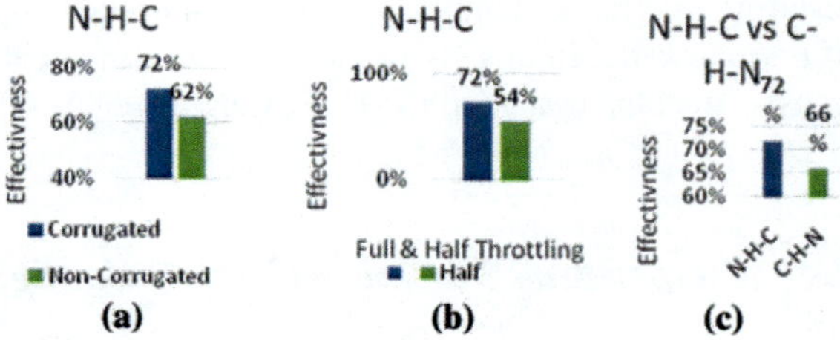

Fig. 8 Effectiveness of heat transfer **a** Corrugated and non-corrugated, **b** Full and Half Throttling, **c** N-H-C versus C-H-N (With corrugated surface)

2.5 *Effectiveness of Heat Exchanger for Different Flow Arrangements*

Due to the presence of corrugated surface, more turbulence could be generated which increased the coefficient of heat transfer and resulted in a higher rate of heat transfer and improved effectiveness (Fig. 8a). It was found that the effectiveness for half throttle arrangement reduced in comparison with full throttle condition in case of hot water flow because in half throttle condition, the fluid flow decreased for which there was a decrease in the mass flow rate and the velocity of flow of the fluid which accounted for a decrease in the coefficient of heat transfer and hence a decrease in the rate of heat transfer. The effectiveness is presented in (Fig. 8b).

Further, the effectiveness of N-H-C and C-H-N is compared and presented above (Fig. 8c), the effectiveness found in more case of N-H-C arrangement due to large surface exposed for the heat transfer and higher differences in temperature between the water above the room temperature and below room temperature in comparison with C-H-N arrangement.

3 Conclusion

Under the adopted experimental conditions and from the experimental observations pertaining to heat transfer in a triple-concentric-pipe heat exchanger with/without corrugations in the middle pipe, the following could be concluded concerning heat transfer between hot, cold and normal water (at room temperature) flowing through the heat exchanger fabricated for the purpose.

(i) N-H-C configuration exhibits better heat transfer capabilities.
(ii) Normal water (water at the temperature of the ambiance) was heated to a greater extent under C-H-N configuration as compared to that under N-H-C configuration.
(iii) Effectiveness of heat transfer was higher in the case pertaining to provision of corrugation on the outer surface of the middle pipe as compared to the case when corrugation was not provided.

(iv) The performance of the machine (heat exchanger) could be reduced by adopting half throttle conditions which reduced the hot water flow rate through the heat exchanger accounting for a reduced driving force for heat exchanges.

References

1. Incropera FP, Dewitt DP (1990) Fundamental of heat and mass transfer, third edn. Wiley, New York
2. Shah RK, Sekulic DP (2007) Fundamental of heat exchanger design. Wiley, Hoboken
3. Zuritz CA (1990) On the design of triple concentric-tube heat exchangers. J Food Process Eng 12:113–130
4. Chalaev D, Silnyagina N, Shmatok O, Nedbailo O (2016) Heat transfer enhancement in a corrugated tube heat exchanger. Ukrainian Food J 5:376–386
5. Wang SM, Li YZ, Wen J, Ma YS (2010) Experimental investigation of header Configuration on two-phase flow distribution in plate-fin heat exchanger. Int Commun Heat Mass Transf 37:116–120
6. Wang CC, Liaw JS, Yang BC (2011) Airside performance of herringbone wavy fin and Tube heat exchangers—data with larger diameter tube. Int J Heat Mass Transf 54:1024–1029
7. Akpinar EK, Bicer Y, Yildiz C, Pehlivan D (2004) Heat transfer enhancements in a concentric Double pipe exchanger equipped with swirl elements. Int Commun Heat Mass Transf 31:857–868
8. Joen CT, De Jaeger P, Huisseune H, Van Herzeele S, Vorst N, De Paepe M (2010) Thermo-hydraulic study of a single row heat exchanger consisting of metal foam covered round tubes. Int J Heat Mass Transf 53:3262–3274
9. Unal A (1998) Theoretical analysis of triple concentric-tube heat exchangers Part 1: mathematical modelling. Int Commun Heat Mass Transfer 25:949–958
10. Unal A (2001) Theoretical analysis of triple concentric-tube heat exchangers Part 2: case studies. Int Commun Heat Mass Transf 28:243–256
11. Ganji DD, Gorji-Bandpy M, Natali M (2016) Numerical study of finned type heat exchangers for ICEs exhaust waste heat recovery

Experimental Analysis on Home-Made Thermal Insulating Material

Ankita Ghosh, Amit Kumar Basu, and Siba Padarbinda Behera

Abstract Intumescent material is a type of material which can be regarded as one of the best types of thermal insulators for passive fire proofing as well as in thermal insulation in air conditioning industries. They can be used as a thermal barrier in various refrigeration and cooling systems. Such a material in refrigeration and cooling system will not allow the heat from the outer system to get conducted within it. This paper tries to justify the capability of the mixture having corn starch, baking soda and PVA glue as stated by NightHawkInLight through experimental setup to find its usage in refrigeration and air conditioning systems as a thermal insulator. The thermal conductivity and dielectric permittivity of the starlite material were found out. Furthermore, a flame test using oxyacetylene flame was conducted to check the maximum range of temperature that the material can sustain.

Keywords Starlite material · Phase change material · PCM material · Thermal insulator · Lee's apparatus

1 Introduction

Starlite [1] is claimed to be the next generation insulating material which has the capability to withstand immense heat. During the 1970s and 1980s as reports suggest, starlite was invented by a British Chemist, Maurice Ward. He revealed the secret recipe and composition of this miraculous material only to few of his close relatives before his death in 2011. The American company Thermashield, LLC, claims to have acquired the rights to starlite and replicated it [2]. A mixture of corn starch, baking soda and PVA glue was used to make a dough like substance. After drying, the sample hardened. When it came in contact with very high amount of heat, a foamy layer of carbon was formed. This layer reportedly barred heat transfer through the sample. When an intumescent comes in contact with a huge amount of heat, a black char type of material is produced [3]. On heating any ablative coating, they get decomposed

A. Ghosh · A. K. Basu · S. P. Behera (✉)
School of Mechanical Engineering, Kalinga Institute of Industrial Technology, Bhubaneswar, Odisha, India
e-mail: siba.behera@kiit.ac.in

M. Ramgopal et al. (eds.), *Advances in Air Conditioning and Refrigeration*, Lecture Notes in Mechanical Engineering, https://doi.org/10.1007/978-981-15-6360-7_14

and release the water vapour henceforth producing a cooling effect. It reduces the rate of heat transfer from the exposed side. Char is a foamy kind of material which has high porosity, and being a poor conductor of heat, it reduces the rate at which heat transfer takes place. The reaction occurs in a matrix of a binder which is molten in nature, for example, vinyl acetate co polymers. Soft char is useful when a light layer or a thin layer of insulation is required. It is used for secondary fire proofing in building structures. Hard char is used in fire proofing externally as they are incapable of fireproofing in the interior [4]. PCM materials or phase change materials are the ones which are often used in industries requiring thermal insulation [5]. They also contribute in storing thermal energy for waste heat recovery systems [6]. It is a type of material which experiences a change in its phase with change in temperature as well as time [7]. In some air conditioning systems, foam insulation is provided to restrict heat conduction through pipes carrying coolants and hot fluids [8] (Fig. 1).

Fig. 1 Intumescent Resin [9]

Fig. 2 Specimen used for Lee's test

2 Experimental Setup

2.1 Test for Thermal Conductivity

The greatest achievement of a good thermal insulator is its low thermal conductivity. Accordingly, the thermal conductivity was tested using Lee's apparatus (Fig. 3). A sample material was created. The sample had corn starch, baking soda and PVA glue as its basic constituents. The corn starch and baking soda were mixed in the ratio 10:1, respectively. After addition of PVA glue, the weight of the sample composite was found to be 48 g (Fig. 2).

Through the Lee's disc experiment, a value for the thermal conductivity k of a bad conductor can be calculated. The disc is made out of the poor conductor material

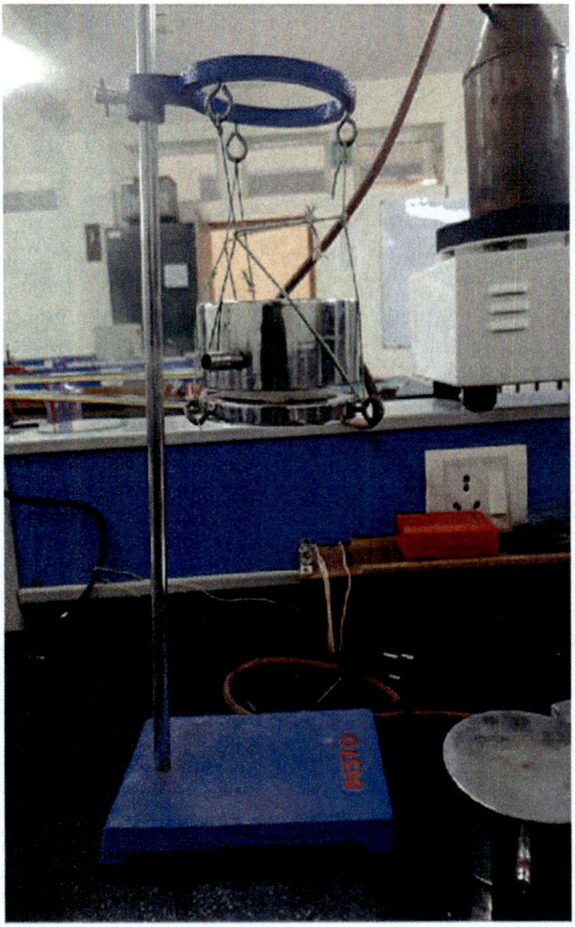

Fig. 3 Lee's apparatus setup for determination of thermal conductivity

having thickness 9 mm and diameter 75.4 mm. It is kept in between two metal discs which are good conductors one of which is connected to a steam chamber.

The approximate thermal conductivity of the sample specimen can be found out using the formula:

$$k = [\{(msd)/A(\Theta_1 - \Theta_2)\} * (\mathrm{d}\Theta/\mathrm{d}t) * \{(r + 2h)/(r - 2h)\}]$$

In the above equation,

k = approximate thermal conductivity

s = Specific heat of the plate material (stainless steel) = 468 J/kg

d = thickness of the specimen = 6.87×10^{-3}m

A = surface area of the specimen = $(41.853 \times 0.001)\ \mathrm{m}^2$

$\Theta 1$ = Temperature of the steam chamber = 98 °C

$\Theta 2$ = Temperature of the lower disc = 55 °C $\mathrm{d}\Theta/\mathrm{d}t$ = slope of the curve = 0.011

r = radius of the insulator specimen = (3.65×0.01) m h = thickness of the lower disc plate = (2.5×0.01) m

2.2 *Test Under High Temperature Oxyacetylene Flame*

The sample was made to dry in sunlight. After that, it was put under high temperature oxyacetylene flame. The oxyacetylene flame has a temperature of more 4000 °C. One side of the specimen was exposed to the flame for some amount of time. The other side which was not completely exposed to the high temperature was immediately touched after the flame was taken back. The temperatures were noted on both the sides (Fig. 4).

Fig. 4 Specimen exposed to oxyacetylene flame

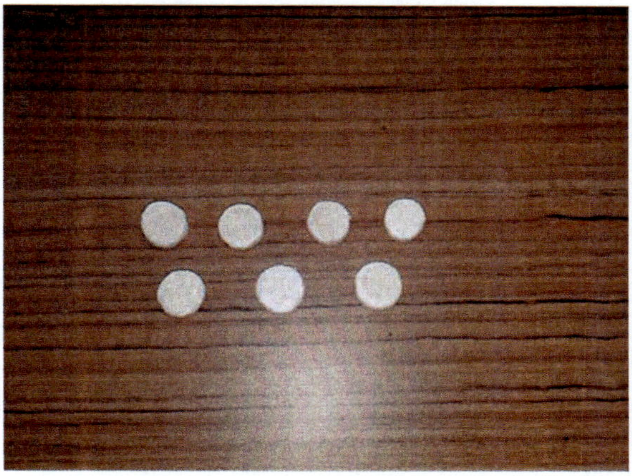

Fig. 5 Button-sized specimen for impedance test

2.3 Test for Dielectric Permittivity

Impedance Test apparatus. The resistance, capacitance and storage capacity were found out from the test. The test for dielectric property was conducted using the sample which was used for the test where of a smaller dimension than the earlier sample. The thickness of the sample was measured to be 3.75 mm and the diameter 13.42 mm (Fig. 5).

3 Results and Discussion

3.1 Test for Thermal Conductivity

The thickness and diameter of the specimen were calculated using vernier callipers and screw gauge, respectively. The thickness and the diameters at different sections of the sample differed slightly. They were measured at different sections, and the mean value was considered for calculation purpose (Tables 1, 2 and 3).

Table 1 i. Thickness of specimen

S. No.	Thickness in mm	Mean thickness in mm (d)
1	7.05	
2	7.03	
3	7.03	
4	6.01	6.87
5	7.01	
6	7.04	

Table 2 Diameter of specimen

S No.	Diameter in mm	Mean Diameter in mm (d)
1	73.04	
2	73.03	
3	72.04	72.43
4	72.04	
5	72.04	

$$\text{Thickness} = (7.05 + 7.03 + 7.03 + 6.01 + 7.01 + 7.04)/6 = 6.87\,\text{mm}$$
$$\text{Diameter} = (73.04 + 73.03 + 72.04 + 72.04 + 72.04)/5 = 72.43\,\text{mm}$$

The setup is allowed to reach a steady state. Here, the objective is to make the heat lost by the lower disc due to convection same as the heat flowing through the sample. The temperatures of the upper disc (Θ1), lower disc (Θ2) were recorded. The sample was removed, and the lower metal disc was allowed to attain the same temperature as the upper disc temperature (Θ2). Later, the upper disc was replaced by the sample which was supposedly an insulator. The setup was allowed to cool down. Throughout the cooling process, the temperature of the metal disc at every instant of time was noted. A curve was plotted using the values. The slope of the curve was measured where the curve passes through Θ1.

From the above values of parameters, the value of thermal conductivity was calculated and found to be **0.01243 W/mK**.

3.2 *Test Under High Temperature Oxyacetylene Flame*

The temperature on this side was recorded to be around 56 °C. Hence, the specimen could actually inhibit the heat from reaching the unexposed side of it. Char type of material was visible on the exposed side after it was separated from the flame (Fig. 6). The foam which is generally used as thermal insulator can sustain a temperature of 100 °C. Furthermore, when a relatively thinner sample having thickness 1.94 cm and diameter 12.8 cm was exposed to that same oxyacetylene flame for more than two minutes, the sample was turned to ashes (Table 4; Figs. 7, 8, 9 and 10).

The experimented material is henceforth found to have a thermal conductivity of 0.01243 W/mK, and it can sustain a temperature of above 4000 °C. Also, it forms char, so it is non-toxic for nature.

Table 3 Cooling characteristics

Time in seconds	Temperature in °C
0	65
30	64.5
60	64
90	63.5
120	63
150	62.5
180	62
210	61.5
240	61
270	60.5
300	60
330	59.5
360	59
390	58.5
420	58
450	58
480	58
510	57.5
540	57
570	56.5
600	56
630	55.5
660	55
690	55
720	54.5
750	54
780	54
810	53.5
840	53.5
870	53
900	53
930	52.5
960	52
990	52
1020	51.5
1050	51.5
1080	51

(continued)

Table 3 (continued)

Time in seconds	Temperature in °C
1110	51
1140	51.5
1170	50.5
1200	50
1250	50
1280	50
1310	49.5
1340	49
1370	49
1400	48.5
1430	48.5
1460	48
1490	48
1520	48
1540	47.5
1570	47
1600	46.5
1630	46.5
1660	46
1690	46
1720	46
1750	45.5
1780	45.5
1810	45
1840	45
1870	45

3.3 Test for Dielectric Permittivity

Using the apparatus, values for resistance, electrical conductivity and impedance were found out. After that, using Origin 8.5 software, the graphs between frequency and impedance were found out for several cases. The first test was carried on at room temperature. After that, the tests were conducted at 40, 60, 80, 100 and 120 °C. For each time, the graphs between impedance and frequency were plotted (Figs. 11, 12, 13, 14, 15, 16 and 17).

From the above test and from the graphs between impedance and frequency at different temperatures, it can be concluded that the impedance changes for every temperature. With change in temperature, positive temperature coefficient is being shown by the sample.

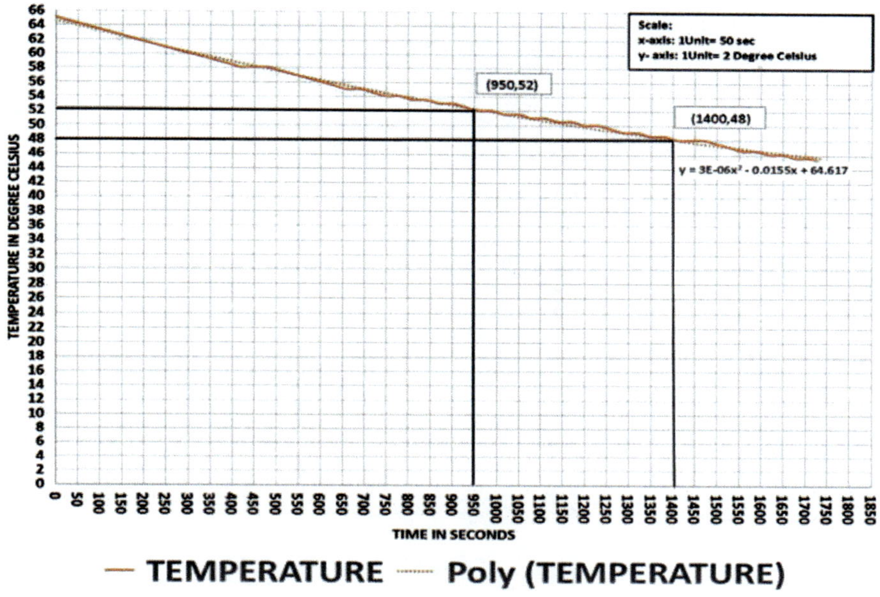

Fig. 6 Cooling Curve (Temperature versus Time)

Table 4 Comparison between conductivities of some thermal insulators

Material	Thermal Conductivity in W/mK	Maximum Temperature material sustains (°C)
Teflon	0.250	326
Extended Polystyrene	0.046	240
Fibreglass	0.045	1200
Mineral Wool	0.04	1000
Cellulose	0.04	380
Polyurethane foam	0.03	100
Silica Aerogel	0.02	650

4 Conclusion

In general, foam is used in refrigerating industries as an insulating material. However, it is toxic in nature. The material tested here is made from ingredients present at home, and hence, it is biodegradable in nature. When completely burnt, it only produces ashes unlike usual foam. Moreover, foam is much more costlier than the experimented material. The material that was tested had a low thermal conductivity, and it could resist the high temperature of the oxyacetylene flame, and it also showed good

Fig. 7 Sample exposed to oxyacetylene flame

Fig. 8 Formation of char like material on thicker specimen after flame test

Fig. 9 Thinner specimen for flame test

Fig. 10 Charred residue of the thinner specimen after flame test

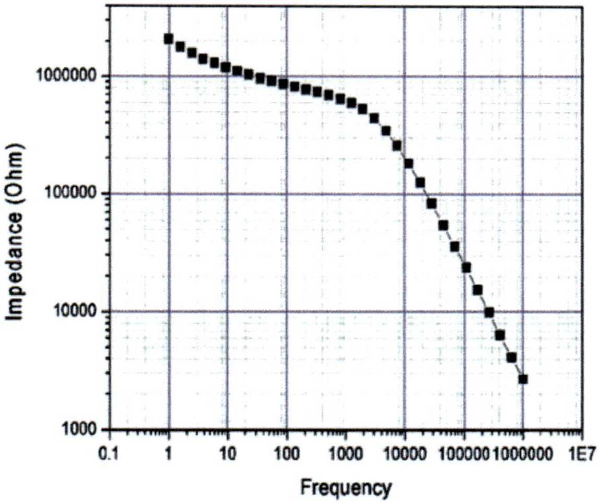

Fig. 11 At room temperature

dielectric property. Hence, in refrigerating industries, this phase changing material can be considered as a replacement for foam in future.

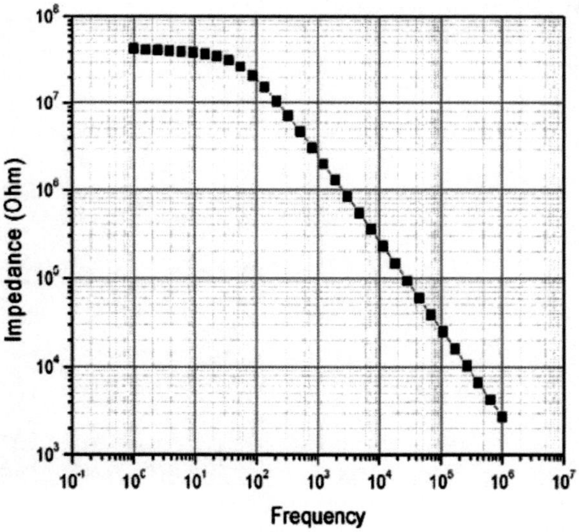

Fig. 12 At 40 °C

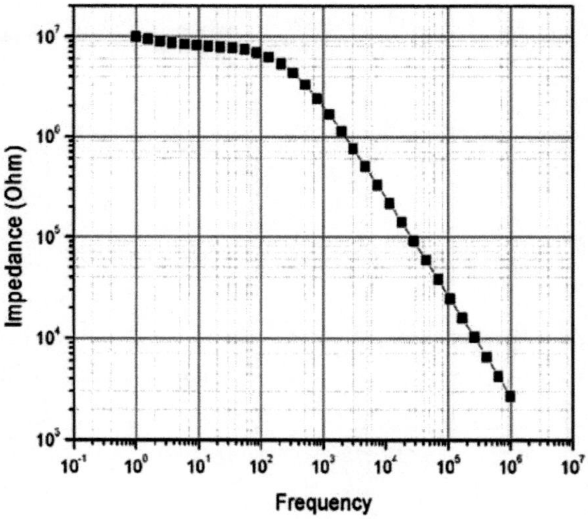

Fig. 13 At 60 °C

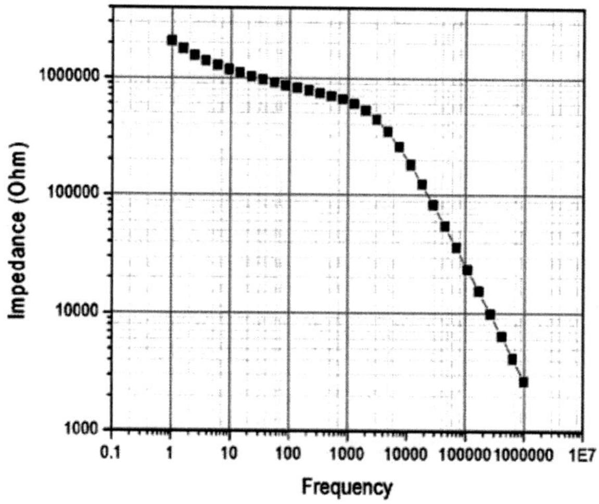

Fig. 14 At 80 °C

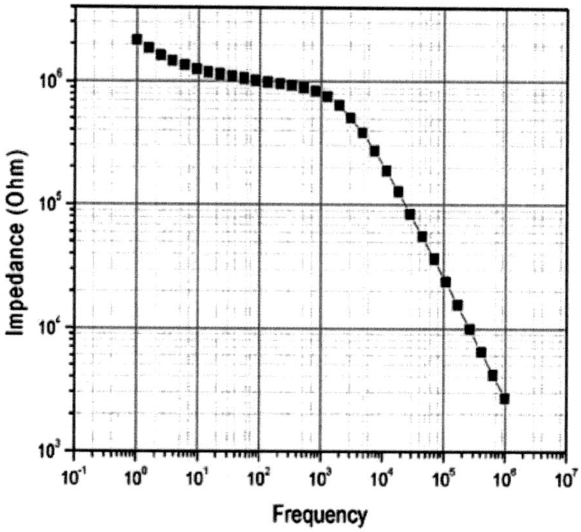

Fig. 15 At 100 °C

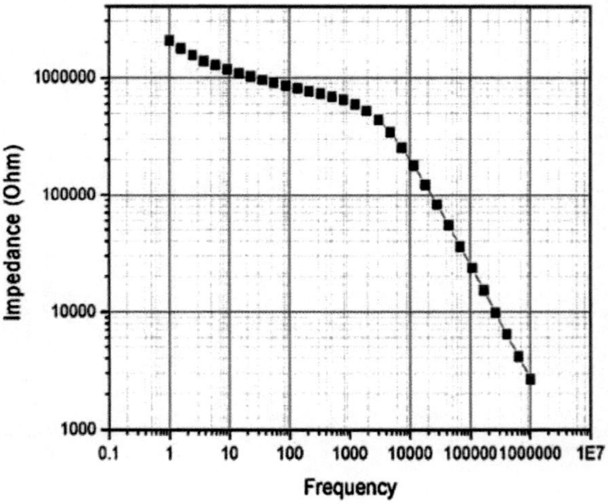

Fig. 16 At 120 °C

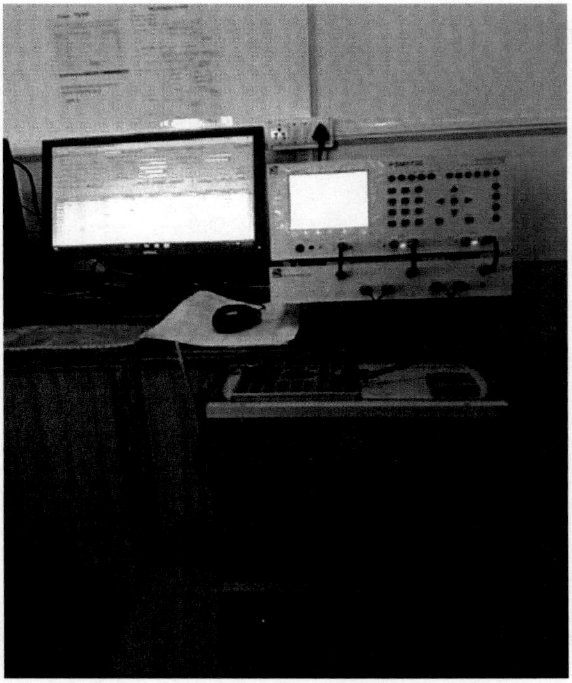

Fig. 17 Test for impedance

References

1. Keene J. Starlite: the miracle material that could be lost forever. Available https://www.theverge.com/2012/5/17/3026074/starlite-maurice-ward-plastic-fireproof
2. Starlite by Thermashield. Available: https://www.starlitethermashield.com/
3. Intumescent-Wikipedia. Available: https://en.wikipedia.org/wiki/Intumescent
4. Miguel AS, Perry JL, Wittman GR (1981) Intumescent material-honeycomb thermal barrier. US Patent 4,299,872
5. Mondal S (2008) Phase change materials for smart textiles—An overview. Appl Therm Eng 28(11–12):1536–1550
6. Pielichowska K (2014) Phase change materials for thermal energy storage. Progr Mater Sci 65:67–123
7. Bajaj P (2001) Thermally sensitive materials. Smart Fibres, Fabrics and Clothing, Woodhead ..., books.google.com
8. Stonitsch LJ (1999) Foam insulation system for pipes. US Patent 5,996,643
9. Achim Hering (1986) https://en.wikipedia.org/wiki/File:Intumescent.jpg. From Wikipedia

Comparative Study of Positioning of Air Conditioner in a Room Using CFD

Manoj Kumar Gopaliya, Madhu Kalyan Reddy Pulagam, and Neha Kumari

Abstract This paper has presented a comparative study of the positioning of an air conditioner in a room using computational fluid dynamics (CFD) simulations. Four different positions namely lower, middle, upper half of wall, and centre of the roof have been studied during this research work, and effect on different thermos-fluid characteristics has been evaluated for these positions. A comparative analysis of these effects has been presented. Considering the overall thermo-fluid flow performance, the current study suggests mounting position at the upper half of the wall as the most suitable among the positions considered.

Keywords CFD · Air conditioner · Positioning · Flow analysis

1 Introduction

An air conditioner is the most power-consuming electrical appliance of any household. Likewise, the overall air conditioning load of any industry or organization is always among the top power-consuming appliances and machines. This very fact is continuously encouraging the researchers across the globe to look for ever-improving air conditioning process. The selection of suitable positioning of an air conditioner in the flow domain is also very critical from power consumption standpoint and hence motivating many types of research in this area.

The design and development of any air conditioning (AC) unit require proper understanding of the dynamics of the airflow in the flow domain. The conventional methods of flow measurement are quite expensive and time-consuming [1]. On the other hand, CFD-based researches give comparable results in a cost and time-effective manner. This has made CFD-based researches increasingly popular in the field of air conditioning. The trend gets further encouraged by ever-increasing computational capabilities which help in extracting both qualitative and quantitative data set for

M. K. Gopaliya (✉) · M. K. Reddy Pulagam · N. Kumari
Department of Mechanical Engineering, CGU-Odisha (Formerly C. V. Raman College of Engineering), Bhubaneswar 752054, Odisha, India
e-mail: manojkumargopaliya@gmail.com

M. Ramgopal et al. (eds.), *Advances in Air Conditioning and Refrigeration*, Lecture Notes in Mechanical Engineering, https://doi.org/10.1007/978-981-15-6360-7_15

important flow parameters; many of these are otherwise difficult or impossible to obtain experimentally.

CFD simulations have helped in analysing and predicting indoor air distribution in air conditioning flow domains [2–7]. Researchers have also undertaken CFD simulations for optimizing the performances of both heating and cooling of an air conditioning system [8–10]. Likewise, researches focusing on indoor thermal comfort and ventilation effectiveness have also roped in CFD simulations for faster results [11–15].

With the growing popularity of CFD-based researches in the field of air conditioning, efforts are also made to improve the performance of different models and also to present the comparative picture of the performance of these models [16–23].

This paper has presented a comparative study of the effect of different positions of air conditioner installation on important flow and thermal characteristics in a room. The room under consideration is having a heat source at the centre equivalent to the body heat of two persons along with constant heat fluxes at walls and an insulated floor. Both qualitative and quantitative analysis has been presented for a comprehensive and better understanding of flow and thermal behaviour of air in the room. The following flow chart describes the work process involved in solving the problem.

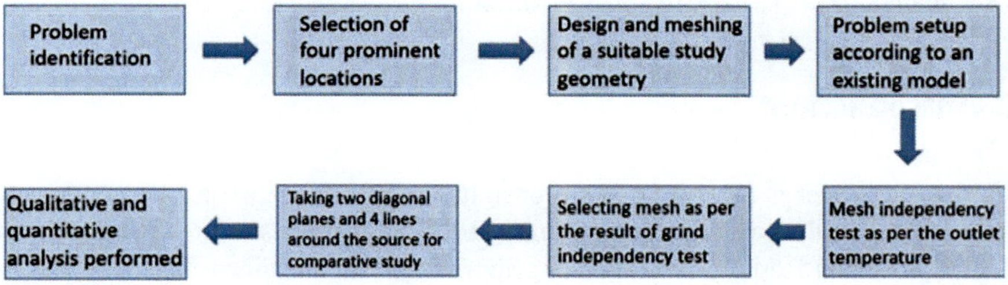

2 Mathematical Model

The selection of appropriate mathematical models is paramount important in any numerical study based on CFD. The selected model also needs to get validated for the same or similar type of flow situations before adopting for simulations.

The mathematical model used by Yao et al. [24] has been selected as the flow situation in the current study is quite similar to what was prevailing in their study. Also, they have duly validated the model against experimental results before adopting it. Details regarding assumptions made, governing equations, wall functions, coupling methodology, etc., can be seen in their research article [24].

3 Geometry and Mesh Generation

The geometry is a simple 10 m × 10 m × 10 m of the room with a sphere of radius 1 m at the centre serving as heat source equivalent to the body heat of two persons. The AC vents are represented by two 1 m × 0.5 m rectangular slots on one of the walls of the room which act as inlet and outlet for the air conditioner. Four different positionings for the AC on this wall have been tried during this study. These positions are very common in practice for room air conditioners (Figs. 1 and 2).

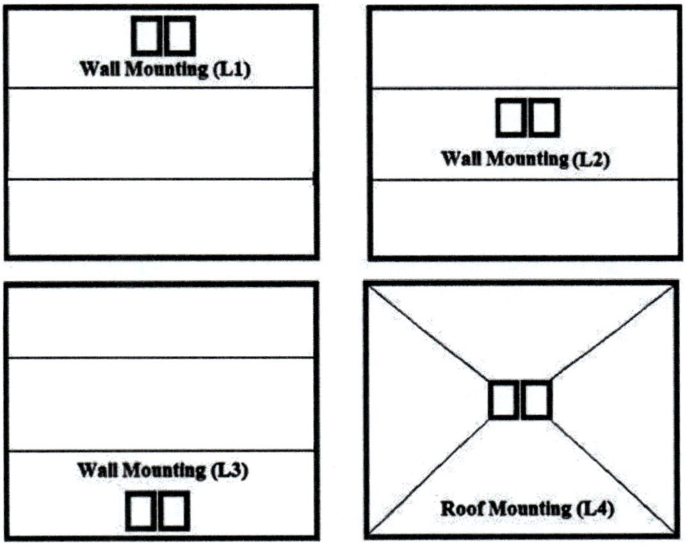

Fig. 1 Schematic layout of mounting positions

Fig. 2 Sample half-section geometry of the four orientations

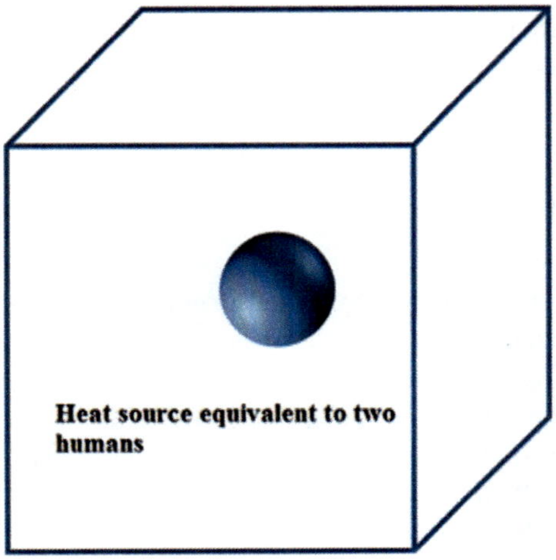

L1—AC vents near to the roof of the room which is the general positioning of a split air conditioner

L2—AC vents at the middle of the room which is the general positioning of a window air conditioner

L3—AC vents near to the floor of the room. This peculiar positioning is sometimes used for a window AC and also for a duct type AC

L4—AC vents on the roof of the room which is the general positioning of a duct type air conditioner.

A grid independence test has been done, and the resulting mesh has approximately 1.8 lacs elements for all the four cases under consideration. Meshing is done by a patch conforming method with a combination of edge and face sizing.

4 Boundary Conditions

Air enters through the inlet with a velocity of 3.5 m/s at a temperature of 290 K at atmospheric pressure (1 atm). The outlet is open to the atmosphere. Walls have a heat flux of 50 W/m^2 each, while the floor is insulated. The heat source (sphere at the centre of the cubical room) has a heat flux of 11.94 W/m^2 which is 150 W in total, the equivalent of body heat emitted by two persons.

5 Solution Process and Convergence Criterion

The selected CFD solver is pressure based on the SIMPLE pressure–velocity coupling. The energy and momentum equations are discretized using a second-order upwind method, while the equations of turbulent kinetic energy and specific dissipation rate are having first-order upwind discretization. A suitable convergence criterion has been adopted for ending the iteration.

6 Results and Analysis

This paper has presented a comparative study of the effect of different positions of air conditioner installation on important flow and thermal characteristics in a room. Four different mounting positions have been analysed during this study. This paper has presented a comparative study of the effect of different positions of air conditioner installation on important flow and thermal characteristics in a room. Important flow and thermal characteristics at selected observation planes and lines in the flow domain have been evaluated for each mounting position. Figure 3 has shown planes and lines

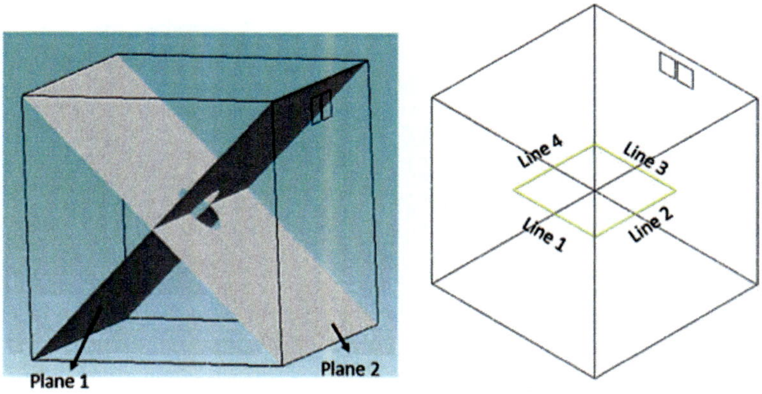

Fig. 3 Sample case to show the orientation of planes and the position of lines in each case

on which observations have been taken.

Both qualitative and quantitative aspects of this thermo-fluid analysis have been presented in this paper. Contours of different thermo-fluid flow characteristics have been obtained in both "Plane 1" and "Plane 2" where quantitative representations have been done across "Line 1 to Line 4".

6.1 Mesh Independency Check

Mesh independency check is one of the important steps in any numerical analysis. This process checks the sensitivity of results towards mesh density. Here, solutions are obtained for progressively finer meshes, and the output at a selected location is captured for comparison. "Mesh Number" after which output becomes constant is selected for further simulations. The above process for the current study has been summarized in Table 1 and Fig. 4. Mesh Number 3 has been selected in this case.

Table 1 Mesh independency data

S. no.	Mesh number	Number of elements ($\times 10^5$)	Outlet temperature
i	1	0.8	291.15
ii	2	1.2	290.92
iii	3	1.8	290.90
iv	4	3.8	290.90

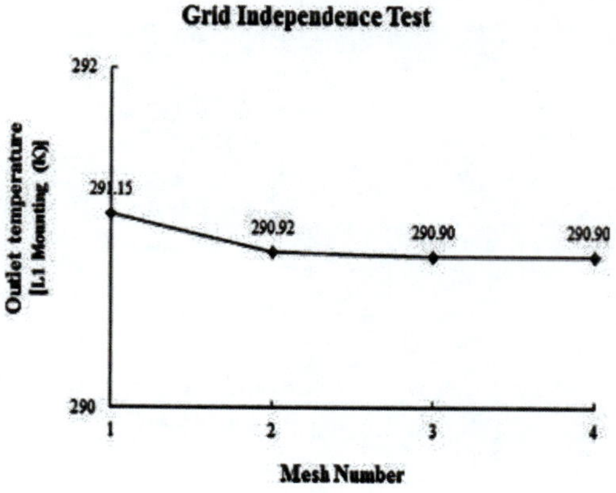

Fig. 4 Mesh independency test data

6.2 *Qualitative Analysis*

a. **Velocity Contours and Velocity Vector Distribution**

Contours of flow velocity have been obtained at "Plane 1" and "Plane 2" for each mounting position and shown in Figs. 5 and 6, respectively. Similarly, velocity vector distributions have also been obtained and presented in Figs. 7 and 8 for "Plane 1" and "Plane 2", respectively.

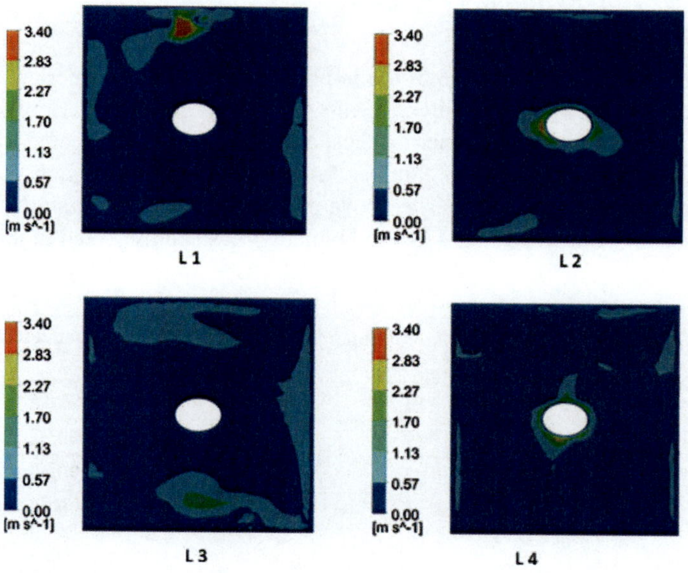

Fig. 5 Velocity contours on plane 1

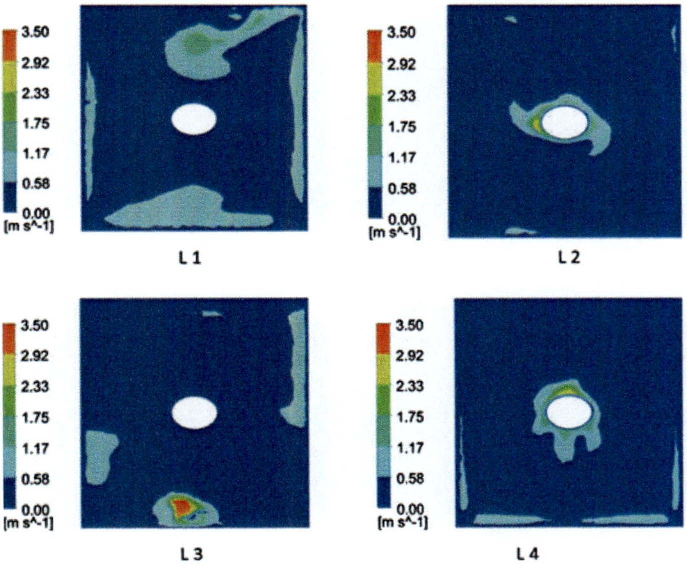

Fig. 6 Velocity contours on plane 2

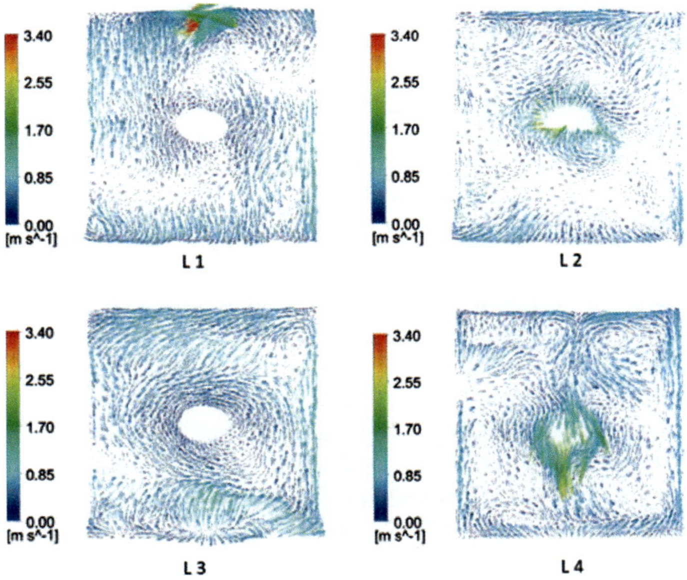

Fig. 7 Velocity vector distribution on plane 1

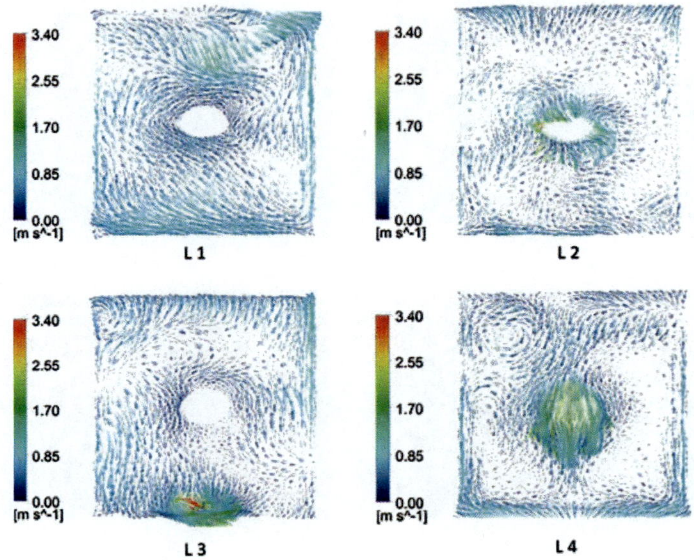

Fig. 8 Velocity vector distribution on plane 2

It has been observed that mounting positions "L2" and "L4" give the most consistent velocity distribution in the flow domain which is confirmed by careful examination of velocity contour and vector distribution on selected planes. This is an important aspect of evaluating human comfort. This is due to equi-distance orientation of these mounting locations from surrounding walls including roof and floor.

b. **Turbulence Contours**

Turbulence contours have been drawn for turbulent kinetic energy and turbulent eddy frequency. Both the contours have been drawn at "Plane 1" and "Plane 2" for all four mounting positions.

Figures 9 and 10 show turbulent kinetic energy contours for "Plane 1" and "Plane 2", respectively. Similarly, Figs. 11 and 12 show turbulent eddy frequency contours for these planes, respectively. Careful observation of these contours confirms that the turbulence distribution has been quite uniform across the flow domain except near the centre heat source for all four mounting positions. Also, its frequency has also been fairly constant across the flow domain for all four cases. These observations may be attributed to AC unit flow speed which is kept mild enough to ensure human comfort. This makes us rely on other factors for deciding the best mounting position among the four positions considered during the present study.

c. **Temperature Contours**

Temperature contours have been plotted for "Plane 1" and "Plane 2" for all four mounting positions to analyse the temperature distribution across the flow domain. These are presented in Figs. 13 and 14, respectively, for "Plane 1" and "Plane 2".

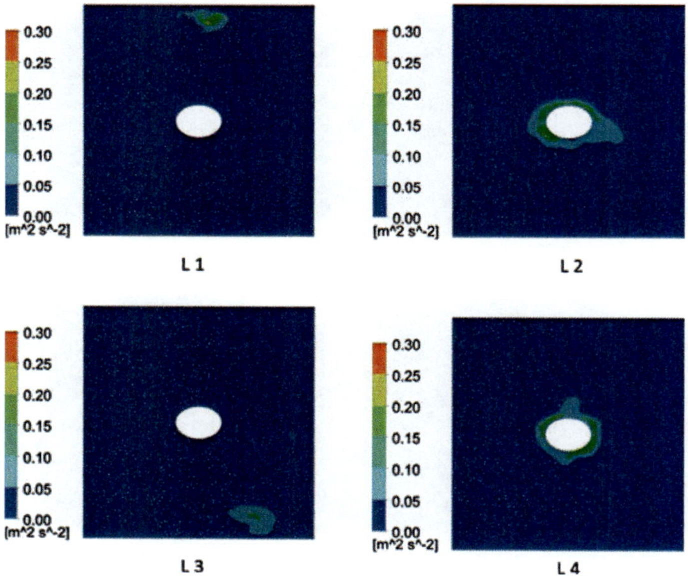

Fig. 9 Turbulence kinetic energy contours on plane 1

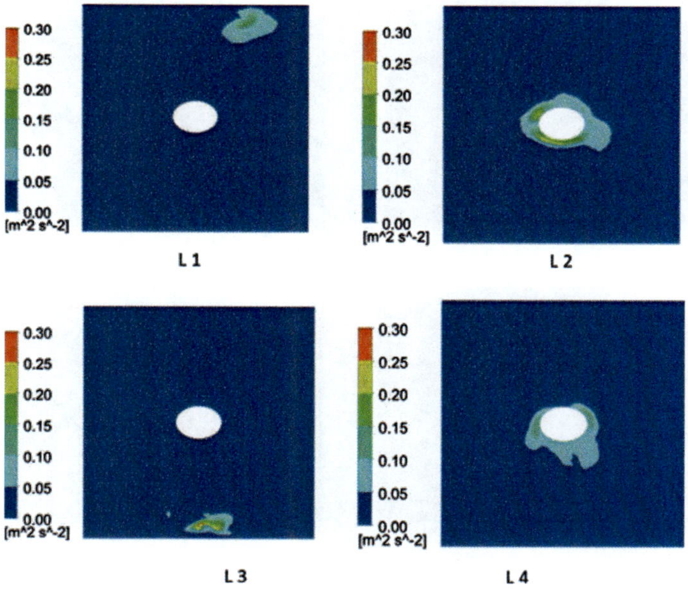

Fig. 10 Turbulence kinetic energy contours on plane 2

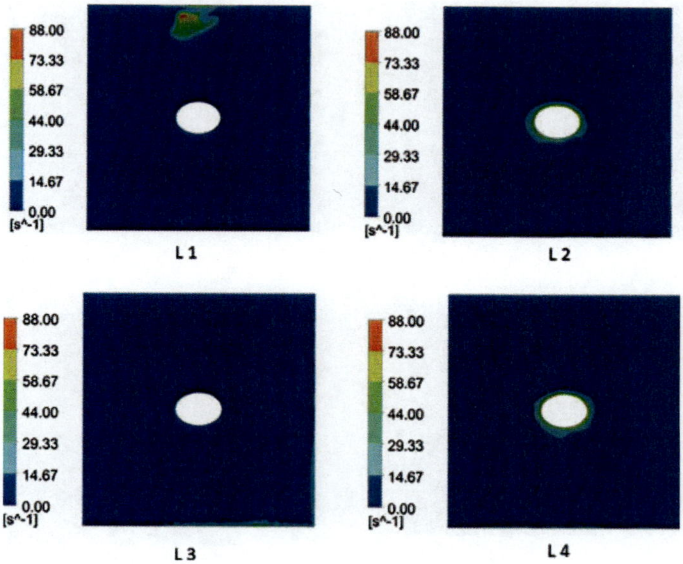

Fig. 11 Turbulence eddy frequency contours on plane 1

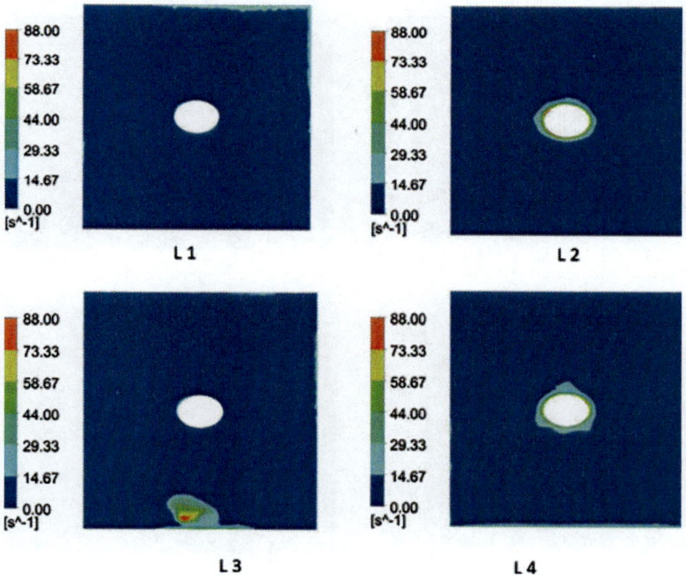

Fig. 12 Turbulence eddy frequency contours on plane 2

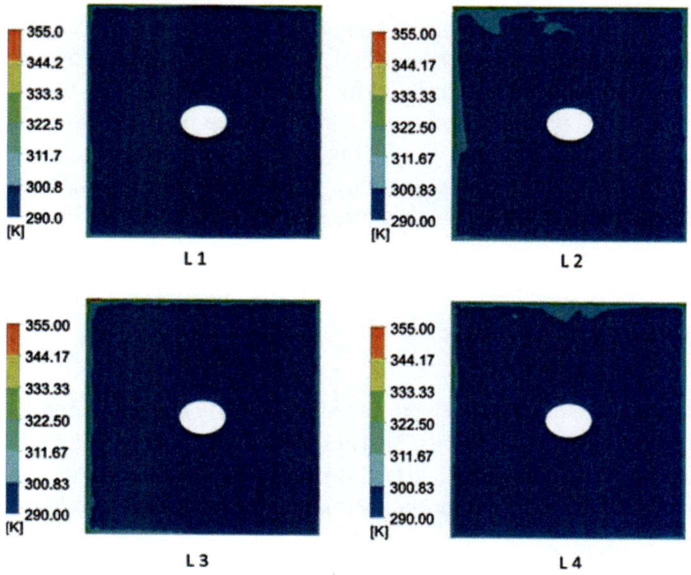

Fig. 13 Temperature contours on plane 1

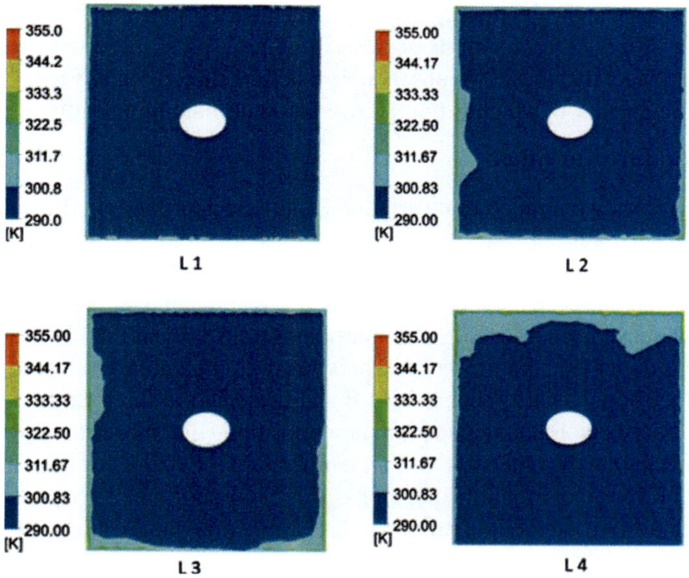

Fig. 14 Temperature contours on plane 2

Temperature distribution has been found to be fairly uniform at "Plane 1" for all mounting positions. However, non-uniformity in temperature distribution near the walls has been noticed at "Plane 2" for all mounting positions except L1. This confirms that the mounting position "L1" has been the best performing among all positions considered during this study. This may be attributed to the fact that being at the upper segment of the wall, AC at this position cools the hot air, locating nears the roof (due to buoyancy effect), most effectively.

6.3 Quantitative Analysis

Quantitative analysis has also been presented for velocity and temperature variations at selected locations (shown in Fig. 3). These locations (lines) are equidistant from surrounding walls including roof and floor so that an overall effect of the thermo-flow characteristics can be reflected through variation drawn across these lines.

a. **Velocity Variation**

Figure 15 has shown velocity variation along selected lines viz. Line 1, Line 2, Line 3, and Line 4. Velocity distributions along these lines are fairly uniform for mounting positions L1, L3, and L4. However, spikes in the distribution along Line 1 and Line 3 have been observed for mounting position L2. These spikes in velocity distribution are due to the fact that these two observatory lines are directly facing the AC inlet in this mounting position; which is not the case in other mounting positions

b. **Temperature Variation**

Figure 16 has shown temperature variation along selected lines viz. Line 1, Line 2, Line 3, and Line 4 for all four mounting positions under consideration. It has been observed that mounting positions L1 and L3 are showing near-constant temperature distribution along selected observatory lines. However, some variations along Line 1 and Line 3 have been noticed for mounting positions L2 and L4.

AC unit's outlet temperature which represents the temperature of air getting drawn out from the room has also been recorded and presented in the form of a bar chart (Fig. 17). It has been found that the mounting position L1 achieves the lowest outlet temperature among the others. The outlet temperature is 290.9 K for L1. The same is around 295 K for other mounting positions. This can be attributed to a fairly uniform cooling achieved throughout the room in the L1 position due to buoyancy effects.

7 Conclusions

This paper has presented a comparative study of the effect of different positions of air conditioner installation on important flow and thermal characteristics in a room. Four different mounting positions have been analysed during this study.

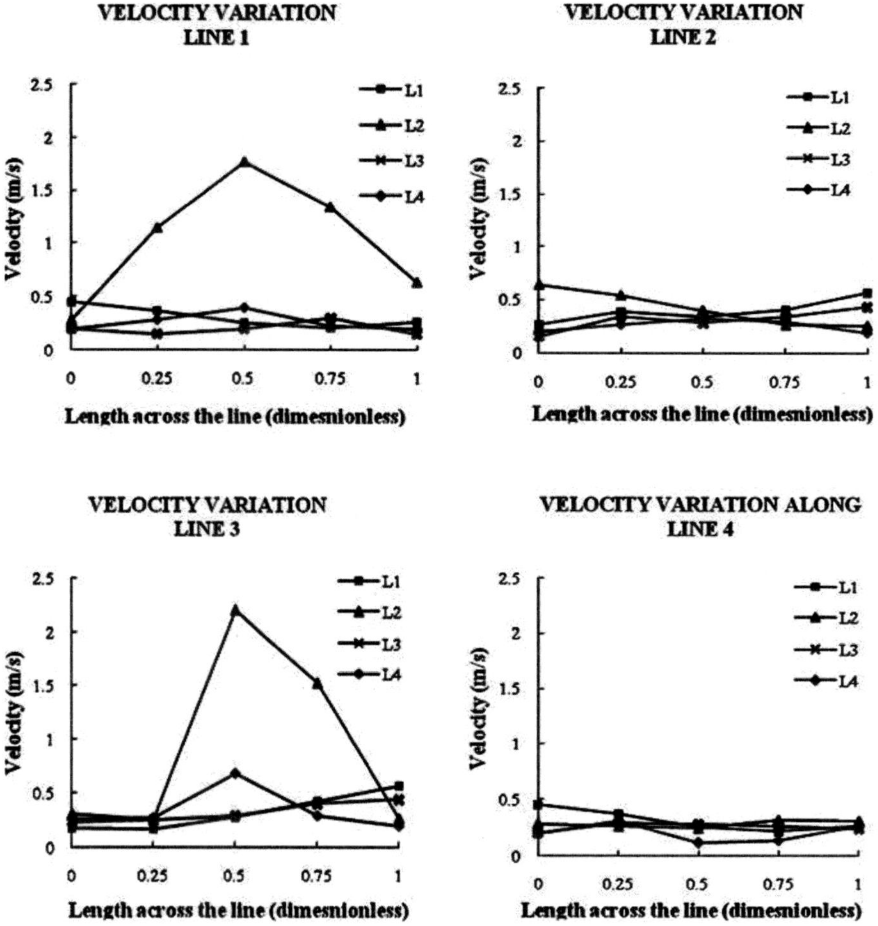

Fig. 15 Velocity variation across selected lines

The following conclusions have been drawn from the current study:

i. Mounting positions "L2" and "L4" give the most consistent velocity distribution in the flow domain.
ii. Flow turbulence and its frequency find to be quite uniform across the flow domain except near the centre heat source for all mounting positions considered during this study.
iii. Mounting position "L1" gives the most uniform temperature distribution across the flow domain among all mounting positions considered during this study.
iv. Velocity distributions along all four observatory lines are fairly uniform for mounting positions L1, L3, and L4.
v. Spikes in velocity distribution along Line 1 and Line 3 have been observed for mounting position L2.

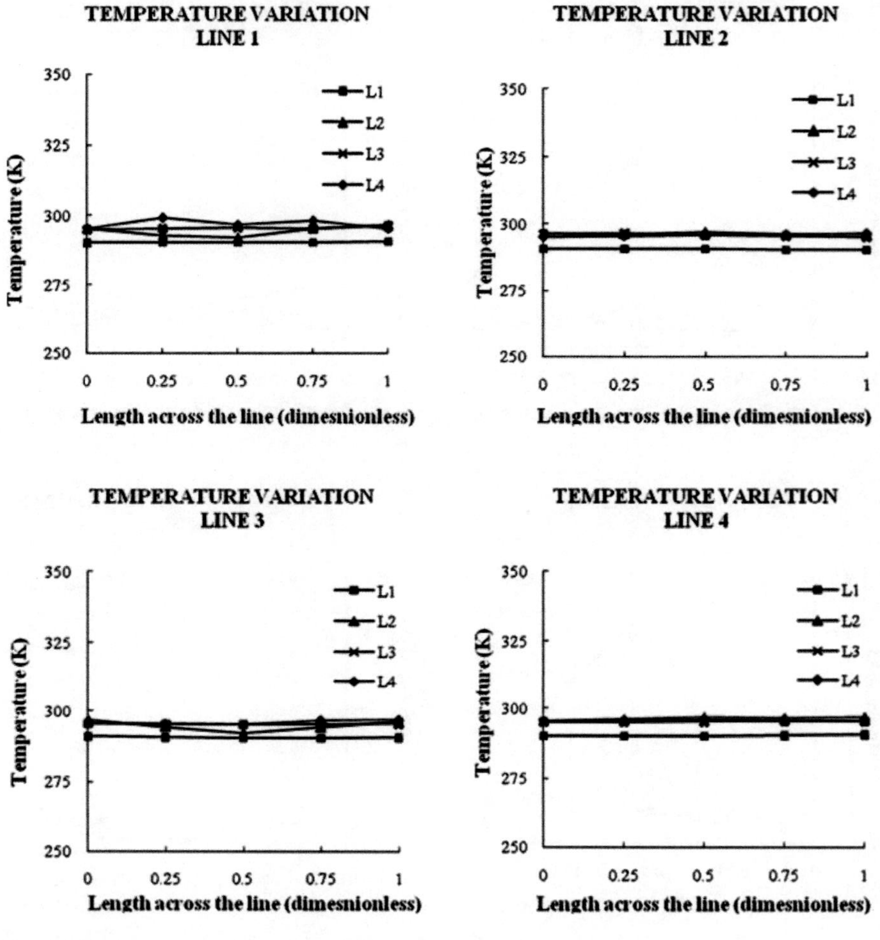

Fig. 16 Temperature variation across the four lines for all the cases

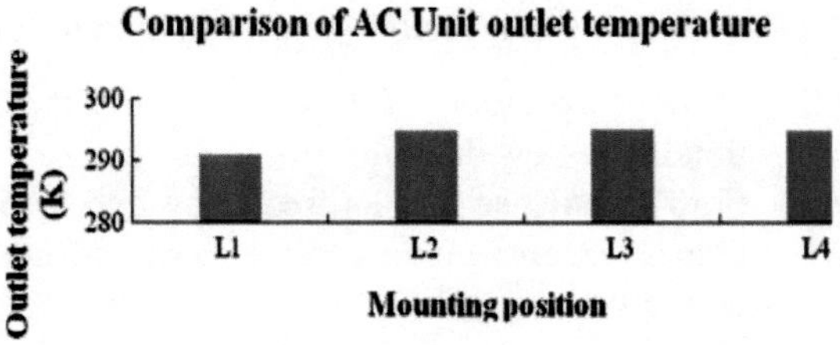

Fig. 17 Comparison of outlet temperature for different mounting positions

vi. Mounting positions L1 and L3 are showing near-constant temperature distribution along all four observatory lines.
vii. Variations along Line 1 and Line 3 have been noticed for mounting positions L2 and L4.

viii. Mounting position "L1" gives the lowest outlet temperature among all mounting positions considered during this study.

Considering the overall thermo-fluid flow performance, the current study suggests mounting position L1 to be the most suitable among the positions considered during this study.

References

1. Huang Y, Shen X, Li J, Li B, Duan R, Lin C, Liu J, Chen Q (2015) A method to optimize sampling locations for measuring indoor air distributions. Atmos Environ 102:355–365
2. Huang C, Lee J, Yu W, Yu W (2011) Study of the temperature flow field change by air-conditioning system in an indoor circular stadium. In: Proceedings of the 2011 international conference on consumer electronics, communications and networks (CECNet), pp 3568–3571, Xianning, China, (2011)
3. Gao CF, Lee WL (2009) Optimized design of floor-based air-conditioners for residential use. Build Environ **44**(10):2080–2088
4. Karlsson JF, Moshfegh B (2005) Investigation of indoor climate and power usage in a data center. Energy Build 37(10):1075–1083
5. Cho J, Lim T, Kim BS (2009) Measurements and predictions of the air distribution systems in high compute density (Internet) data centers. Energy Build 41(10):1107–1115
6. Yang L, Ye M, He BJ (2014) CFD simulation research on residential indoor air quality. Sci Total Environ 472:1137–1144
7. Cehlin M, Karimipanh T, Larsson U (2014) Unsteady CFD simulations for prediction of airflow close to a supply device for displacement ventilation. In: Proceedings of the 13th international conference on indoor air quality and climate, IndoorAir 2014, pp 47–54, Hong Kong
8. Li K, Xue W, Xu C, Su H (2013) Optimization of ventilation system operation in office environment using POD model reduction and genetic algorithm. Energy Build 67:34–43
9. Qi R, Lu L (2014) Energy consumption and optimization of internally cooled/heated liquid desiccant air-conditioning system: a case study in Hong Kong. Energy 73:801–808
10. Cui X, Chua KJ, Yang WM (2014) Numerical simulation of a novel energy-efficient dew-point evaporative air cooler. Appl Energy 136:979–988
11. Cehlin M, Taghi K, Ulf L (2014) Unsteady CFD simulations for prediction of airflow close to a supply device for displacement ventilation. In 13th International Conference on Indoor Air Quality and Climate, Indoor Air 2014, 7–12 July 2014, Hong Kong, pp 47–54
12. Wang D, Zhang Z, Han L, Li L (2011) Numerical simulation on air conditioning system and flow field of passenger compartment of type—a subway. Pain 152(6):1317–1326
13. Ascione F, Bellia L, Capozzoli A (2013) A coupled numerical approach on museum air conditioning: energy and fluid dynamic analysis. Appl Energy 103:416–427
14. Norton T, Grant J, Fallon R, Sun D-W (2009) Assessing the ventilation effectiveness of naturally ventilated livestock buildings under wind dominated conditions using computational fluid dynamics. Biosys Eng 103(1):78–99
15. Yang Li, Ye Miao (2014) CFD simulation research on residential indoor air quality. Sci Total Environ 472:1137–1144
16. Li B, Liu J, Luo F, Man X (2015) Evaluation of CFD simulation using various turbulence models for wind pressure on buildings based on wind tunnel experiments. In: Proceedingsof the 9th international symposium on heating, ventilation and air conditioning, ISHVAC 2015 Joint with the 3rd international conference on building energy and environment, COBEE 2015, pp 2209–2216, China
17. Fariborz H, Cherif MA (1996) A comprehensive validation of two airflow models—COMIS and CONTAM. Indoor Air 6(4):278–288

18. Madyira DM, Muaaz B (2010) Comparative study of the performance of displacement vs conventional ventilation using CFD. In: Proceedings of the Third IASTED African Conference, pp 056–149 (2010)
19. Chen Q (1995) Comparison of different k-models for indoor air flow computations. Numer Heat Transf Part B: Fundam 28(3):353–369
20. Shrivastava S, Sammakia B, Schmidt R, Iyengar M (2009) Comparative analysis of different data center airflow management configurations. In ASME 2005 pacific rim technical conference and exhibition on integration and packaging of MEMS, NEMS, and electronic systems collocated with the ASME 2005 heat transfer summer conference, pp 329–336. American Society of Mechanical Engineers Digital Collection (2009)
21. Li Y, Nielsen PV (2011) Commemorating 20 years of Indoor Air: CFD and ventilation research. Indoor Air 21(6):442–453
22. Zhai ZJ, Zhang Z, Zhang W, Chen QY (2007) Evaluation of various turbulence models in predicting airflow and turbulence in enclosed environments by CFD: Part 1-summary of prevalent turbulence models. HVAC&R Res 13(6):853–870
23. Zhang Z, Zhang W, Zhai ZJ, Chen QY (2007) Evaluation of various turbulence models in predicting airflow and turbulence in enclosed environments by CFD: Part 2-comparison with experimental data from literature. HVAC&R Res 13(6):871–886
24. Yao Q, Bai H, Kwan TH, Kase K (2018) A parametric study and optimization of an air conditioning system for a heat-loaded room. Mathematical Problems in Engineering, Hindawi, Article ID 2385691, 10p, https://doi.org/10.1155/2018/2385691

Experimental Study on an Inclined Pyramid-Type Single Basin Solar Pond for Water Distillation

Dibya Padhi and S. Kumar

Abstract The major problem for most developing countries is the shortage of good and clean drinking water. In the present experimental study, the conventional basin type has been modified to inclined pyramid single basin solar pond and produced distillate from the still. Tests were carried out for different water samples, namely saline water and mud water with a quantity of 750 g of salt in 10 L of freshwater and 10 L of mud water, respectively. To conduct the experiment, solar radiation has been checked by using a pyranometer. In order to enhance the distillate efficiency, the glass roof inclined angle was maintained as 61.9°. Both for desalination and mud water purification performances were observed. Both the experiments were conducted on inclined-type single basin solar pond set-up. The laboratory results showed that the pH values for desalination and mud water purification were 7.4 and 7.2, respectively. The total desalination and mud water purification efficiencies were found 15 and 16%, respectively, also the total solar radiation on glass cover was 6 kWh/m^2/day and the average ambient temperature was 26 °C.

Keywords Solar distillation · Desalination · Mud water purification

1 Introduction

A solar pond is a reservoir of solar energy, with a large shape and size that looks like a pond. The ponds that exist are natural or either man-made but solar ponds are artificial (manufactured and designed by the human). Natural process of desalination of water with the help of direct sunrays is a very accessible way. Their basic functions are to collect the incoming solar radiation from the sun, act as a warehouse; in the heated core of the pond, due to convection current, it leads to heat loss at the boundaries [1]. By far, the most common approach is the establishment of salt density gradient [2].

D. Padhi (✉) · S. Kumar
Alliance University, Chandapura, Bangalore 562106, India
e-mail: dibyadarshini.padhi@gmail.com

M. Ramgopal et al. (eds.), *Advances in Air Conditioning and Refrigeration*, Lecture Notes in Mechanical Engineering, https://doi.org/10.1007/978-981-15-6360-7_16

A solar pond construction majorly lies on a non-convective zone (NCZ] [3]. A handful of developed methods such as Royal Melbourne Information of Technology(RMIT) solar pond was established in 2001 with a 15 m^2 surface area, and the reason for which the salinity decreases with height is elaborated in this experiment with saturated brine to start lower convective zone (LCZ) as well as the middle insulating layer [4]. Karakilcik et al. [5] have proposed that the level for highest thermal efficiency of the pond was found as 28.1, 13.8 and 4.5% for the heat storage zone (HSZ), convective zone (NCZ) and upper convective zone (UCZ) in the month of August. It has been found by some of the researchers initiated by proceedings that before and after each of the experiment the continuation of the density profile was taken [6].

The higher value of temperature potential is defined as ΔT_{higher} is equal to 70 °C. It is instinctive that various zones give in different ways to the temperature gradients [7]. Due to the greater amount of the salt present in the lower convective zone (LCZ), heat will be collected for a longer time period and will contribute thermal energy at any time which is required to have a temperature difference from 50 to 90 °C [8]. The 24 h aggregate distillate output of a modified still was 9.38 kg/m^2, which was 56.92% higher than the conventional still 5.98 kg/m^2 [9]. As discovered by some researchers, some part of the low-temperature water required for the process of mineral distillation permitting a huge reduction in the quantity of fuel oil utilized by the system is generated by the solar pond [10].

Nayi et al. have investigated that for the pyramid pond, the daily aggregate yearly incident and absorbed radiation were 4% more but also has a 1% higher daily average loss of solar radiation [11]. Experimentally noticed by the researcher that the results indicated by the per day efficiency for modified stepped and conventional solar pond are nearly 61, 42, 55 and 37%, respectively, for sea and saltwater at black absorber and also per day efficiency at cotton absorber for sea and saltwater are nearly 61, 40, 70 and 48%, respectively [12]. The solar pond which is normally referred to as salt gradient solar pond (SGSP) contains three different zones such as upper convective zone(surface zone), non convective zone(insulation zone) and lower convective zone(storage zone), and the upper convective zone (UCZ) is less effective which has low and approximately uniform salt concentration [13].

"Saxena et al." investigated that the effect of water level depth on solar pond affects thermal performance [14]. In solar pond where solar energy is stored which is a basin of water is called a salt gradient solar pond (SGSP). A temperature gradient which shows bottom hotter and top cooler can be established due to solar radiation absorption at the surface and the salt concentration [15]. Three zones are LCZ, NCZ and UCZ with salt concentration are shown in Fig. 1 [16].

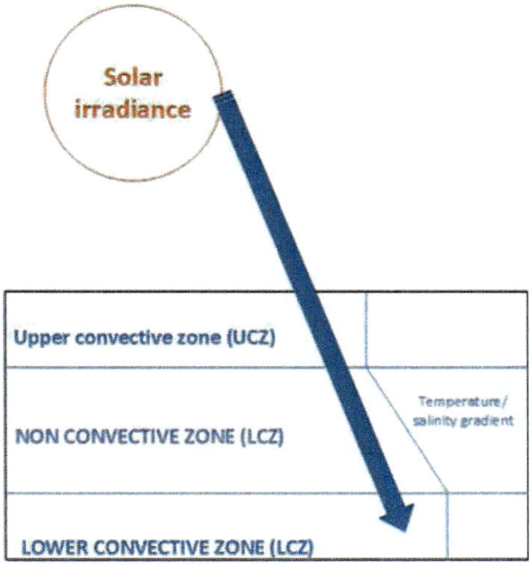

Fig. 1 Simple illustration of various zones in solar pond

2 Experimental

2.1 Experimental Set-up

The set-up has a square aluminium box as shown in Fig. 2, which is placed on a vertical mild steel stand. In order to avoid re-evaporation of desalinated water, PVC made red-coated funnel was used at the bottom side of aluminium box. Solar pond with pyramid structure consisted of four slope single basin sections to achieve more water temperature and less condensing cover temperature. To absorb more heat, black painted 1.5 mm thickness and 400 mm square aluminium box was used. The top surface of the solar pond was covered with four transparent toughened glasses glued with 789 silicon weather proofing sealants with a dimension of 4 mm thickness.

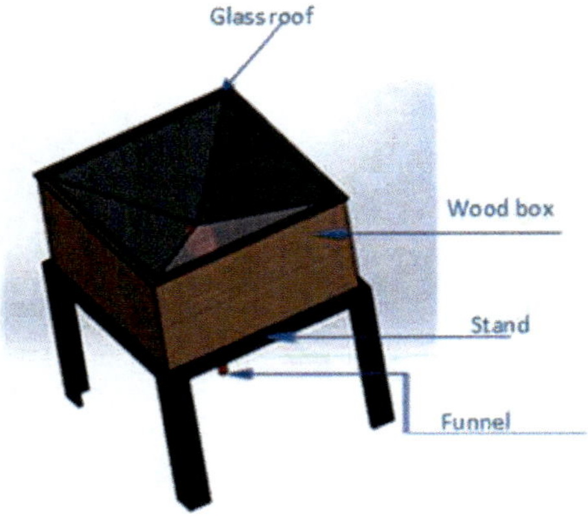

Fig. 2 Inclined pyramid-type solar pond

The aluminium box was covered with wooden sheets of thickness of 70 mm because of high insulation due to air pockets within its cellular structure.

The solar radiation on glass cover was measured by the CMP3 pyranometer. The "K-type" thermocouples were used to measure the temperature of water and various locations of the solar pond. The water quality of saline water and mud water was repeatedly measured by measuring their pH throughout the experimental work. The pH value was verified by the below formula:

$$\mathrm{pH} = -\log\left[H^{+}\right]$$

where

H^{+} Hydrogen ions in a solution;

$[H^{+}]$ Concentration of H^{+}.

2.2 *Experimental Procedure*

Solar Desalination

The pictorial view of the pyramid-type solar pond for desalination is shown in Fig. 3. The experiments were performed at Alliance University in Bangalore, India, (12.9716° N latitude, 77.5946° E longitude) by keeping the device in shadow-free

Fig. 3 Saline to desalination

Fig. 4 Mud water purification

area. The solar pond was filled with 750 grams of salt in 10 L of freshwater because 30–50% of salt in freshwater is equivalent to seawater [1]. After an hour, evaporation was started. It has been observed after 2 h, the water droplets started falling into the collecting tank. The dry patches of salt were fermented on the bottom surface of the basin.

Mud Water Purification

The pictorial view of solar pond during mud water purification is shown in Fig. 4. The set-up was placed in maximum sunlight area without any shadow effects. Ten litres of mud water were filled in pyramid-type solar pond, and the top glasses were sealed properly so that there is no way of leakage of temperature. The sidewall of aluminium box was coloured with black paint to absorb more heat as a black body.

2.3 Design of Pyramid-Type Solar Pond

By considering factors like solar radiation available at various times, pyramid-type solar pond is designed and fabricated as shown in Fig. 4. The design of inclined pyramid single basin solar pond with dimension is shown in Table 1 and in Fig. 5. The software package "ENOVIA V5 VPM" of version 5 is accomplished for the present experiment. As per the latitude of Bangalore and atmospheric temperature in the month of May, the angle of the glass roof was set as 61.9°. The aluminium basin area of 447 mm × 298 mm × 1.5 mm was kept for the present experiment. For better insulation, the outer cover of aluminium has covered by a wooden box area of 405 mm × 280 mm × 12 mm.

Table 1 Details of pyramid-type solar pond

Si no.	Component	Material	Dimension
1.	Aluminium box	Aluminium sheet	1.5 mm × 447 mm × 298.5 mm
2.	Glass roof	Toughened glass	119.53 mm × 447 mm × 4 mm
3.	Funnel	PVC	Ø63 mm × Ø18 mm × 300 mm
4.	Wooden box	Wood	12 mm × 449 mm × 280 mm
5.	Stand	Mild steel	451 mm × 300 mm, 90°

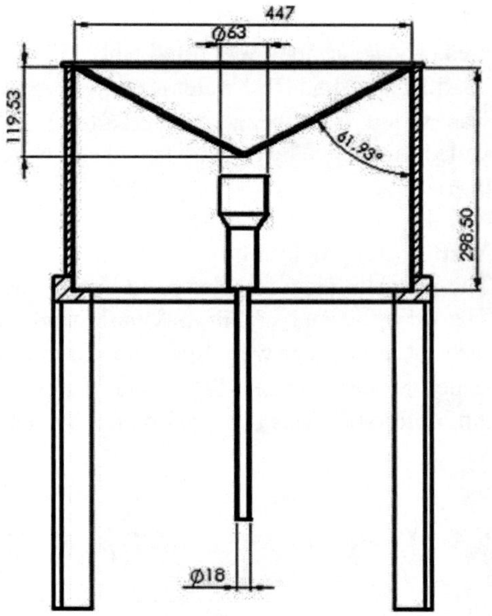

Fig. 5 Design of inclined pyramid-type solar pond

3 Result and Discussion

The experiments were performed for both desalination and mud water purification during May 2019. The test results of experiments taken in May 2019 are reported here. Figures 6 and 8 show the variation of ambient temperature with time. The cumulative solar radiation of 6 kWh/m^2/day and an average ambient temperature of 26 °C were tested during day period. The peak solar radiation occurred at 12:30 pm and the peak ambient temperature reached at 1:00 pm.

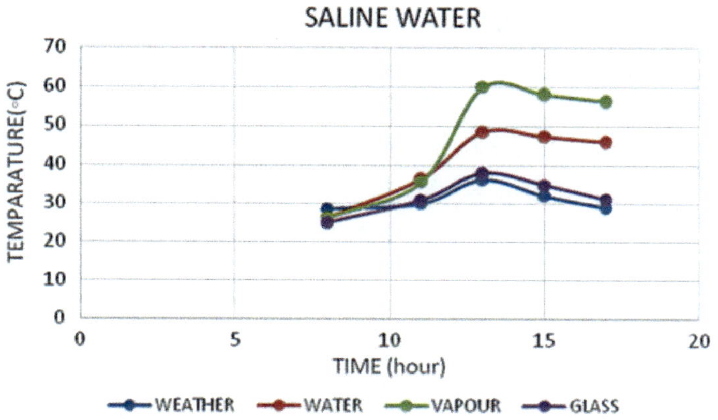

Fig. 6 Variation of ambient temperature with time for desalination

3.1 *Desalination Output*

The output for desalination was measured. To complete the whole experiment and achieve the desired output, it has taken 9:20 h. The experiment was started at morning 8 am, and the desired output was achieved in the afternoon at 3 pm on the same day. The desalinated water temperature was collected in a container. The total quantity of desalinated water after purification was collected of 260 ml. The pH value test was carried out in a laboratory, and it has found that the pH value of desalinated water is 7.4. It has further validated with pure water pH value, i.e. 7.

3.2 *Component Temperature of Solar Pond for Desalination*

Figure 6 represents the variation of temperature in °C with time in an hour. At 1:00 pm, the temperature of water vapour is found maximum. It is also observed that the heating of water at the same time was high due to the maximum solar radiation energy. Pyranometer has shown the value of that period was 5.63 kWh/m^2. At high temperature, it has been observed that high-density salt was patched on the bottom wall of the box and desalinated water was collected in a jar at its bottom.

3.3 *Mud Water Purification Output*

The output of mud water purification is shown in Fig. 7. After a 9 h complete observation, a pure distilled water quantity of 210 ml was collected in a container at the

Fig. 7 Distilled water after purification

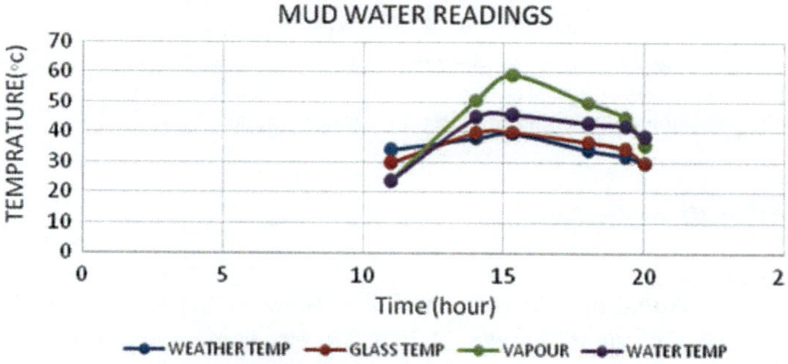

Fig. 8 Variation of ambient temperature with time for mud water purification

bottom of the box through the PVC funnel. It is observed that the pH value of purified water found 7.2 was nearly the value of pure water, i.e. 7.

3.4 *Component Temperature of Solar Pond for Purification of Mud Water*

Figure 8 represented the variation of ambient temperature with time. The increase of water vapour was observed around 3:00 pm, and the temperature was 60 °C. The temperature reading was noted after every one hour. From day to till night, the temperature was effective due to some energy reserved in an insulated box.

4 Conclusion

The conclusions can be drawn from the present experimental work as follows:

(1) The use of a pyramid-type solar pond for the supply of desalination water and purification of mud water was found effective.

(2) Passive solar pond with 4 mm thickness of glass is found better compared with 5 and 6 mm thickness. It enhanced the distillate output by 12–27% compared with 5 and 6 mm thick glass, and the total radiation on glass cover was 6 kWh/m^2/day.
(3) Inclined pyramid single basin passive solar still received 260 ml desalinate water and 210 ml pure water after mud water purification efficiency is increased by 15 and 16%.

References

1. Akbarzadeh A, Andrews J, Ponds S (2018) Solar energy conversion and photoenergy system 1:1–10
2. Kaushika ND (1986) Solar ponds. Adv Energy Syst Technol 5:75–165
3. Abdullah AA, Lindsay KA, AbdelGawad AF (2016) Construction of sustainable heat extraction system and a new scheme of temperature measurement in an experimental solar pond for performance enhancement. Sol Energy 130:10–24
4. Faqehaa H, Bawahaba M, Veta QL, Abhijit Datea AF, Akbarzadaha A (2019) An experimental study to establish a salt gradient solar pond (SGSP). Energy Proc 160:239–245
5. Karakilcik M, Erden M, Cilogulları M, Dincer I (2018) Investigation of hydrogen production performance of a reactor assisted by a solar pond via photoelectrochemical process. Int J Hydrogen Energy 43:10549–10554
6. Faqehaa H, Bawahaba M (2019) Setting up salinity gradient in an experimental solar pond (SGSP). Energy Procedia 156:115–121
7. Kumar A, Singh K, Verma S, Das R (2018) Inverse prediction and optimization analysis of a solar pond powering a thermoelectric generator. Sol Energy 169:658–672
8. Sathish D, Jegadheeswaran S (2018) Relative study of steel solar pond with sodium chloride and pebbles. Mater Sci Energy Technol 1:171–174
9. Dhindsa GS, Mittal MK (2018) Experimental study of basin type vertical multiple effect diffusion solar still integrated with mini solar pond to generate nocturnal distillate. Energy Convers Manag 165:669–680
10. Alcaraz A, Montalà M, Cortina JL, Akbarzadeh A, Aladjem C, Farran A, Valderrama C (2018) Design, construction, and operation of the first industrial salinity-gradient solar pond in Europe. Solar Energy 164:316–326
11. Nayi KH, Modi KV (2018) Pyramid solar still: a comprehensive review. Renew Sustain Energy Rev 81:136–148
12. El Agouz SA (2014) Experimental investigation of stepped solar still with continuous water circulation. Energy Convers Manag 86:186–193
13. Sifuna DB, Kinyanjui TK, Ndiritu FG, Ngubu RG (2014) Comparison of thermal storage efficiency of solar pondwith and without a polyethene membrane. Int J Sci Res Publ 4:2250–3153
14. Saxena AK, Sugandhi S, Husain M (2009) Significant depth of ground water table for thermal performance of salt gradient solar pond. Renew Energy 34:790–793
15. Giestasa MC, Milhazesb JP, Pina HL (2014) Numerical modeling of solar ponds. Energy Procedia 57:2416–2425
16. Kadi KE, Elagroudy S, Janajre I (2019) Flow simulation and assessment of a salinity gradient solar ponddevelopmen. Energy Procedia 158:911–917

Development of Indigenous Technology for Large Cooling Capacity GM Cryorefrigerator for Application to High Tc Superconducting Magnets—Prospects and Problems

Sachindra Kumar Rout, Balaji Kumar Choudhury, Suraj Kr Behera, and Sunil Kr Sarangi

Abstract Superconducting magnets have traditionally been used in MRI medical diagnostic equipment, particle accelerators and a variety of industrial applications. Although most magnets made in our country so far have used low Tc superconductors operating at liquid helium temperatures, there is a distinct possibility that magnets of tomorrow will be made of high Tc materials that offer zero resistance around the temperature of liquid nitrogen (77 K). The optimum temperature for operating these magnets will be around 40–50 K, which is achievable by using GM-type cryocoolers. Unfortunately, in spite of nearly a century of experience in cryogenic technology, our country continues to import basic cryogenic cooling equipment. It is time for taking up indigenous development of such important and easy-to-achieve equipment. We are happy to report that such a development project has been taken up by the authors at the C. V. Raman College of Engineering, Bhubaneswar. The target of the development project is to design and build a GM-type closed-cycle cryocooler delivering about 100–150 W of refrigeration at 45 K. The paper presents the design methodology and means of construction. The complete design, fabrication and testing process are being documented so that other workers in India, both researchers and entrepreneurs, can build their own systems and even improve on our design, without additional help from any quarter. The objective of the present activity is to develop the complete end-to-end indigenous technology so that a techno-entrepreneur can initiate a business start-up basing on our experience and our recommendations..

Keywords Cryocoolers · Superconductors · Indigenous

S. K. Rout (✉) · S. K. Sarangi
C. V. Raman Global University, Bhubaneswar 752054, Odisha, India
e-mail: sachindra106@gmail.com

B. K. Choudhury
Parala Maharaja Engineering College, Berhampur, Odisha, India

S. K. Behera
National Institute of Technology, Rourkela 769008, India

M. Ramgopal et al. (eds.), *Advances in Air Conditioning and Refrigeration*, Lecture Notes in Mechanical Engineering, https://doi.org/10.1007/978-981-15-6360-7_17

1 Introduction

Superconducting magnets have traditionally been made of NbTi or Nb3Sn wires (low Tc) and cooled either by liquid helium or by two-stage GM/pulse tube cryocoolers [1] through conduction. A major change, however, is in the horizon. After three decades of development, companies are now selling high Tc (YBCO/BSSCO) tapes of sufficient length to build viable superconducting magnets. High Tc superconducting magnets of small, but still useful, size are being constructed by researchers. Time is probably not far when practically useful high Tc SC magnets will be in the international market. Although the Tc of the concerned materials (YBC0, BSCCO) are well above 77 K, their current-carrying capacity at these temperatures is small, which forbids practical application. Using liquid helium to operate these magnets at lower temperature defeats the advantage of using high Tc tapes. The optimal solution is to operate high Tc SC magnets at a temperature of 40–50 K, where the tapes have adequate current-carrying capacity (JC), accept large magnetic field (HC) and are thermally economical, i.e. the corresponding refrigerating power input is moderate. This refrigeration requirement can be met comfortably by Gifford-McMahon cycle-based closed-cycle cryocoolers [2–4], i.e. the temperature and cooling power necessary for small SC magnets fall within the zone of operation of GM and GM-type pulse tube refrigerators.

Till recently, there was little market demand for such refrigerators. Therefore, commercial development of GM refrigerators in the Western world has focused mostly on two-stage 10 K refrigerators for cryopumps applications, 4 K refrigerators for superconducting magnets and helium re-liquefactions and to a lesser extent on single-stage 75 K machines for laboratory scale LN_2 generators. The intermediate temperature level (40–50 K) and high cooling capacity (>100 W) combination is thus a new requirement. In fact, major international firms like Cryomech, USA, and Sumitomo, Japan, have already started developing this class of cryorefrigerators.

In our country, in spite of nearly a century of industrial and research experience in several laboratories, there is no commercial production of GM cryocoolers. In all cryocooler-based equipment—scientific experiments, cryopumps, liquid nitrogen generators, helium re-liquefiers, conduction-cooled SC magnets—every device continues to be imported; a situation that is both unaffordable and avoidable.

The objective of the present effort at the C. V. Raman College of Engineering, Bhubaneswar, is to develop the complete end-to-end indigenous technology for a closed-cycle cryocooler based on GM cycle delivering 100–150 W cooling power at 45 K for application to cooling of high Tc superconducting magnets and similar devices. In this paper, we present the development target, the design methodology and the fabrication methodology being followed at the C. V. Raman College.

2 The Technology

A GM-type cryocooler consists of three main subsystems—(i) the helium compressor operating at room temperature and delivering around 100 nm^3/h of helium with 10 bar inlet and 20 bar discharge pressure along with necessary instrumentation, (ii) the cold head that contains the rotary valve, the oscillating displacer holding the regenerator in its belly and the cold end heat exchanger that receives the heat load from the application and (iii) a pair of flexible hoses connecting the two. The compressor subsystem contains not only the mechanical compressing unit but also a set of heat exchangers and lubricating oil separators. While the techniques under current use in imported machines are time tested, we have conceived several innovative new ideas which are more suitable to Indian conditions. Those concepts when implemented will improve reliability and reduce cost of operation well below that of current imports. Figure 1 gives a schematic of the GM cryorefrigerator system showing the three subsystems.

The Helium Compressor

The helium compressor system used for GM cryocoolers has some special features:

a. The inlet pressure is around 10 bar and the discharge pressure around 20 bar.
b. Even minor leak of helium is not admissible, because the system has to work for 10 years or longer without refilling of helium gas. This demands a hermetically sealed arrangement and leak-proof plumbing.
c. The peak temperature in the presence of a lubricant must be within 1500 °C to avoid cracking of lubricating oil and carbon deposit.

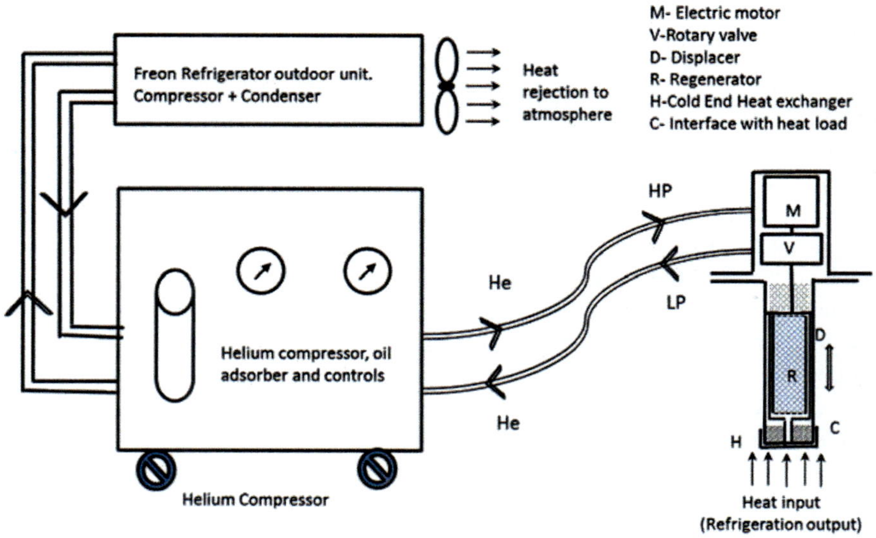

Fig. 1 Schematic of GM cryorefrigerator

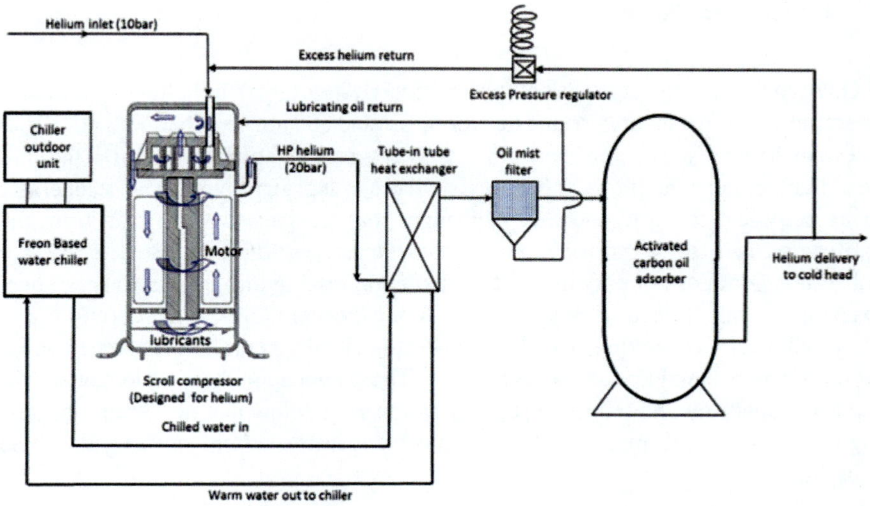

Fig. 2 Helium compressor subsystem (conventional design)

d. The discharged gas must be free from traces of lubricating oil to avoid freezing in the cold zone of the refrigerator. Desirable oil vapour level is about 1 ppm.

Figure 2 shows the schematic of the helium compressor subsystem as delivered by international manufacturers. It consists of a basic compressor unit followed by an after cooler, a coalescing oil filter (with oil return to LP side), an activated carbon oil adsorber and an upstream pressure regulator or differential pressure regulator. The compressor body as well as the outlet helium stream can be either air cooled or water cooled.

More recent compressors come with specially designed and marketed helium compressors based on basic scroll-type compressor normally used for air conditioning. These compressors need flooding of the compression space with lubricating oil which is achieved by draining oil from the bottom of the compressor shell, cooling by chilled water in a plate-type heat exchanger and returning to the suction line. The helium gas and the lubricating oil are compressed together, the mass ratio being about 1:15 and the volume ratio 50:1 between helium and oil. That keeps the temperature of the discharge helium within limits (Fig. 3).

The helium compressor used for GM or GM-type pulse tube refrigerator is a rather complex subsystem containing a basic compressing unit, a conventional tube fin or plate heat exchanger serving as an after cooler for the discharge helium stream, a lubricating oil coalescing unit and an oil vapour adsorption unit, along with some simple instrumentation and control devices. The basic compressing unit for helium can be purchased from international firms (Hitachi and Copeland); but they are expensive and importing one will not be in sync with the "indigenous product" approach that we have taken. We are yet to take up the fabrication of the helium compressor subsystem. When we do, we propose to follow the traditional approach of

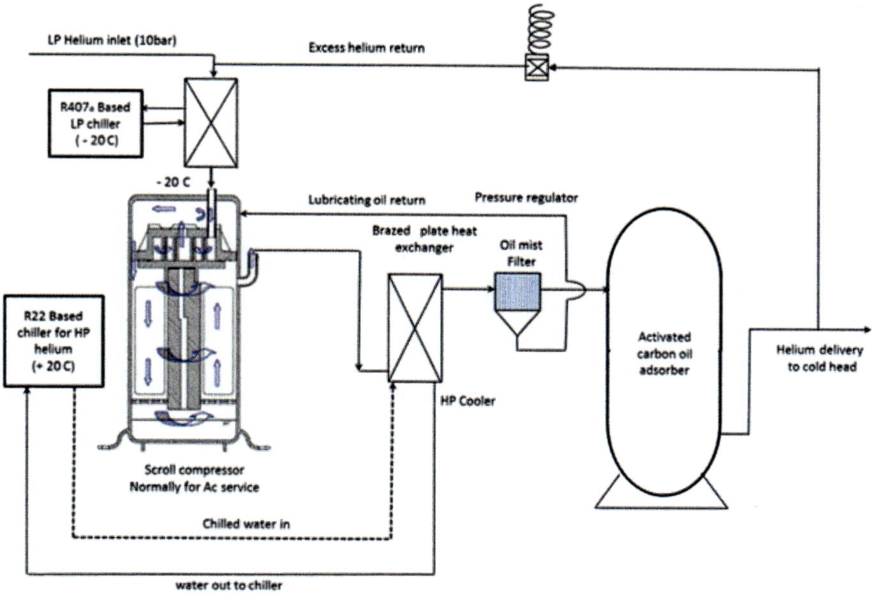

Fig. 3 Helium subsystem (proposed configuration, employing cheap refrigeration compressor)

converting an ordinary off-the-shelf hermetically sealed air conditioning compressor for helium service. For this purpose, we propose to take two approaches, distinct from each other.

Our first approach shall be to use a Hitachi make basic compressor unit with high pressure in the hermetic shell. We shall carefully drill a hole in the bottom of the shell and connect a tube to drain lube oil during operation, cool it in a plate heat exchanger and feed the oil to the compressor suction. We hope to succeed, though we are yet to work on the compressor module. The second approach is to use a Copeland type of refrigeration compressor where the shell is maintained at the suction pressure. In this arrangement, unlike the existing compressor modules obtained from abroad that are either air or water cooled (European or American laboratory temperature), we propose to incorporate a Freon-based chiller to carry the heat of compression from the compressor unit and dissipate it to the atmosphere through its outdoor condensing unit. A unique feature of the proposed configuration shall be chilling of the inlet helium stream to about 200 °C for keeping temperature rise within the compression chamber under control. This is expected to offer several benefits and may be seen as a positive innovation even in international standards. Buy a regular low-cost Freon compressor and convert it for helium service. We propose to drill a hole at the bottom of the unit and braze a tube that will drain a portion of the lubricating oil, coot it in a plate-type heat exchanger and feed it to the compressor suction. And (2) in Copeland type of compressor, where the shell is maintained at the suction pressure, we propose to pre-chill the helium stream to approximately 20 °C by using an appropriate Freon-based chiller before admitting the gas to the

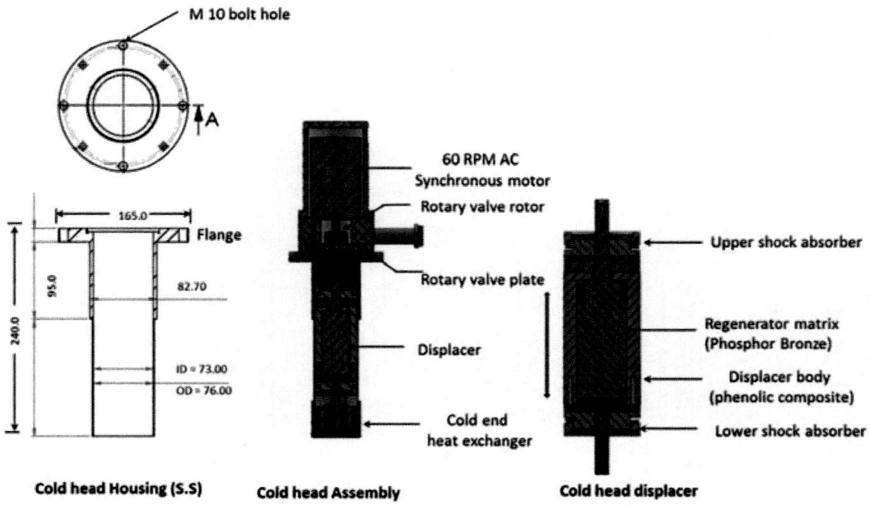

Fig. 4 Cryorefrigerator cold head and displacer

compressor suction. The exact configuration shall be decided during the course of the project depending on the result of initial trials.

The Cold Head and the Connecting Hoses

The cryorefrigerator cold head is the heart of the proposed GM cryocooler. Figure 4 shows a schematic of the cold head along with some details of the displacer and the displacer housing. As a part of the institutional research work, we have already made a moderately detailed design of the cold head unit and worked out most of the fabrication procedures. Fabrication of the components is ongoing. A few example of the design and drawing is shown in Figs. 5 and 6 for the helium gas inlet/outlet unit, the main cylinder and the rotary valve, all critical components of the cryorefrigerator. Similar drawings are available for about 30 other components, ten subassemblies and ten tools and fixtures for supporting machining and assembly.

3 Conclusion

An effort is being made to develop the complete end-to-end technology (with full documentation and working prototype) for single-stage GM cryocooler of capacity 100–150 W at 45 K. While the system has been designed for 150 W of refrigeration at 45 K temperature, it may not be achieved in the first prototype. However, sustained testing, analysis and rebuilding are expected to deliver the desired product. Apart from its application as a cooler for superconducting magnets, the GM refrigerator can be used as the heart of the small liquid nitrogen generator delivering about 200 W of refrigeration at 70 K or about 50 L of liquid nitrogen per day.

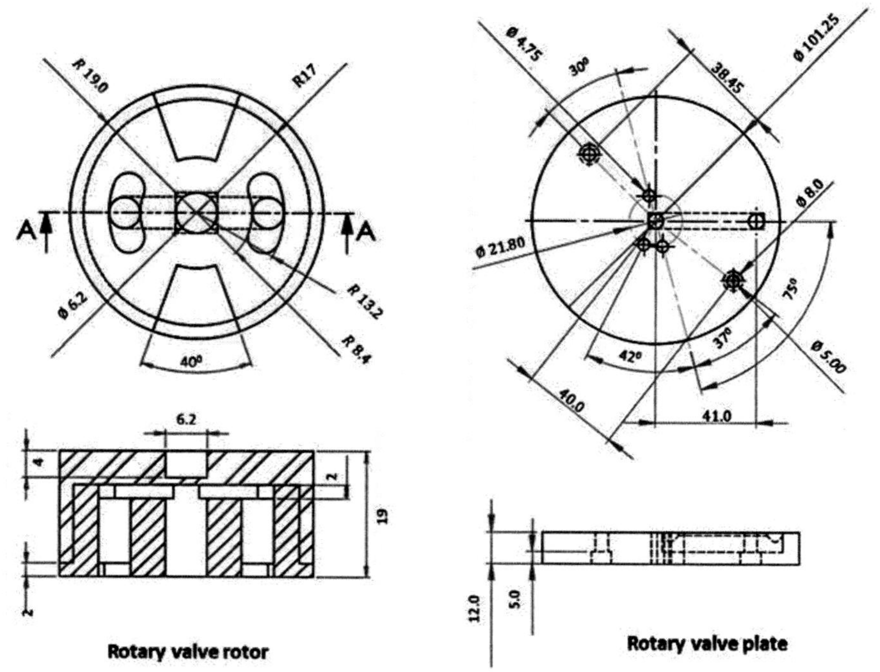

Fig. 5 Proposed details of the rotary valve

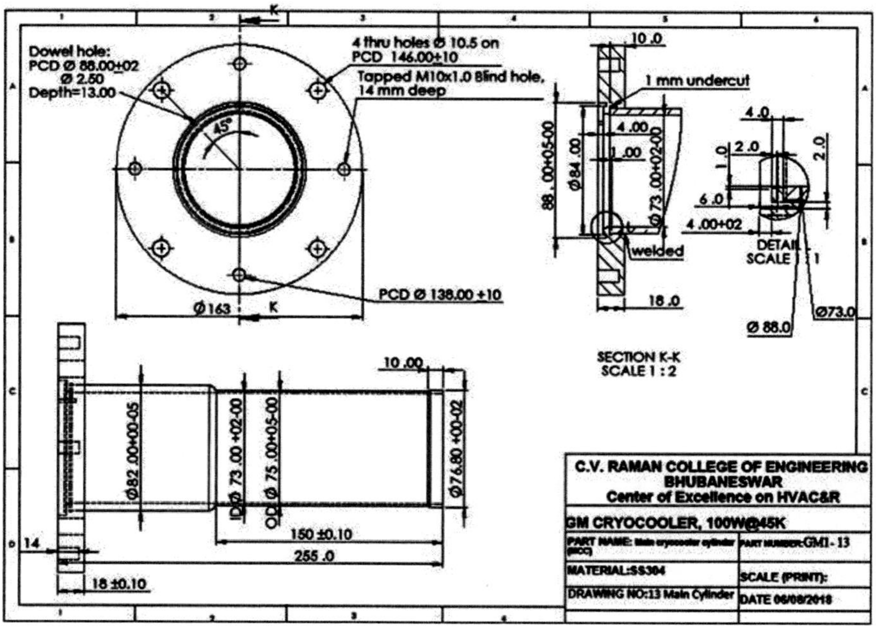

Fig. 6 Main cylinder of GM cryocooler

Acknowledgements The authors greatly acknowledge the support from the BPUT-TEQIP III under CRIS scheme (BPUT-XIX-TEQIP-III/17/19/117).

References

1. Rout SK, Choudhury BK, Sahoo Ranjit K, Sarangi Sunil K (2014) Multi-objective parametric optimization of Inertance type pulse tube refrigerator using response surface methodology and non-dominated sorting genetic algorithm. Cryogenics 62:71–83
2. Radebaugh R (2009) Cryocoolers: the state of the art and recent developments. J Phys: Condens Matter 21(16):164219
3. De Waele ATAM (2011) Basic operation of cryocoolers and related thermal machines. J Low Temp Phys 164(5-6):179
4. Haran KS et al (2017) High power density superconducting rotating machines—development status and technology roadmap. Supercond Sci Technol 30(12):123002

Numerical Investigation of a Shell and Coil Tube Heat Exchanger used in Solar Domestic Hot Water System

Ashutosh Rout, Taraprasad Mohapatra, Sachindra Kumar Rout, and Dillip Kumar Biswal

Abstract In this study, a shell and coil tube heat exchanger used in a solar domestic hot water system is numerically studied. A 3-D model of the SCTHE is prepared using the commercial software package ANSYS 16.1. The result of the current approach is verified and validated with the results available in the literature with good conformity. Simulation runs are performed to determine the effect of three major parameters, i.e. the mass flow rate of the coil-side fluid, the mass flow rate of the outer annulus-side fluid and the inlet temperature of the coil-side fluid on the thermal performance of the SCTHE, i.e. overall heat transfer coefficient and heat transfer effectiveness. It is found out that with the increase in inner Dean number, outer Dean number and inlet temperature of the coil-side fluid, the overall heat transfer coefficient increases in every case; however, the effectiveness decreases with increment in inner Dean number and inlet temperature of the coil-side fluid and increases with increment in outer Dean number.

Keywords Shell and coil tube heat exchanger · CFD modelling · Numerical computation · Heat transfer analysis

Nomenclature

A	Surface area, m^2
C	Heat capacity, J/K
d_c	Diameter of helical tube, m
D_c	Coil diameter, m
De	Dean number
h	Convective heat transfer coefficient, W/m^2 K
K	Thermal conductivity, W/m K
L	Length, m

A. Rout · T. Mohapatra (✉) · S. K. Rout · D. K. Biswal
Department of Mechanical Engineering, C. V. Raman Global University, Bhubaneswar 752054, Odisha, India
e-mail: tpmohapatra@cvrce.edu.in

M. Ramgopal et al. (eds.), *Advances in Air Conditioning and Refrigeration*, Lecture Notes in Mechanical Engineering, https://doi.org/10.1007/978-981-15-6360-7_18

p	Pitch, m
R	Capacity ratio
Re	Reynolds number
U	Overall heat transfer coefficient, W/m^2 K

Greek symbols

ε	Effectiveness

Subscripts

c	Coil
cu	Copper
CFC	Counterflow configuration
h	Hot water
i	Inner
min	Minimum
n	Normal water
o	Outer
PFC	Parallel flow configuration

Abbreviations

CFD	Computational fluid Dynamics
LPH	Litre per hour
NTU	Numbers of transfer unit
SCTHE	Shell and coil tube heat exchanger

1 Introduction

Owing to the shortage of non-renewable energy, the demand for renewable energy is increasing day by day. In the current scenario, people are aware of the energy crisis and thinking differently towards saving and efficient use of energy in our everyday activities. Out of so many options, solar energy is the better choice and can be widely used for residential and industrial applications. Solar water heating is one of them, where heat energy converted from sunlight is utilized through a solar thermal collector. Varieties of configurations of solar domestic hot water systems (SDHW) are nowadays available worldwide to provide better solutions to these water heating problems at different climates and latitudes. However, in some configuration of SDHW, shell and coil tube heat exchanger (SCTHE) is commonly used for efficient recovery of solar thermal energy. In this study, a small attempt is made to investigate the effect of three major parameters on the heat transfer performance of the SCTHE after acquiring some basic information from the detailed literature review. Shokouhmand et al. [1] experimentally investigated the performance of three shell and coiled tube heat exchangers for different coil pitches, curvature ratios and flow configurations. The coils with larger pitches and counterflow configuration affect shell-side heat transfer coefficients and Nusselt numbers more as resulted. A coil-in-shell heat exchanger was studied by Ghorbani et al. [2] to determine the heat transfer behaviour of fluid flow in mixed convection condition. Decreased LMTD and effectiveness were noted with an increment in mass flow rate ratio. Jamshidi et al. [3] attempted to enhance the heat transfer rate in shell and coiled tube heat exchangers experimentally. Increment in coil diameter, pitch and rate of flow mass in helical coil improves heat transfer rate which was noticed. Pawar and Sunnapwar [4] studied heat transfer characteristics of Newtonian and non-Newtonian fluids in helical coil experimentally for different types of flows and curvature ratios. From results, it was found that glycerol–water mixture and non-Newtonian fluids possess low overall heat transfer coefficient and Nusselt numbers than water. Thermal performance and irreversibility effect of steady-state flow were studied by Alimoradi [5] for a shell and coiled tube heat exchangers. Shell-side heat transfer rate decreases with increase in shell diameter as it increases the gap between coil and shell. Etghani and Baboli [6] numerically investigated the thermal performance of a shell and helical tube heat exchanger and optimized the performance for maximum heat transfer rate and minimum exergy loss. Higher heat transfer rate was observed for increased cold and hot fluid flow rates. Kharat et al. [7] experimentally and numerically studied the effect of various geometric parameters of a concentric helical coil heat exchanger on heat transfer. Numerical and experimental results were analyzed by a mathematical model, optimized and the new correlation for heat transfer coefficient were developed. Shell-side thermal performances were studied by Huminic and Huminic [8] for different heat exchangers with helical tube. It was found out that the shell-side hydraulic diameter is a major parameter for determining the shell-side heat transfer characteristics. The heat transfer characteristic of fluid flow inside a helical coil was experimentally and numerically investigated by Jayakumar et al. [9] for various boundary conditions. It is resulted that in an actual heat exchanger, the constant temperature or constant heat flux boundary condition does not yield proper modelling as constant fluid properties are considered instead of temperature-dependent fluid properties. Genić et al. [10] experimentally studied the shell-side heat transfer coefficient and fouling factors of eight shell-and-tube heat exchangers with parallel helical coils. The fouling resistances calculated in eight heat exchangers were resulted slightly lesser than usual values with straight tubes. Pimenta and Campos [11] experimentally determined the heat transfer coefficients of Newtonian and non-Newtonian fluids for fully developed laminar flow inside a helical coil at constant wall temperature boundary condition. For identical Prandtl numbers and Dean numbers, the Nusselt numbers for the CMC solutions are slightly higher and the XG solutions are significantly lower than the Newtonian fluids which were observed. The heat transfer coefficient and pressure drop characteristics of R-134a condensing inside a horizontal helical coil tube in counterflow direction were experimentally investigated by Gupta et al. [12] for different mass flux, vapour quality and saturation temperature. For mixed convection condition, the Nusselt number and pressure drop correlation for condensation of R-134a inside horizontal helical coils were presented.

2 Objective

The objectives are:

2.1 Numerical modelling of the SCTHE
2.2 Validation of numerical model
2.3 Performance investigation of SCTHE for mass flow rate variations of different fluids and inlet temperature of hot water.

3 Numerical Simulation

3.1 Mathematical Modelling

The SCTHE test section and its geometry are schematically represented in Fig. 1.

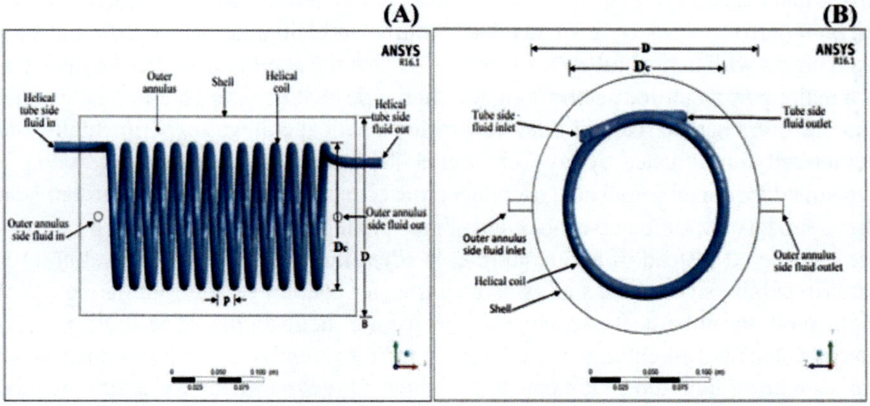

Fig. 1 **a, b** Numerical modelling of the SCTHE

SCTHE test section simply contains an outer shell and a helical coil. For modelling, length, inner diameter and thickness of the shell are chosen as 0.3, 0.19 and 0.005 m, respectively. Similarly, coil diameter, coil pitch, thickness and length of the helical coil are considered for the modelling as 0.13, 0.018, 0.0095 and 0.235 m, respectively. It is considered during the study that normal water is flowing through the outer shell and hot water flows are flowing through the helical coil. The adiabatic boundary is assumed at outside of the shell to not permit any kind of heat interaction with outside environment. The geometry and process parameters for the SCTHE are mentioned in Table 1.

The helical coil inside the SCTHE consists of 13 numbers of turns and fluid flow presumed to be turbulent when passing through it.

3.2 Numerical Computation

3.2.1 Grid Generation

The computational domain has been meshed by automatic mesh generation code using ANSYS ICEM CFD. The shell conduction [13] approach is considered in this study for the wall of helical coil and shell, however these walls are not physically modeled and meshed. Advanced size function on curvature is used for mesh refinement due to the cylindrical geometry of the computational model. The assembly meshing method has been kept as default (tetrahedrons). For the current study, grid with 828,564 elements and 160,411 nodes has been chosen. The mesh preview of both the shell and the helical tube is shown in Fig. 2.

3.2.2 Boundary Conditions and Solver Inputs

Finite volume iterative method (FVM) is used for solution of the 3-D computational domain of the SCTHE. The tube and shell walls are specified by no-slip boundary conditions. At the inlet of the helical tube and shell, the velocity inlet (V_{in}) boundary condition has been opted. Temperature at the inlet (T_{in}) for hot water and cold water has been specified as a thermal boundary condition. The pressure outlet boundary condition has been specified at the outlet of the helical tube and shell. Heat transfer between the shell and nearby environment is restricted by using adiabatic boundary condition. The shell conduction model has been opted with appropriate wall thickness and material properties.

The thermal performance of fluid flow in the SCTHE is studied with respect to variation in the helical tube-side fluid (100–200) LPH, shell-side fluid (450–650) LPH and inlet temperatures of the helical tube-side fluid (338–358 K) in counterflow configurations. In this study, during the variation in volumetric flow rate of one fluid, other fluids volumetric flow rates are taken constant. Similar method is also adopted for inlet temperature variation condition by maintaining shell-side fluid inlet

Table 1 Geometry and process condition for the SCTHE

Tube type	Inner diameter, m	Thickness, m	Outer diameter, m	Length, m	Pitch, m	Coil diameter, m	Number of turn	Fluid type	Velocity, m/s
Shell	0.19	0.005	0.195	0.3	–	–	–	Normal water	0.006184
Coil	0.0092	0.005	0.0142	0.235	0.018	0.13	13	Hot water	0.5846

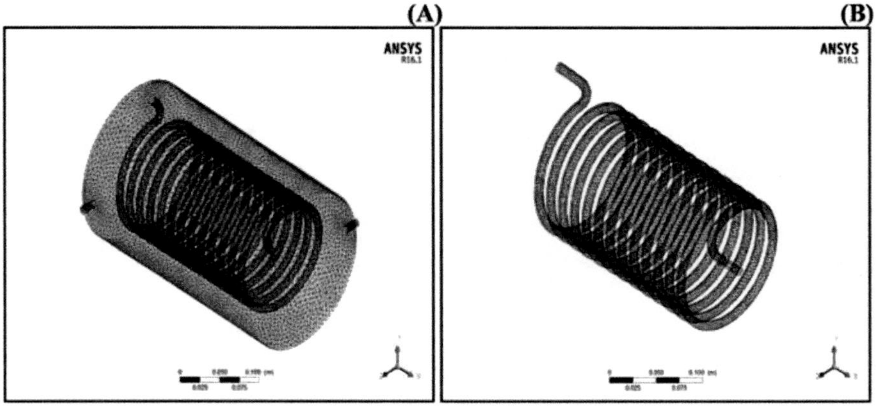

Fig. 2 Mesh distribution of **a** shell and **b** coil of SCTHE model

temperature constant at 308 K. The equivalent diameter is used for calculation of Reynolds number for fluid flow in helical tube and shell. The average temperature is used in this study to predict different fluid properties.

3.2.3 Dean Number

It is a dimensionless number normally calculated for the fluid flow in helical coils, which is the ratio of the viscous force to the centrifugal force. Dean number usually used to predict the flow characteristics of fluids flowing inside and outside of the helical coils.

$$De = Re\sqrt{d_{\mathrm{c,i}}/D_{\mathrm{c}}} \tag{1}$$

In this study performance of the SCTHE is predicted with respect to inner and outer Dean number, where

$$\text{Inner Dean number, } De_i = Re_i\sqrt{d_{\mathrm{c,i}}/D_{\mathrm{c}}} \tag{2}$$

$$\text{And Outer Dean number, } De_{\mathrm{o}} = Re_{\mathrm{o}}\sqrt{d_{\mathrm{c,o}}/D_{\mathrm{c}}} \tag{3}$$

3.2.4 Calculation of Overall Heat Transfer Coefficients

The overall heat transfer coefficient, U_{o}, is calculated using the following formula for energy interaction between helical tube-side fluid and the shell-side fluid.

$$\frac{1}{U_o} = \frac{d_{c,o}}{h_h.d_{c,i}} + \frac{A_{c,o}.\ln\left(\frac{d_{c,o}}{d_{c,i}}\right)}{2\pi K_{cu} L_c} + \frac{1}{h_n} \quad (4)$$

where h_h and h_n are the convective heat transfer coefficients of tube-side and shell-side fluid which are determined numerically.

3.2.5 Calculation of Effectiveness

Effectiveness, ε is calculated for heat transfer from the helical tube side fluid to the shell side fluid using the NTU method. Shell side NTU is calculated using

$$\text{NTU} = \frac{U_o A_{c,o}}{C_{\min}} \quad (5)$$

where U_o is the overall heat transfer coefficient outside of the helical tube, $A_{c,o}$ is the outside heat transfer area and $C_{\min}$ is the minimum heat capacity.

For parallel flow configuration, effectiveness, ε, of the TFHE is determined

$$\varepsilon_{\text{PFC}} = \frac{1 - \exp(-\text{NTU}(1+R))}{1+R} \quad (6)$$

Similarly, for counterflow configuration, effectiveness, ε, of the TFHE is determined

$$\varepsilon_{\text{CFC}} = \frac{1 - \exp[-\text{NTU}(1-R)]}{1 - C.\exp(-\text{NTU}(1-R))} \quad (7)$$

3.3 *Verification and Validation of the Model*

For validation of the model, the current results, i.e. inner Nusselt number, Nu_i, are verified with the literature for counterflow configuration as shown in Fig. 3. The Nusselt number correlations presented by Seban and McLaughlin, Xin and Ebadian, and Kalb and Seader are considered in this study. There is a good concurrence observed between the results and the model which is validated.

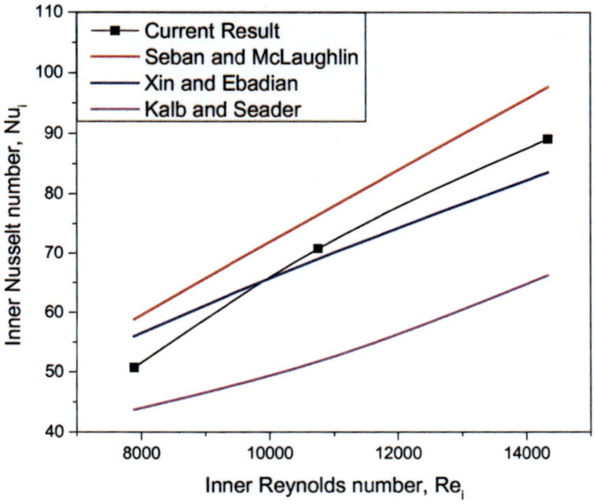

Fig. 3 Validation of the current approach

4 Results and Discussions

4.1 Effect of Inner Dean Number on Thermal Performance of the SCTHE

Figure 4 illustrates the effect of inner Dean number, De_i, on thermal performance of the SCTHE for counterflow configuration. The overall heat transfer coefficient U_o and effectiveness, ε, are calculated as the thermal performance parameter for this study for three different inner Dean numbers, i.e. 2135.66, 2911.37 and 3881.06, from the validated numerical model. The outer annulus-side mass flow rate is taken as constant during the study. From Fig. 4, it is observed that overall heat transfer

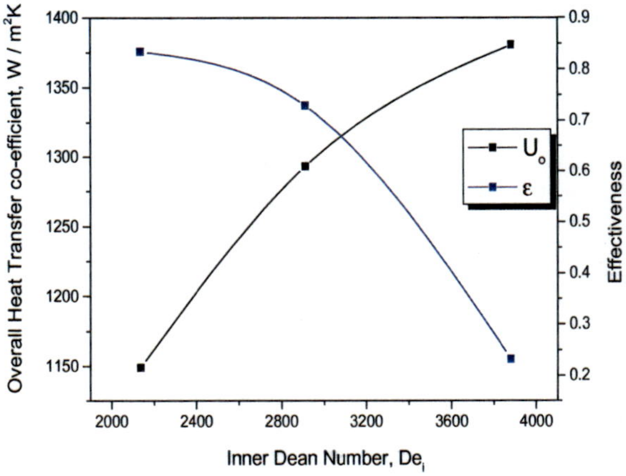

Fig. 4 Effect of inner Dean number on heat transfer performance of the SCTHE

coefficient increases and the heat transfer effectiveness decreases with increment in inner Dean number. As inner Dean number is dependent on fluid velocity, so increment in fluid velocity of fluid flow inside the helical tube reduces the heat rejection rate and increases corresponding Reynolds number, mean temperature and Prandtl number. So, tube-side Nusselt number, convective heat transfer coefficients and overall heat transfer coefficients increases as these are dependent on tube-side Reynolds number, Prandtl number and coil curvature ratio. Reduced residence time and heat transfer rate decrease the heat transfer effectiveness.

4.2 *Effect of Outer Dean Number on Thermal Performance of the SCTHE*

Figure 5 illustrates the effect of outer Dean number, De_o, on thermal performance of the SCTHE for counterflow configuration. The overall heat transfer coefficient U_o and effectiveness, ε, are calculated as the thermal performance parameter for this study for three different outer Dean numbers, i.e. 3841.84, 3232.03 and 2944.32, from the validated numerical model. The tube-side mass flow rate is taken as constant during the study. From Fig. 5, it is observed that overall heat transfer coefficient increases and the heat transfer effectiveness initially decreases and later increases with increment in outer Dean number. Dean number is a function of Reynolds number and dependent on fluid velocity, so increment in outer Dean number is associated with rise in fluid velocity and decrement in residence time. Decreased residence time reduces the rate of heat rejection from tube-side fluid to outer annulus-side fluid, whereas outer annulus-side fluid Reynolds number, mean temperature of heat rejection and Prandtl number increase. So, Nusselt number and overall heat transfer coefficients for fluid flow in outer annulus side increase as these are dependent on outer annulus-side

Fig. 5 Effect of outer Dean number on heat transfer performance of the SCTHE

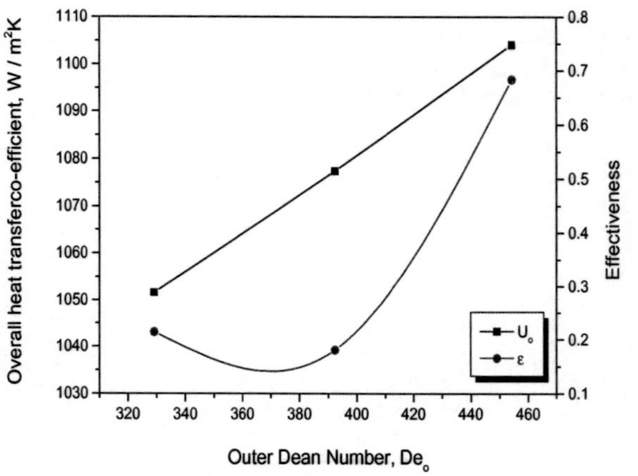

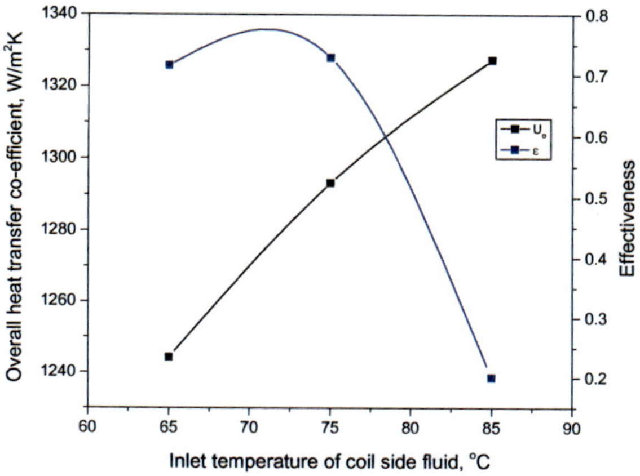

Fig. 6 Effect of variation of inlet temperature of the coil-side fluid on heat transfer performance of the SCTHE

Reynolds number, Prandtl number and coil curvature ratio. Reduced residence time and heat transfer rate decrease the heat transfer effectiveness initially.

4.3 Effect of Variation of Inlet Temperature of the Tube-Side Fluid on the Thermal Performance of the SCTHE

Figure 6 illustrates the variation in inlet temperature for fluid flow inside the helical tube on thermal performance of the SCTHE for counterflow configuration. The overall heat transfer coefficient U_0 and effectiveness, ε, are calculated as the thermal performance parameter in this study for 65, 75 and 85 °C from the validated numerical model. Mass flow rate of fluid flow inside the helical tube and outer annulus side is kept constant for this study.

From Fig. 6, it is observed that by increasing inlet temperature of tube-side fluid, the Reynolds number, Nusselt number, convective heat transfer coefficient and overall heat transfer coefficient at the outside of the helical tube increase. As a result, the heat transfer rate from the helical tube-side fluid to the outer annulus-side fluid increases. Hence, the exit temperatures of the outer annulus side increase. This results in the partial increase of the effectiveness, and gradually decrease in the effectiveness is observed.

5 Conclusion

In this study, a numerical investigation of the SCTHE was performed. SCTHE is numerically modelled and validated with the results available in the literature. Out of several parameters, the mass flow rate and inlet temperature for fluid flow inside and outside of the helical coil were considered for the thermal analysis of the SCTHE. The results of this work are presented as follows.

i. Keeping the fluid flow mass flow rate at outside of the helical tube constant, the effect of inner Dean number on the overall heat transfer and effectiveness was determined. It was found that with the increase in inner Dean number, the overall heat transfer coefficient increases while the effectiveness decreases.
ii. Similarly, keeping the fluid flow mass flow rate inside helical coil constant, the effect of outer Dean number on the overall heat transfer and effectiveness is determined. It was found that with the increase in outer Dean number, the overall heat transfer coefficient and the effectiveness increases.
iii. Again, while keeping the mass flow rate of both inner coil-side and the outer annulus-side fluid as constant, the inlet temperature of the coil-side fluid was varied on a range (65–85 °C) and the effect of outer Dean number on the overall heat transfer and effectiveness was plotted on a graph. And, it was found that with the increase in outer Dean number, the overall heat transfer coefficient increases while the effectiveness decreases.

References

1. Shokouhmand H, Salimpour M, Akhavan-Behabadi M (2008) Experimental investigation of shell and coiled tube heat exchangers using Wilson plots. Int Commun Heat Mass Transfer 35:84–92
2. Ghorbani N, Taherian H, Gorji M, Mirgolbabaei H (2010) Experimental study of mixed convection heat transfer in vertical helically coiled tube heat exchangers. Exp Thermal Fluid Sci 34:900–905
3. Jamshidi N, Farhadi M, Ganji DD, Sedighi K (2013) Experimental analysis of heat transfer enhancement in shell and helical tube heat exchangers. Appl Therm Eng 51:644–652
4. Pawar S, Sunnapwar VK (2013) Experimental studies on heat transfer to Newtonian and non-Newtonian fluids in helical coils with laminar and turbulent flow. Exp Thermal Fluid Sci 44:792–804
5. Alimoradi A (2017) Investigation of exergy efficiency in shell and helically coiled tube heat exchangers. Case Stud Therm Eng 10:1–8
6. Etghani MM, Baboli SAH (2017) Numerical investigation and optimization of heat transfer and exergy loss in shell and helical tube heat exchanger. Appl Therm Eng 121:294–301
7. Kharat R, Bhardwaj N, Jha R (2009) Development of heat transfer coefficient correlation for concentric helical coil heat exchanger. Int J Therm Sci 48:2300–2308
8. Huminic G, Huminic A (2011) Heat transfer characteristics in double tube helical heat exchangers using nanofluids. Int J Heat Mass Transf 54:4280–4287
9. Jayakumar J, Mahajani S, Mandal J, Vijayan P, Bhoi R (2008) Experimental and CFD estimation of heat transfer in helically coiled heat exchangers. Chem Eng Res Des 86:221–232
10. Genić SB, Jaćimović BM, Jarić MS, Budimir NJ (2013) Analysis of fouling factor in district heating heat exchangers with parallel helical tube coils. Int J Heat Mass Transf 57:9–15
11. Pimenta TA, Campos J (2013) Heat transfer coefficients from Newtonian and non-Newtonian fluids flowing in laminar regime in a helical coil. Int J Heat Mass Transf 58:676–690

12. Gupta A, Kumar R, Gupta A (2014) Condensation of R-134a inside a helically coiled tube-in-shell heat exchanger. Exp Therm Fluid Sci 54:279–289
13. ANSYS Fluent 16. Tutorial Guide. Canonsburg, PA: ANSYS, Inc; 2016

Exergetic Study of a Three-Fluid Heat Exchanger used in Solar Flat Plate Collector System

Taraprasad Mohapatra, Sudhansu S. Sahoo, and Biranchi N. Padhi

Abstract An exergetic investigation for heat transfer among different fluids in a combined three-fluid heat exchanger and solar flat plate collector system is carried out here. Present three-fluid heat exchanger (TFHE) is a modification of double pipe exchanger, where an additional helical coil is included between two concentric straight pipes for enhanced performance. The present study is an extension of previous published experimental work, where the effects of flow and thermodynamic parameter on exergetic characteristics of the TFHE are assessed in counter flow configuration. For this study, mass flow rate of different fluids and inlet temperature of hot fluid are chosen as the input parameters, whereas exergy loss and dimensionless exergy loss are considered as the performance parameters for this study. It is found out from the result that the exergetic performance of the TFHE is significantly affected by hot water and normal water mass flow rate. Inlet temperature of hot water is also other major contributing factor towards prediction of exergy loss in TFHE. It is noticed that exergy loss increases with rise in inlet temperature of hot water through helical coil in TFHE. Maximum exergy loss observed at higher mass flow rate and inlet temperature of hot water flow in TFHE.

Keywords Solar energy · Three fluid heat exchanger · Helical coil · Exergy

T. Mohapatra (✉)
Department of Mechanical Engineering, C. V. Raman Global University, Bhubaneswar, Odisha, India
e-mail: tpmohapatra@cvrce.edu.in

S. S. Sahoo
Department of Mechanical Engineering, CET, Bhubaneswar, Odisha, India

B. N. Padhi
Department of Mechanical Engineering, IIIT, Bhubaneswar, India

M. Ramgopal et al. (eds.), *Advances in Air Conditioning and Refrigeration*, Lecture Notes in Mechanical Engineering, https://doi.org/10.1007/978-981-15-6360-7_19

Nomenclature

c_p	Specific heat, J/kg-K
e	Dimension exergy loss
E	Exergy loss, W
$\dot{m}$	Mass flow rate, kg/sec
s	Specific entropy, J/kg-K
T	Temperature, K
T_o	Ambient temperature, K

Subscripts

in	Inlet
out	Outlet
h	Hot water
n	Normal water
a	Air
min	Minimum

1 Introduction

Demand for energy is rising day by day due to socioeconomic development of human being. Everywhere energy is required, i.e. from fulfilment of basic needs to serve productive processes. Till date the major portion of the energy requirement is fulfilled by coal, oil and gas. But the use of these fuels has a significant contribution on greenhouse gas (GHG) emission and CO_2 concentration, which are the key factors of global warming and climatic change. Therefore, reduction in GHG emission is essential, and there are multiple options are available to lower GHG emission such as efficient energy conservation, switching of fossil fuel and progressive use of renewable energy (RE), etc. Out of so many options, renewable energy (RE) and its efficient use are the keen interest of current researchers. Solar energy is an important source of renewable energy (RE) can have significant contribution towards today's world energy crisis. According to National Renewable Energy Laboratory (NREL), US Department of Energy, solar energy received by earth in one hour is enough to meet the annual energy needs of all people worldwide. Therefore, efficient use of solar energy is vital and necessary. Current work is also dedicated towards exergetic analysis on heat transfer characteristics of a combined three-fluid heat exchanger and solar flat plate collector system intended for air and water heating applications during winter.

These days' heat exchangers are the vital components of several industrial processes. Recently, importance is given towards production of new proficient heat exchangers by incorporating different heat transfer augmentation method in conventional heat exchangers for better heat transfer. Different researchers used numerous techniques in conventional heat exchangers towards their performance enhancement. Some of them are reviewed for the current study. Sekulić and Shah [1] presented a review on the thermal design theory of three-fluid and multi-fluid heat exchangers. Comprehensive information on the thermal design theory of existing three-fluid heat exchangers was presented. Ünal [2] conducted a theoretical study on triple concentric-tube heat exchangers with mathematical modelling, where the derivation and possible solutions of the governing differential equations for both counter-flow and parallel-flow arrangements were presented. An experimental and numerical investigation of heat transfer characteristics of a triple concentric-tube heat exchanger was carried out by Gomaa et al. [3] with inserted ribs. The performance was measured for different mass flow rate, temperature, rib height and rib pitch. Appreciable increment in convective heat transfer was observed for the insertion of ribs with higher rib pitches and lower rib height to the fluid flow at inner annulus side. Touatit and Bougriou [4] numerically determined the temperature distribution, heat transfer coefficients and total frictional power expenditure for fluid flow in a triple concentric-tube heat exchanger using FORTRAN code. The optimal diameters of the heat exchanger were determined from techno-economic analysis. Nema and Datta [5] numerically studied the milk fouling characteristics for milk and steam flow inside a helical triple tube heat exchanger. Steam temperature and pressure were predicted as the major parameter which controls milk outlet temperature affected by fouling. Using Biot numbers, the performance of the HTTHE can be optimized was suggested. Mohapatra et al. [6] experimentally investigated the convective heat transfer characteristics of fluid flow in a three-fluid heat exchanger (TFHE) for different mass flow rate. The overall heat transfer coefficients increases with increment in volumetric flow rate of fluid flow were observed, and optimum performance of the TFHE was examined for highest volumetric flow rate of normal water at counter flow configuration.

Numerous literatures were studied for better understanding of exergy analysis in heat exchangers with helical coil. Various heat transfer enhancement techniques and their relative merits were studied towards reduction of irreversibility in heat exchangers by Oullette and Bejan [7]. The exergetic characteristics for fully developed laminar flow inside helical coil was investigated by Ko and Ting [8]. Performance optimization was carried out for minimum entropy generation with respect to three non-dimensional design parameters, i.e. Re, δ and λ. A helically coiled heat exchanger was studied, and its performance was optimized for minimum entropy generation by Farzaneh-Gord et al. [9] for four different dimensionless parameters, i.e. De, Pr, δ and the duty parameter. Dizaji et al. [10] studied numerically the effect of CuO and TiO_2 nanofluids flow in a helically coiled tube-in-tube heat exchanger on heat transfer and entropy generation. It was observed that performance of the said heat exchanger increases with the use of nanofluids.

From literature review, very good information about all existing three-fluid heat exchangers and their application area are obtained apart from the TFHE presented here. So, in this paper, an exergetic analysis of the proposed TFHE is aimed for determination of its heating performance. The current TFHE may be used in heat recovery, HVAC, food and dairy applications, etc.

2 Material and Method

2.1 *Experimental Procedure*

An experimental setup of a combined three-fluid heat exchanger and a solar flat plate collector system shown in Fig. 1 is tested for prediction of its heating characteristics. As the current work is an extension of previous experimental work [7], so once more explanation about the TFHE test section and its specification in detail is not required. However, a small description about the experimental setup is given below.

The setup prepared for experiment mainly consists of a three-fluid heat exchanger and a solar flat plate collector. The heat transfer fluid (HTF) is supplied from the solar flat plate collector, and anticipated heating takes place inside the TFHE. The heating fluid, i.e. hot water is allowed in a closed loop through the helical coil of the TFHE. Normal tap water is flowing through the outer annulus of the helical coil in an opposite direction to the flow hot water. Similarly, air is supplied from a fan is allowed to flow through the innermost tube of the TFHE in an opposite direction to the flow hot water. The normal water and air get heated due to effective heat transfer from hot water during the fluid flow along the TFHE test section. The hot

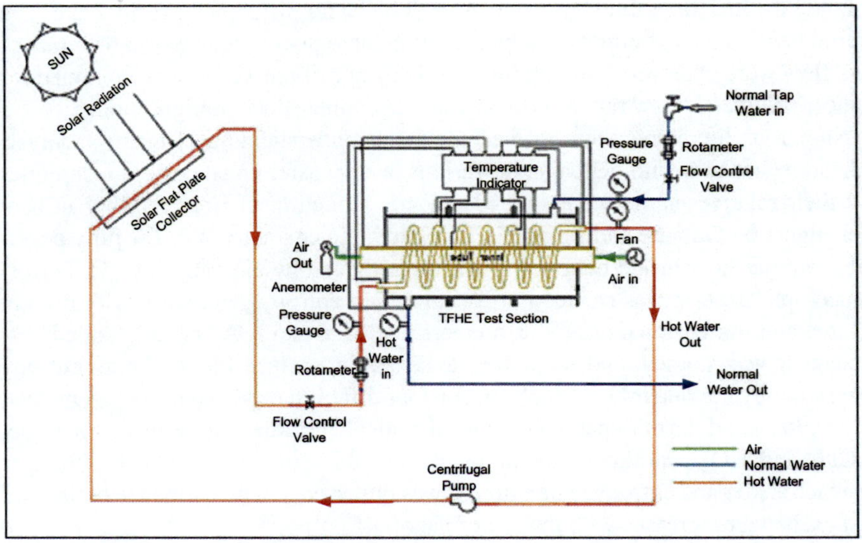

Fig. 1 Schematic diagram of the experimental setup

water is supplied from a solar flat plate collector (2120 mm × 1040 mm × 100 mm) through a ½ HP centrifugal pump. Flow control valves and rotameters are connected along the hot water and normal water loop to control and measure the flow rate. Four pressure gauges (0–60 psi) are connected at the inlet and outlet of hot and normal water passage of the TFHE test section to measure fluid pressure. Anemometer is used to measure the air velocity. The inlet, intermediate and outlet temperatures of three different fluids are measured using 4 K-type thermocouples located at different position of the TFHE test section.

2.2 *Uncertainty in Experimental Data*

The measuring instruments used for evaluation of various parameters such as volumetric flow rate, temperature and air velocity during the experimental investigation were minor error assisted, which results uncertainty in output performance data. The precision of the thermocouples is ±2.2 °C, rotameters are ±0.804 L/min and ±0.0176 m/s. These instruments's preciseness affected the exergy analysis of TFHE, i.e. the calculated uncertainty associated with exergy loss, *E* is 12.32%.

3 Exergy Analysis

The exergetic performance of the TFHE is assessed for different mass flow rate and inlet temperature of fluid flow in counter flow configuration. Exergy loss, *E* and dimensionless exergy loss, *e* are the parameters chosen for the exergetic study of the TFHE can be calculated as follows.

3.1 *Exergy Loss,* E

Exergy is the maximum useful work obtained from a reversible system is well known. For certain control volume, exergy balance can be written as

$$\sum E_{\text{in}} = \sum E_{\text{out}} - \sum E_{\text{product}} \tag{1}$$

Exergy loss for the current heat exchanger (TFHE) under steady and adiabatic condition can be expressed as follows.

Exergy loss,

$$E = E_{\text{h}} + E_{\text{n}} + E_{\text{a}} \tag{2}$$

where E_{h}, E_{n} and E_{a} are exergetic changes take place in hot water, normal water and air can be calculated as follows.

$$E_{\mathrm{h}} = T_{\mathrm{o}}\left[\dot{m}_{\mathrm{h}}\left(s_{\mathrm{h,out}} - s_{\mathrm{h,in}}\right)\right] \quad (3)$$

$$E_{\mathrm{n}} = T_{\mathrm{o}}\left[\dot{m}_{\mathrm{n}}\left(s_{\mathrm{n,out}} - s_{\mathrm{n,in}}\right)\right] \quad (4)$$

$$E_{\mathrm{a}} = T_{\mathrm{o}}\left[\dot{m}_{\mathrm{a}}\left(s_{\mathrm{a,out}} - s_{\mathrm{a,in}}\right)\right] \quad (5)$$

For three different fluids, $(S_{\mathrm{out}} - S_{\mathrm{in}})$ can be calculated as follows.

$$(s_{\mathrm{out}} - s_{\mathrm{in}}) = c_{\mathrm{p}} \ln\left(\frac{T_{\mathrm{out}}}{T_{\mathrm{in}}}\right) \quad (6)$$

$$E_{\mathrm{h}} = T_{\mathrm{o}}\left[\dot{m}_{\mathrm{h}}.c_{\mathrm{p,h}} \ln\left(\frac{T_{\mathrm{h,out}}}{T_{\mathrm{h,in}}}\right)\right] \quad (7)$$

$$E_{\mathrm{n}} = T_{\mathrm{o}}\left[\dot{m}_{\mathrm{n}}.c_{\mathrm{p,n}} \ln\left(\frac{T_{\mathrm{n,out}}}{T_{\mathrm{n,in}}}\right)\right] \quad (8)$$

$$E_{\mathrm{a}} = T_{\mathrm{o}}\left[\dot{m}_{\mathrm{a}}.c_{\mathrm{p,a}} \ln\left(\frac{T_{\mathrm{a,out}}}{T_{\mathrm{a,in}}}\right)\right] \quad (9)$$

3.2 Dimensionless Exergy Loss, E

Dimensionless exergy loss,

$$e = \frac{E}{T_{\mathrm{o}}.C_{\mathrm{min}}} \quad (10)$$

4 Results and Discussions

In this paper, the TFHE is experimented for deviation in mass flow rate of three different fluids and inlet temperature of hot water. However, during the test, normal water and air temperature are kept constant, i.e. 32 and 35 °C, respectively. Similarly, geometric parameters of the TFHE are kept constant during the test. The effect of above-said parameters on exergy loss and dimensionless exergy loss is depicted as follows.

4.1 Effect of Hot Water Mass Flow Rate on Exergetic Performance of the TFHE

Figure 2 illustrates the consequence of hot water mass flow rate deviation on exergetic performance of the TFHE. From Fig. 2a, it is observed that increment in mass flow rate of hot water increases the exergy loss in TFHE. Minimum and maximum exergy loss takes place at 0.016 and 0.082 kg/s mass flow rate of hot water, respectively. It may be due to the enhancement of Reynolds number and heat transfer rates corresponding to the increment of hot water mass flow rate in TFHE. It is obvious that higher heat transfer rate always associated with more finite temperature difference and more

Fig. 2 Mass flow rate variation of hot water on exergetic performance

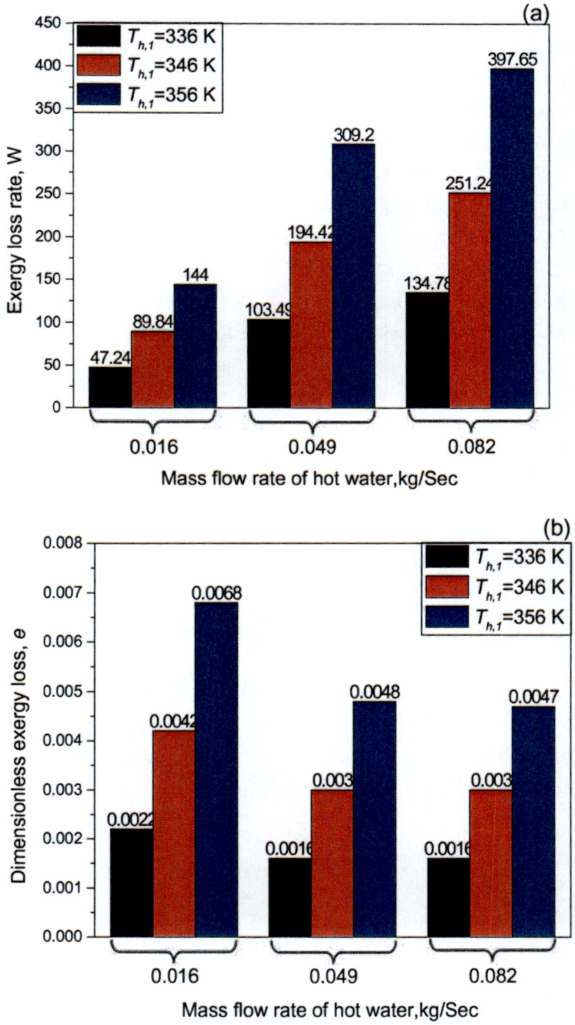

exergy loss. From Fig. 2b, it is observed that dimensionless exergy loss decreases with the increase in mass flow rate. The reason of this phenomenon can be well understood from Eq. (10). It may be noted increment in mass flow rate of hot water increases both numerator and denominator of Eq. (10). As a result, the dimension less exergy loss decreases.

4.2 Effect of Normal Water Mass Flow Rate on Exergetic Performance of the TFHE

Figure 3 illustrates the consequence of normal water mass flow rate deviation on exergetic performance of the TFHE. From Fig. 3a, it is observed that exergy loss increases with the rise in mass flow rate of normal water in TFHE. Minimum and maximum exergy loss takes place at 0.033 and 0.099 kg/s mass flow rate of normal water, respectively. It may be due to the enhancement of Reynolds number and heat transfer rates corresponding to the increment of normal water flow rate in TFHE. It is obvious that higher heat transfer rate always associated with more finite temperature difference and more exergy loss. From Fig. 3b, it observed that dimensionless exergy loss increases with the increase in mass flow rate of normal water in TFHE. The reason of this phenomenon can be well understood from Eq. (10). It may be noted that the denominator of Eq. (10) does not influence by normal water flow rate. As a result, the dimension less exergy loss increases.

4.3 Effect of Air Mass Flow Rate on Exergetic Performance of the TFHE

Figure 4 illustrates the consequence of air mass flow rate deviation on exergetic performance of the TFHE. From Fig. 4a, it is observed that increment in mass flow rate of air, exergy loss in TFHE negligibly changes. It may be due to the insignificant enhancement in magnitude of E_{a} in Eq. (10), as air side convective heat transfer coefficient is too small compared to water side. Similar trend is also observed in Fig. 4b that dimensionless exergy loss negligibly changes with the increase in air mass flow rate. It may be due to the little changes in the numerator of Eq. (10) corresponding to increment in mass flow rate of air.

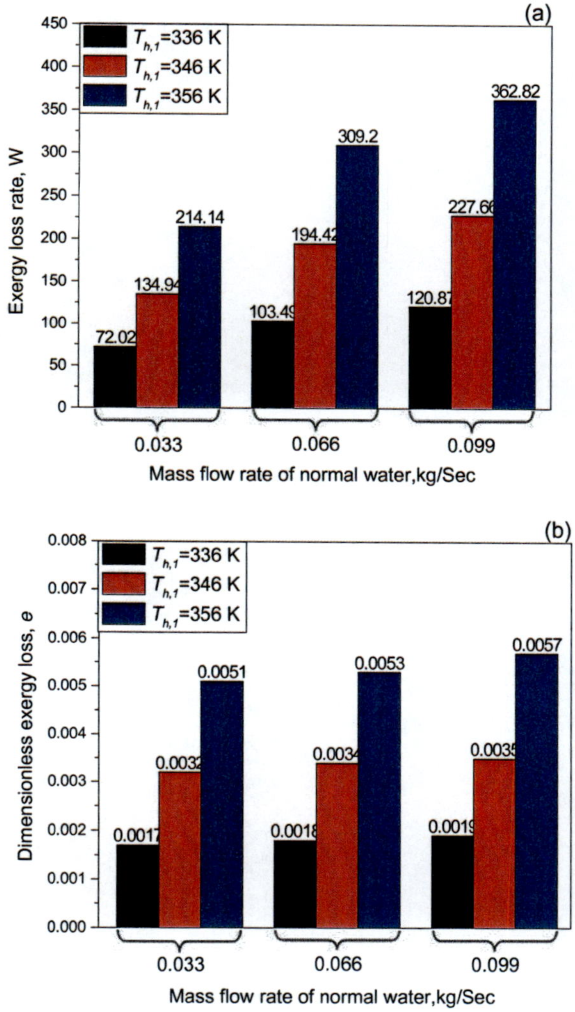

Fig. 3 Mass flow rate variation of normal water on exergetic performance

4.4 Effect of Hot Water Inlet Temperature on Exergetic Performance of the TFHE

The effect of hot water inlet temperature of on exergetic performance of the TFHE is presented in Figs. 2, 3 and 4. A good observation is noticed that exergetic performance decreases with rise in hot water inlet temperature. It may be due to the considerable increment in temperature difference associated with different fluids.

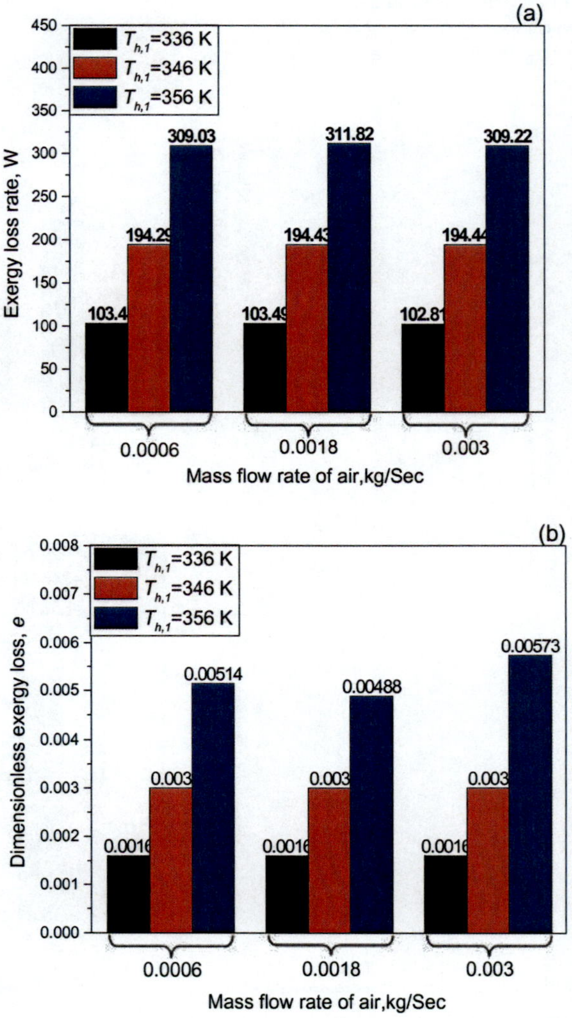

Fig. 4 Mass flow rate variation of air on exergetic performance

5 Conclusion

An experimental investigation was performed to predict the exergetic characteristics of TFHE for different mass flow rate of various fluids and inlet temperature of hot water. The experiments were conducted for the range of Reynolds number of 9000–54000 within the helical coil for coil curvature ratio, $\delta = 0.509$. As the current work is an extension of previous experimental work, so the effect of mass flow rate and inlet temperature on exrgetic performance of the TFHE was determined from prior validated experimental work. Following results were concluded from this study.

i. Hot water and normal water mass flow rate are the major parameters which affect exergy loss and dimension exergy loss of the TFHE significantly; however, these said performance parameters are little affected by the mass flow rate of air.
ii. Inlet temperatures of hot water affect significantly the exergetic performance of the TFHE in each case of mass flow rate variation. It is noticed that exergy loss increases with rise in inlet temperature of hot water through helical coil in TFHE.
iii. Maximum exergy loss is associated with higher mass flow rate and higher inlet temperature of hot water flow in TFHE.

References

1. Sekulić D, Shah R (1995) Thermal design theory of three-fluid heat exchangers. Adv Heat Transf 26:219–328
2. Ünal A (1998) Theoretical analysis of triple concentric-tube heat exchangers Part 1: Mathematical modelling. Int Commun Heat Mass Transf 25:949–958
3. Gomaa A, Halim M, Elsaid AM (2017) Enhancement of cooling characteristics and optimization of a triple concentric-tube heat exchanger with inserted ribs. Int J Therm Sci 120:106–120
4. Touatit A, Bougriou C (2018) Optimal diameters of triple concentric-tube heat exchangers. Int J Heat Technology 36:367–375
5. Nema P, Datta A (2005) A computer based solution to check the drop in milk outlet temperature due to fouling in a tubular heat exchanger. J Food Eng 71:133–142
6. Mohapatra T, Padhi BN, Sahoo SS (2017) Experimental investigation of convective heat transfer in an inserted coiled tube type three fluid heat exchanger. Appl Therm Eng 117:297–307
7. Oullette WR, Bejan A (1980) Conservation of available work (exergy) by using promoters of swirl flow in forced convection heat transfer. Energy 5:587–596
8. Ko T, Ting K (2005) Entropy generation and thermodynamic optimization of fully developed laminar convection in a helical coil. Int Commun Heat Mass Transf 32:214–223
9. Farzaneh-Gord M, Ameri H, Arabkoohsar A (2016) Tube-in-tube helical heat exchangers performance optimization by entropy generation minimization approach. Appl Therm Eng 108:1279–1287
10. Dizaji HS, Khalilarya S, Jafarmadar S, Hashemian M, Khezri M (2016) A comprehensive second law analysis for tube-in-tube helically coiled heat exchangers. Exp Therm Fluid Sci 76:118–125

Comparative Energetic and Exergetic Analyses of a Cascade Refrigeration System Pairing R744 with R134a, R717, R1234yf, R600, R1234ze, R290

Ipsita Das and Samiran Samanta

Abstract A comparative analysis of cascade refrigeration system is presented in this paper. Different pairs of refrigerants are used in this simulation. Both energy and exergy analyses have been conducted. Since a cascade system uses two different refrigerants, R744 is used in the low-temperature unit and R134a, R717, R1234yf, R290, R1234ze, R600 are used alternatively in the high-temperature unit for the purpose of comparative study. Values of COP and the second law efficiency are evaluated and compared with each different pairs of refrigerants used. This paper shows whether the use of refrigerants with low ODP and GWP values is viable or not. The sole purpose of this paper serves in identifying an appropriate refrigerant pair with lower environmental risk which can be really efficient in operating a cascade system. It was observed that R744-R717 attained the highest COP value equivalent to 7.848, and the highest ECOP value of 0.9838 was also attained by R744-R717. R744-R600 also obtained COP value of 7.741 and ECOP value of 0.9833 which is quite good.

Keywords Cascade refrigeration system · COP · GWP · Exergy analysis

Nomenclature

COP	Coefficient of Performance
VCR	Vapor Compression Refrigeration
VAR	Vapor Absorption Refrigeration
GHG	Greenhouse Gas
ODP	Ozone Depletion Potential
GWP	Global Warming Potential
CHX	Cascade Heat Exchanger
LTE	Low-Temperature Evaporator

I. Das · S. Samanta (✉)
School of Mechanical Engineering, KIIT Deemed to be University, Bhubaneswar, Odisha, India
e-mail: samirankgec@gmail.com

M. Ramgopal et al. (eds.), *Advances in Air Conditioning and Refrigeration*, Lecture Notes in Mechanical Engineering, https://doi.org/10.1007/978-981-15-6360-7_20

LTC	Low-Temperature Condenser
HTE	High-Temperature Evaporator
HTC	High-Temperature Condenser
HTX	High-Temperature Expansion Valve
LTX	Low-Temperature Expansion Valve
LTET	Low-Temperature Evaporator Temperature
HTET	High-Temperature Evaporator Temperature
HTCT	High-Temperature Condenser Temperature
LTCT	Low-Temperature Condenser Temperature
HT	High Temperature
LT	Low Temperature
EXD	Exergy Destruction Rate
EX	Exergy Rate
ECOP	Exergetic Coefficient of Performance

1 Introduction

In this era of rapid advancements in technology and progressive developments in research, controlled energy consumption and lesser environmental degradation are major concerns. Among many other usages, one such thermal application is refrigeration. Vapor-compression refrigeration (VCR) systems are common name for industrial, commercial and domestic purposes. However, incorporation of auxiliary components like ejectors, expanders or use of different types of compressors or introduction of different system configurations or use of different energy sources or variation in the operating parameters like evaporator and condenser temperatures have resulted in different observations which show the way of performance enhancement. Different concepts have evolved with different analyses conducted by various researchers, which are integrated in the literature. The VCRs are effective for domestic purposes. But for the applications where low temperature is required, these conventional VCR systems cannot function effectively. It is very difficult for a simple VCR system to achieve such low temperature, due to high compression ratios leading to a low volumetric efficiency and high discharge problem. Cascade refrigeration system is a low-temperature application system [1], where the temperature range varies from -40 to -130 °C. It is a combination of two refrigerating units known as lower temperature unit and higher temperature unit coupled with a cascade heat exchanger (CHX). Energy analysis deals with the quantity of energy, which reveals the way energy is used in that particular operation, whereas exergy is the maximum work potential of the system. Exergy analysis helps in identifying the irreversibility occurred in a system. Lower the irreversibility, higher is the second law efficiency.

Gholamin et al. [2] conducted an advanced exergy analysis of a cascade system using R744-R717 where it was analyzed that CO_2-throttling valve and compressor and CHX are the components where improvement is necessary. One more research [3] stated that the systems, using mixed refrigerants, are thermally efficient. But properties like no combustibility, reasonable physical and thermodynamic properties must be always considered while choosing appropriate refrigerant for a system. Another experiment related to cascade [4] system was conducted using a vapor absorption (VAR) and a VCR system, which is a hybrid cascade system where Li-Br was used in VAR at high-temperature cycle and R134a, R32, R1234yf were used in VCR at low-temperature cycle. Shah et al. [5] also did a comparative assessment of a cascade refrigeration using different refrigerant pairs by varying parameters like subcooling, superheating and condenser and evaporator temperatures. Domestic refrigerators contribute to greenhouse gas (GHG) emissions through refrigerant leakage and conventional energy consumption. To counterattack this problem, Zhaohua et al. introduced [6] high-efficiency linear compressor with a capacity modulator incorporated in the refrigerators. To reduce environmental impact, refrigerants with low global warming potential (GWP) values were used. Yataganbaba et al. conducted [7] an exergy analysis of a multi-evaporator system using R1234yf and R1234ze as an alternative to R134a. Multi-evaporator systems reduce the compressor work and result in efficient system performances. Alhamid et al. [8] conducted an exergy analysis of a cascade system purposely used for biomedical cold storage application through using azeotropic mixture of carbon dioxide, and ethane–propane as refrigerant was utilized. To tackle excessive exploitation of environment by harsh usage of refrigerants, much attention is given to promote utilization of eco-friendly products. An energy analysis of a cascade system was performed with refrigerants having low GWP. This was conducted by varying certain operating parameters. Outcomes of this analysis revealed that maximum exergy destruction occurred at the condenser. Usage of low GWP is also extended to supermarkets, where two multi-ejector based CO_2 refrigeration plants as well as two CO_2/R1234ze indirect refrigerating configurations were energetically compared to a R404A multiplex direct expansion system. A comparative analysis of a simple VCR and VAR system was done. Both energy and exergy analyses of both the systems were analyzed. It was noted that COP of a VCR cycle is greater than COP of VAR cycle, whereas exergetic performance of a VAR cycle at dead environmental state is better than the VCR cycle [9–11]. Messineo evaluated the performance [12] of a R744/R717-based cascade system and compared its results with 404A. The results discovered that R744/R717 will be a good alternative to R404A for low evaporating temperatures. The second law analysis of a mobile air conditioning system was performed using low GWP refrigerants. R1234yf has been used for testing the performance of a refrigeration system for automobile sectors. In the literatures studied, the aim of replacing conventional hydrofluorocarbons with refrigerants having low GWP is abundant due to environmental risks harmful refrigerants carry [13–15].

In this paper, a comparative first law and second law analyses of a cascade VCR system have been conducted using different refrigerants pairing with R744

2 System Description and Modeling

The cascade system is comprised of two circuits. One is the high-temperature unit and other is the low-temperature unit. Figure 1 shows a schematic circuit of the proposed cascade refrigeration system. Two vapor compression refrigeration (VCR) cycles are thermally coupled with the help of a cascade heat exchanger (CHX). The CHX is comprised of high-temperature evaporator (HTE) and low-temperature condenser (LTC). Each VCR cycle comprises a compressor, condenser, expansion valve and an evaporator. The cascade heat exchanger acts as a condenser for the low-temperature cycle and as an evaporator for the high-temperature cycle. Beneficial refrigerating effect is obtained along the low-temperature evaporator (LTE). To simulate the system theoretically, certain assumptions were made which are given below.

1. System operates at steady condition [5].
2. Heat losses are negligible in the cascade heat exchanger [2].
3. Compression is adiabatic in nature with the isentropic efficiency (η_i) = 1, for both the compressors.
4. Refrigerants undergo an isenthalpic expansion in the expansion valves [5].
5. Pressure losses along the connected pipes are negligible [5].
6. Changes in K.E and P.E are negligible.

2.1 Thermodynamic Modeling

Thermodynamic analysis of the system is done by using Engineering Equation Solver software package. Steady flow energy equation and mass flow equations for the system analysis are given by Refs. [2, 5].

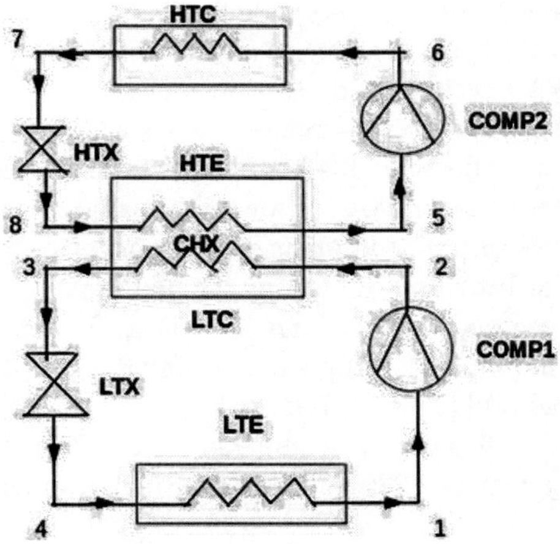

Fig. 1 Schematic diagram of the proposed cascade refrigeration system

Rate of heat absorbed by LTE (Q_L) is given by the equation given below.

$$Q_L = m_l\ (h_1 - h_4) \tag{1}$$

where m_l is the mass flow rate of the refrigerant in the low-temperature cycle, h denotes the enthalpy at the respective state points. Work done by the LT compressor is given by the following equation

$$W_{compl} = m_l(h_2 - h_1) \tag{2}$$

$$W_{comp2} = m_h\ (h_6 - h_5) \tag{3}$$

$$COP = Q_L / W_{net} \tag{4}$$

where

$$W_{net} = W_{comp1} + W_{comp2} \tag{5}$$

2.2 *Exergy Balance*

Correct utilization of the system lies in its exergy analysis [2]. For the exergy analysis of the system, here in the paper, only the physical exergy is considered and kinetic and potential exergy is neglected. Expression for exergy (EX) value for any stream is given as follows

$$EX_i = m\ [h_i - h_0] - T_0\ [s_i - s_0] \tag{6}$$

where i represents the state points, m represents the mass flow rate, T_0 is the reference temperature, s_i is the entropy of that sate point and s_0 is the entropy at the reference condition. Exergy destruction of each component of the cascade system is calculated as per the following Eqs. (9)–(14).

$$EXD_{compl} = EX_1 - EX_2 + W_{comv1} \tag{7}$$

$$EXD_{comp2} = EX_5 - EX_6 + W_{comp2} \tag{8}$$

$$EXD_{LTX} = EX_3 - EX_4 \tag{9}$$

$$EXD_{HTX} = EX_7 + EX_8 \tag{10}$$

$$EXD_{CHX} = (EX_2 + EX_8) - (EX_3 + EX_5) \quad (11)$$

$$EXD_{LTE} = (1 - T_0/T_E) \times Q_L + EX_4 - EX_1 \quad (12)$$

$$EXD_{HTC} = (T_0/T_C - 1)\, Q_H + EX_6 - EX_7 \quad (13)$$

Exergetic efficiency or the second law efficiency is expressed as net exergy output divided by net exergy input.

$$ECOP = Q_L\ \{(T_0 - T_E/T_E\}/W_{net} \quad (14)$$

where

EXD_{comp1}, EXD_{comp2}, EXD_{LTX}, EXD_{HTX}, EXD_{CHX}, EXD_{LTE}, EXD_{HTC} represents the exergy destruction at the compressor 1 & 2, low-temperature expander, high-temperature expander, cascade heat exchanger, low-temperature evaporator, high-temperature condenser, respectively. T_E is the low-temperature evaporator temperature

3 Results and Discussions

System simulation is performed to analyze the trend of change in the COP and ECOP with the change in a particular operating parameter of the system, keeping the other parameters fixed. For the analysis, the ambient pressure and temperature are taken as 1.013 bar and 25 °C and a fixed cooling load (Q_L) of 9 kW. The following are the range of the operating parameters which are varied to obtain the change in COP and ECOP. The low-temperature cycle evaporator temperature (LTET) varies from −15 to 15 °C. The high-temperature cycle evaporator temperature (HTET) varies from 0 to 25 °C. The low-temperature cycle condenser temperature (LTCT) varies from 5 to 30 °C. The high-temperature cycle condenser temperature (HTCT) varies from 25 to 45 °C. The GWP value of R134 is 1430 which is comparatively higher than the other refrigerant used at the high-temperature cycle in the simulation of this model.

3.1 Effect on COP Due to Change in Different Operating Parameters

It is observed from Table 1 and Fig. 2 that while varying the temperature range of LTET from −15 to 15 °C, maximum COP was attained by R744-R717 refrigerant pair as compared to other pairs such as R744-R1234yf, R744-1234ze, R744-R134a

Table 1 Values of COP at different LTE

LTET	R744-R717	R744-R134a	R744-R1234yf	R744-R290	R744-R600	R744-R1234ze
−15	3.472	3.354	3.257	3.311	3.448	3.345
−10	3.913	3.768	3.648	3.715	3.883	3.757
−5	4.432	4.251	4.103	4.186	4.395	4.237
0	5.048	4.82	4.634	4.738	5.001	4.803
5	5.785	5.492	5.258	5.389	5.724	5.471
10	6.694	6.313	6.011	6.178	6.614	6.264
15	7.848	7.339	6.941	7.161	7.741	7.301

Fig. 2 Effect of LTET on COP

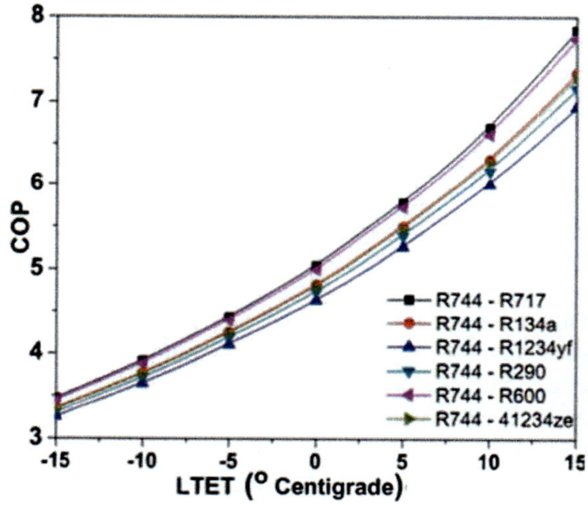

and R744-R290 for other constant thermodynamic parameters. The COP of R744-R600 is much near to the R744-R717 pair. The value of COP increases gradually as the LTET increases. With rise in evaporator temperature, the compressor work decreases due to fall in pressure ratio. At the same time, the cooling effect per unit mass flow rate increases. So, the COP of the low-temperature cycle and the whole cascade system increases simultaneously. From Table 2 and Fig. 3, it can be noticed

Table 2 Values of COP at different HTC

HTCT	R744-R717	R744-R134a	R744-R1234yf	R744-R290	R744-R600	R744-R1234ze
25	4.769	4.718	4.67	4.685	4.475	4.725
30	4.26	4.19	4.127	4.154	4.259	4.192
35	3.834	3.742	3.662	3.702	3.823	3.738
40	3.472	3.354	3.257	3.311	3.448	3.345
45	3.16	3.014	2.898	2.968	3.121	3

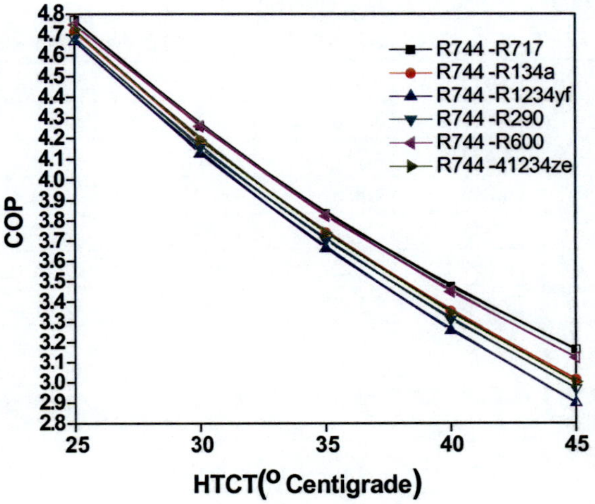

Fig. 3 Effect of HTCT on COP

that the HTCT temperature varies from 25 to 45 °C. It is observed that under this range, the COP of R744-R717 is the highest with a value of 4.769. This also illustrates that as the temperature increases, the COP decreases. R744-R134a and R744-290 also hold a good value of COP. Reversely, when the HTCT increases, the pressure ratio of the compressor of the high-temperature cycle increases leading to increase in total compressor work. At the same time, the refrigerating effect per unit mass flow rate decreases. As a result, the COP of the high-temperature cycle as well the whole cascade system decreases with the increase in HTCT. The change in the value of COP with the change in the LTCT is given in Table 3 and Fig. 4. With the increase in LTCT, the COP of system decreases gradually. The COP values of R744-R717 and R744-R600 pairs have a very minimal difference. R744-R717 achieved the highest COP. Figure 5 shows the variation of COP with respect to the change in HTET. It shows that with the increase in HTET, the COP of the system increases as shown in Table 4. R744-R717 achieves the highest COP of 6.345. Nearest to this value is 6.342, attained by R744-R600. R744-R134a has also obtained a feasible COP, but, since R134a has a higher GWP value, it will not be suitable from environmental

Table 3 Values of COP at different LTC

LTCT	R744-R717	R744-R134a	R744-R1234yf	R744-R290	R744-R600	R744-R1234ze
5	3.472	3.354	3.257	3.311	3.448	3.345
10	3.074	2.979	2.899	2.944	3.054	2.971
15	2.714	2.537	2.573	2.609	2.698	2.631
20	2.378	2.317	2.265	2.294	2.366	2.312
25	2.05	2.002	1.962	1.984	2.04	1.998
30	1.668	1.634	1.605	1.621	1.661	1.631

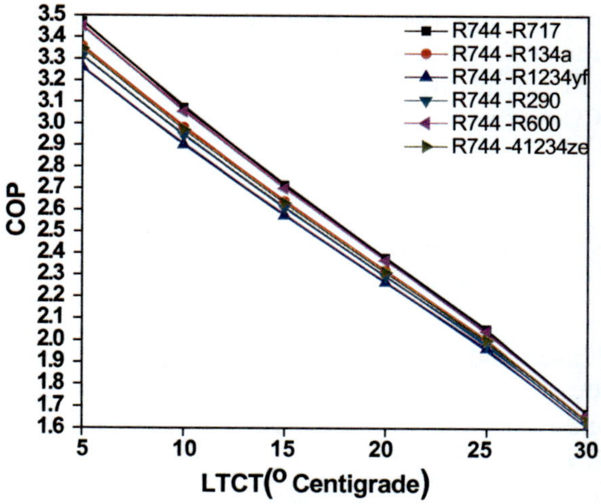

Fig. 4 Effect of LTCT on COP

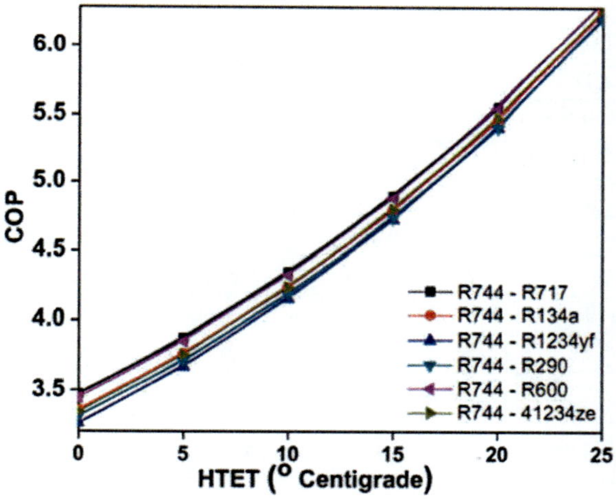

Fig. 5 Effect of HTET on COP

Table 4 Values of COP at different HTE

HTET	R744-R717	R744-R134a	R744-R1234yf	R744-R290	R744-R600	R744-R1234ze
0	3.472	3.354	3.257	3.311	3.448	3.345
5	3.874	3.76	3.672	3.713	3.854	3.757
10	4.343	4.233	4.155	4.183	4.326	4.235
15	4.893	4.79	4.724	4.738	4.882	4.798
20	5.55	5.456	5.403	5.403	5.543	5.469
25	6.345	6.264	6.225	6	6.342	6.281

point of view. The reason behind the trend of change of COP with LTET is explained earlier. For the same reason, the trend of change in COP with HTET remains the same. Similarly, the trend of the graph of COP with respect to LTCT is similar as the COP versus HTCT graph, and the technical reason is explained in the above section.

3.2 *Effect on ECOP Due to Change in Different Operating Parameters*

Table 5 and Fig. 6 show the variation of ECOP with the change in LTET. It shows that with the increase in LTET, the ECOP value decreases. The reason for this is clear from the exergy analysis and equation number (14) that ECOP and LTET are inversely correlated. So keeping other parameters fixed is seen that as the LTET increases, the availability of the cooling load decreases (numerator part of Eq. 14).

Table 5 Values of ECOP at different LTE

LTET	R744-R717	R744-R134a	R744-R1234yf	R744-R290	R744-R600	R744-R1234ze
−15	0.5383	0.5201	0.5049	0.5134	0.5346	0.5187
−10	0.5208	0.5014	0.4855	0.4944	0.5168	0.5
−5	0.4961	0.4759	0.4593	0.4686	0.492	0.4743
0	0.4623	0.4414	0.4244	0.4339	0.458	0.4398
5	0.4162	0.3951	0.3783	0.3877	0.4118	0.3936
10	0.3548	0.3346	0.3186	0.3275	0.3506	0.3331
15	0.2725	0.2548	0.241	0.2487	0.2688	0.2535

Fig. 6 Effect of LTET on ECOP

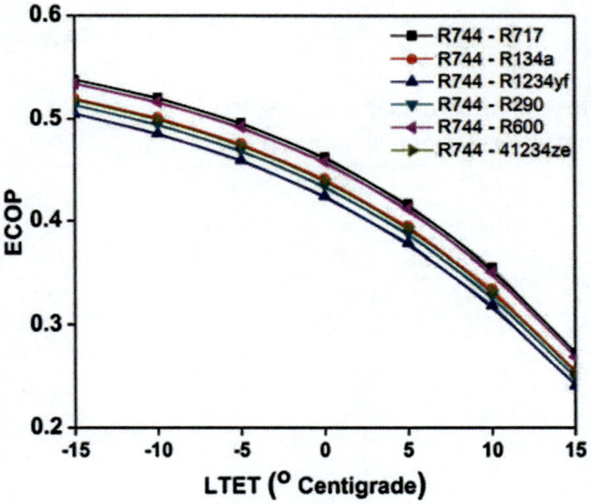

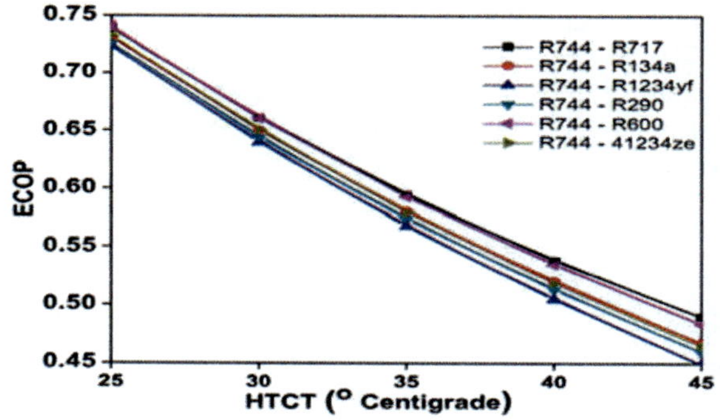

Fig. 7 Effect of HTCT on ECOP

Table 6 Values of ECOP at different HTC

HTCT	R744-R717	R744-R134a	R744-R1234yf	R744-R290	R744-R600	R744-R1234ze
25	0.7394	0.7315	0.724	0.7263	0.7404	0.7326
30	0.6605	0.6497	0.6398	0.644	0.6604	0.6499
35	0.5945	0.5801	0.5677	0.574	0.5927	0.5796
40	0.5383	0.5201	0.5049	0.5134	0.5346	0.5187
45	0.4899	0.4673	0.4493	0.4601	0.4838	0.4651

As a result, the ECOP of the system also decreases. It is observed that R744-R717 refrigerant pair achieves maximum ECOP equivalent to 0.5383. It is clear from Fig. 7 and Table 6 that as the HTCT increases, ECOP decreases. As we have seen from Fig. 3 that the COP of the system decreases with increase in HTCT because of increase of the compressor work, keeping other parameters fixed. So, here, also as the HTCT increases, keeping others parameter fixed, the ECOP decreases. From Table 6, it is quite obvious that maximum ECOP is attained by R744-R600 with a value of 0.7404. ECOP value of R744-R717 is very close to the value of R744-R600. Lowest ECOP is obtained with R744-R1234ze. From Fig. 8 and Table 7, it is observed that with the increase in LTCT the ECOP decreases. As the LTCT increases, the work input into the compressors increases. So, it is clear from Eq. (14) keeping other parameters fixed, if the work input to the system increases, then the ECOP will decrease. So, as the LTCT increases, the ECOP decreases. The ECOP of R744-R717 is highest among other refrigerant pairs. The second best results are achieved by R744-R600. It is seen from Fig. 9 that as the HTET increases, the ECOP also increases. As we know when the HTET increases, the compressor work for the high-temperature cycle decreases. It happens due to the decrease in pressure ratio of the compressor of the high-temperature cycle (COMP2 of Fig. 1). So, the total compressor work input of the whole cascade system decreases but other parameters remain unaltered. As a result, the ECOP of the system increases with the increase in HTET. It is observed from Table 8 that the ECOP value reaches up to a very high value at the temperature

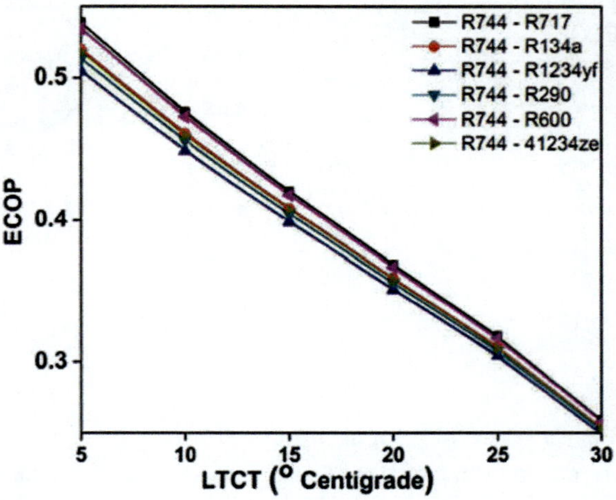

Fig. 8 Effect of LTCT on ECOP

Table 7 Values of ECOP at different LTC

LTCT	R744-R717	R744-R134a	R744-R1234yf	R744-R290	R744-R600	R744-R1234ze
5	0.5383	0.5201	0.5049	0.5134	0.5346	0.5187
10	0.4765	0.4618	0.4495	0.4564	0.4735	0.4607
15	0.4207	0.4088	0.3988	0.4044	0.4183	0.4079
20	0.3687	0.3592	0.3512	0.3557	0.3668	0.3585
25	0.3178	0.3104	0.3041	0.3077	0.3163	0.3098
30	0.2568	0.2533	0.2488	0.2513	0.2575	0.2529

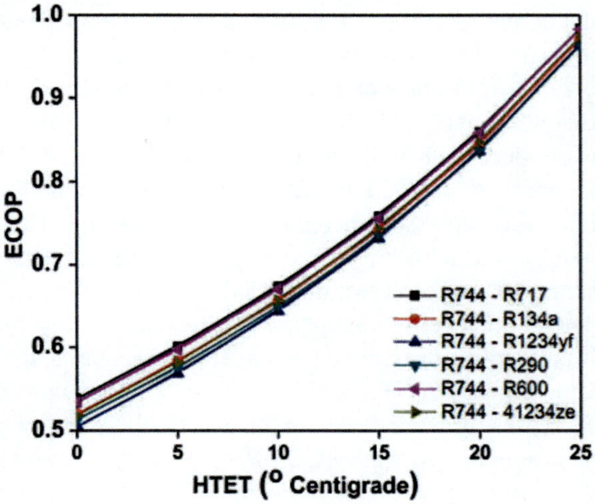

Fig. 9 Effect of HTET on ECOP

Table 8 Values of ECOP at different HTE

HTET	R744-R717	R744-R134a	R744-R1234yf	R744-R290	R744-R600	R744-R1234ze
0	0.5383	0.5201	0.5049	0.5134	0.5346	0.5187
5	0.6007	0.5829	0.5692	0.5757	0.5975	0.5824
10	0.6733	0.6562	0.6441	0.6485	0.6708	0.6567
15	0.7587	0.7427	0.7324	0.7345	0.7564	0.7439
20	0.8605	0.8459	0.8377	0.8376	0.8593	0.8479
25	0.9838	0.9712	0.9651	0.9632	0.9833	0.9738

of 25 °C. The highest value obtained is 0.9838 for R744-R717 combination pair. The second best result is achieved by R744-R600.

4 Conclusion

Among all the refrigerants used in the simulation of the cascade system, R744-R717 achieved the maximum COP as well ECOP. This pair produces the maximum overall COP of 7.848 and the maximum overall ECOP of 0.9838. Other than R744-R717 pair, R744-R600 pair also showed good results with an overall COP value of 7.741. Results show that for functioning of a cascade system, R744-R717 is the best working pair. R717 is eco-friendly and has a low GWP. Opting for this refrigerant is quite viable. Other alternatives like R600 and R1234yf can be used since it promotes the safety of the environment. This study shows that efficient and environment friendly refrigeration system can be considered without using conventional refrigerants like R134a. R744-R717 and R744-R600 refrigerant pair may be opted as an alternative to operate cascade system efficiently with expectation of lower environmental risks.

5 Future Scope

Researches can be taken forward by experimenting the cascade systems with more combinations of refrigerants with different energy sources, and its effect on the system can be simulated and can be presented in the form of advancements in this field. Also, energy efficient systems with lower environmental risks are required to be optimized for better understanding of the failure points of the system. Cascade system is vast research area, and testing the system with different temperature limits may help in analyzing the system in a better way.

References

1. Naik RH, Manohar KR (2017) The performance analysis of cascade refrigeration system with and without phase change material. IOSR J Mech Civil Eng 4:7–16
2. Gholamian E, Hanafizadeh P, Ahmadi P (2018) Advanced exergy analysis of a carbon dioxide ammonia cascade refrigeration system. Appl Therm Eng 137:689–699
3. Pilla TS, Sunkari PKG, Padmanabhuni SL, Nair SS, Dondapati RS (2017) Experimental evaluation Mechanical performance of the compressor with mixed refrigerants R-290 and R-600a. Energy Procedia 109:113–121
4. Ahmad N, Arora A, Manjunath K (2018) Optimum performance analysis of a hybrid cascade refrigeration system using alternative refrigerants. Mater Today: Proc 5(14):374–383
5. Shah AHA, Kapadia BRG (2011) Comparative assessment of a cascade refrigeration cycle with different refrigerant pair. In: International conference on current trends in technology, 'NUiCONE—2011 proceedings, pp 8–10, Institute of Technology, Nirma University, Ahmedabad, 08–10 Dec 2011
6. Li Z, Jiang H, Chen X, Liang K (2019) Energy and buildings comparative study on energy efficiency of low GWP refrigerants in domestic refrigerators with capacity modulation. Energy Build 192:93–100
7. Yataganbaba A, Kilicarslan A (2015) Exergy analysis of R1234yf and R1234ze as R134a replacements in a two evaporator vapor. Int J Refrig 60:26–37
8. Alhamid MI, Syaka DRB (2010) Nasruddin: exergy and energy analysis of a cascade refrigeration system using R744 + R170 for low temperature applications. Int J Mech Mech Eng 10(6):1–8
9. Sun Z, Wang Q, Xie Z, Liu S, Su D, Cui Q (2018) Energy and exergy analysis of low GWP refrigerants in cascade refrigeration system. Energy 170:1170–1180
10. Gullo P, Hafner A (2017) Comparative assessment of supermarket refrigeration systems using ultra low-GWP refrigerants—case study of selected American cities. In: ECOS 2017—30th international conference on efficiency, cost, optimization, simulation and environmental impact of energy systems proceedings, pp 1–13, San Diego (USA)
11. Chattopadhyay S, Roy D, Ghosh S (2017) Comparative energetic and exergetic studies of vapour compression and vapor absorption refrigeration cycles. Int J Renew Energy Technol 8(3/4):222–238
12. Getu HM, Bansal PK (2006) Simulation model of a low temperature supermarket refrigeration system. Int J HVAC & R Res 12(4):1117–1139
13. Messineo A (2012) R744-R717 cascade refrigeration system: performance evaluation compared with a HFC two-stage system. Energy Procedia 14:56–65
14. Vicente P, Belman-Flores J, Rodriguez-Munoz JL, Victor R, Gallegos-Munoz A (2017) Second law analysis of a mobile air conditioning system with internal heat exchanger using low GWP refrigerants. Entropy 19(4):175–188
15. Patel V, Panchal D, Prajapati A, Mudgal A, Davies P (2019) An efficient optimization and comparative analysis of cascade refrigeration system using NH_3/CO_2 and C_3H_8/CO_2 refrigerant pairs. Int J Refrig 102:62–76

Modeling of Frosting on Fin-and-Tube Heat Exchanger of a Domestic Refrigerator

A. Saikiran Pegallapati and Maddali Ramgopal

Abstract Domestic refrigerators typically use fin-and-tube type heat exchangers with variable fin pitch across the tube rows to accommodate for the reduction in the flow area due to frost formation. In the current study, a numerical model is developed in Modelica language to predict the heat and mass transfer across a typical tube–fin heat exchanger employed in a domestic refrigerator. Results obtained from the numerical model are validated using a purpose-built experimental setup. A comparison between the experimental and numerical results showed good agreement. Frost formation rate observed is constant, and the rate of heat transfer is higher in the bottom rows and decreases along the direction of the airflow.

Keywords Domestic refrigerator · Frost modeling · Heat and mass transfer

1 Introduction

Freezer compartment of a domestic two-door frost free refrigerator has a typical operating temperatures of −18 °C which requires the evaporator of the domestic refrigerator to operate well below freezing point of water (about −23 °C). This results in frosting on the evaporator surface due to simultaneous transfer of both heat and water vapor from moist-air in the cabinets. Frosting on heat exchanger is detrimental to its performance as the capacity of the heat exchanger reduces with time, primarily by reduction in the flow rate of air in case of a fan-supplied heat exchanger and secondarily due to formation of the frost layer which adds resistance to the flow of heat. Hence, understanding the phenomenon of frosting on heat exchanger surface and its effect on the performance is essential for efficient design of the refrigeration system and the selection of proper defrosting methods.

A detailed review on the models for frost properties and frost formation on simple geometries is presented by Iragorry et. al. [1]. Frost formation is characterized into three stages consisting of crystal growth period followed by frost layer growth period

A. S. Pegallapati (✉) · M. Ramgopal
Mechanical Engineering Department, Indian Institute of Technology Kharagpur, Kharagpur, WB 721302, India
e-mail: aaditya.sai.kiran@gmail.com

M. Ramgopal et al. (eds.), *Advances in Air Conditioning and Refrigeration*, Lecture Notes in Mechanical Engineering, https://doi.org/10.1007/978-981-15-6360-7_21

and frost layer full growth period by Hayashi et. al. [2]. Several numerical models are presented for prediction of frost growth on simple geometries. Sami and Duong [3] provided a lumped (integral) analysis whereas Tao et. al. [4] developed a distributed (differential) analysis. Lee et. al. [5] developed a simplified model which assumes a homogeneous behavior of the frost layer with uniform properties to obtain ordinary quasi-steady state differential equations. Na and Webb [6] used one dimensional, transient model based on differential approach considering porous structure of frost and using local volume averaging technique. Cheng and Cheng [7] developed a semi-empirical model that uses the correlation introduced by Hayashi et al. [2] for frost density, making frost thickness as the only dependent variable in the model.

Several studies have been conducted for understanding the phenomenon and effect of frosting on the heat exchangers. Seker et al. [9] and Kondepudi and O'Neal [8] investigated the effect of frosting on tube–fin type heat exchangers both numerically and experimentally. Padhmanabhan et al. [10] developed a semi-empirical model to estimate the non-uniform growth of frost on heat exchangers which is experimentally validated. Ribeiro and Hermes [11] presented and validated an algebraic model based on first principles to simulate the thermo-hydraulic behavior of tube–fin evaporators undergoing frost formation. Da Silva et al. [12] performed experiments on fan-supplied tube–fin evaporators undergoing frosting. Barbosa et al. [13] studied experimentally the heat transfer and pressure drop behavior on the air-side of a discrete flat plate-finned tube heat exchangers used in domestic refrigerators. More recently, Ribeiro et al. [14] performed optimization using quasi-steady algebraic models to reduce the system-level energy consumption during frosting of fan-supplied tube evaporators. Zhang et al. [15] performed experiments to study the frost formation on tube–fin heat exchangers with different fin pitches.

In the present paper, a numerical model for prediction of frost formation on a typical domestic refrigerator evaporator is presented. The results obtained from the analysis are compared to that of the experiments conducted on a purpose-built in-house experimental setup.

2 Geometry and Mathematical Modeling

2.1 Geometry of Heat Exchanger

Figure 1 shows the schematic of the tube–fin heat exchanger of a domestic two-door refrigerator that is employed in the current study. The geometrical details of the heat exchanger are provided in Table 1.

The heat exchanger consists of seven tube rows with varying fin pitch and is divided into six partitions across the tube length giving a total of 42 control volumes. Balance equations for mass and energy are invoked across each of the control volumes as detailed in the subsequent sections.

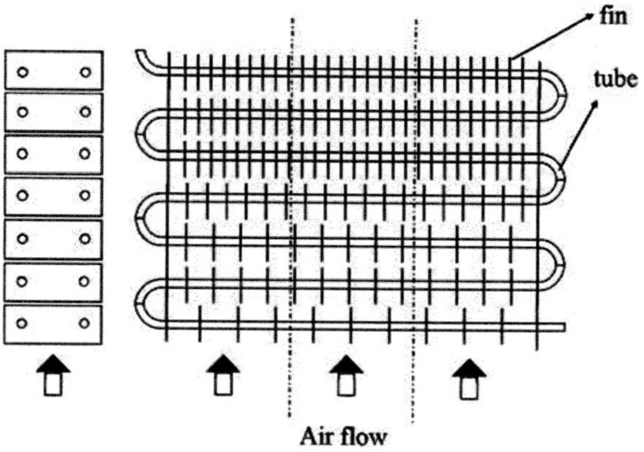

Fig. 1 Schematic of fin-and-tube type heat exchanger

Table 1 Details of fin-and-tube heat exchanger geometry

Parameter	Value
Tube length per row (mm)	270
Outer diameter of tube (mm)	8.7
Width of fin (mm)	80
Thickness of fin (mm)	0.15
Height of fin (mm)	27
Number of fins (respectively) Tube rows 1, 2, 3, 4, 5, 6, 7	6, 9, 9, 12, 18, 18, 18
Tube spacing-Transversal (mm)	25
Tube spacing-Longitudinal (mm)	30

2.2 Air Model

A one-dimensional quasi-steady approach is followed for modeling of air flow across the heat exchanger. Heat conduction in the air flow direction is assumed to be negligible, and Lewis analogy is applicable during the simultaneous heat and mass transfer. Also, contact resistance between the tube and fin is neglected.

At the outlet of the control volume, temperature and humidity ratio of the exit air are obtained using the following equations which are obtained by invoking the energy and mass balance of air across the control volume.

$$T_{\text{a,out}} = T_{\text{a,in}} + \left(T_{\text{w}} - T_{\text{a,in}}\right)\left\{1 - \exp\left[-\frac{h\phi A_{\text{tot}}}{\dot{m}_{\text{a}} c_{\text{p,a}}}\right]\right\} \tag{1}$$

$$\omega_{\text{a,out}} = \omega_{\text{a,in}} + \left(\omega_{\text{a,s}} - \omega_{\text{a,in}}\right)\left\{1 - \exp\left[-\frac{h\phi A_{\text{tot}}}{\dot{m} c_{\text{p,a}} Le^{2/3}}\right]\right\} \tag{2}$$

$$\Delta p_{\mathrm{a}} = \frac{G_{\mathrm{a.max}}^2}{2\rho_{\mathrm{a,in}}}\left\{f\frac{A_{\mathrm{tot}}}{A_{\mathrm{ff,min}}}\frac{\rho_{\mathrm{a,in}}}{\rho_{\mathrm{avg}}} + \left(\frac{\rho_{\mathrm{a,in}}}{\rho_{\mathrm{a,out}}} - 1\right)\left(1 + \frac{A_{\mathrm{ff,min}}^2}{A_{\mathrm{fc}}^2}\right)\right\} \tag{3}$$

The values of $h\phi$ and friction factor f are obtained from the correlation given by Barbosa et. al. [13].

2.3 Frost Growth and Densification Model

Model developed by Lee et. al. [5] and Cheng and Cheng [7] is employed in the present work with the following simplifying assumptions to model frosting. (i) The frost layer is characterized by average properties, and the thickness is assumed to be same on both fin and tube. (ii) Quasi-steady conditions are assumed, and the model is one dimensional. (iii) Lewis analogy is applicable. (iv) No energy storage within frost layer. In the total water vapor transferred from the air, a portion of water vapor goes into increasing the frost average density, and the rest goes into increasing the frost thickness. Determining the percentage contribution of these two portions in the total mass transferred is the essence of modeling the frosting process.

The sensible heat and the total mass transfer to the frost layer from moist air can be obtained by applying conservation of energy and mass at interface of air and frost as.

$$q''_{\mathrm{sen}} = \frac{\dot{m}_{\mathrm{a}} c_{\mathrm{p,a}}\left(T_{\mathrm{a,in}} - T_{\mathrm{w}}\right)}{A_{\mathrm{tube}} + A_{\mathrm{fin}}}\left\{1 - \exp\left[-\frac{h\phi A_{\mathrm{tot}}}{\dot{m}_{\mathrm{a}} c_{\mathrm{p,a}}}\right]\right\} \tag{4}$$

$$m''_{\mathrm{v}} = \frac{\dot{m}_{\mathrm{a}}\left(\omega_{\mathrm{a,in}} - \omega_{\mathrm{f,s}}\right)}{A_{\mathrm{tube}} + A_{\mathrm{fin}}}\left\{1 - \exp\left[-\frac{h\phi A_{\mathrm{tot}}}{\dot{m}_{\mathrm{a}} c_{\mathrm{p,a}} Le^{2/3}}\right]\right\} \tag{5}$$

The total mass flux into the frost consists of densification mass flux and thickness mass flux defined as

$$m''_{\mathrm{v}} = \bar{\rho}_{\mathrm{f}}\frac{\mathrm{d}\delta_{\mathrm{f}}}{\mathrm{d}t} + \delta_{\mathrm{f}}\frac{\mathrm{d}\bar{\rho}_{\mathrm{f}}}{\mathrm{d}t} = m''_{\rho} + m''_{\delta} \tag{6}$$

The densification mass flux at the interface of air and the frost layer can be obtained by applying Fick's law as

$$m''_{\rho} = -D_{\mathrm{eff}}\left.\frac{\partial \rho_{\mathrm{v}}}{\partial x}\right|_{x=\delta_{\mathrm{f}}} \tag{7}$$

Lee et al. [5] assumed that the rate at which water vapor is absorbed by the frost layer is proportional to the local water vapor density within the frost layer to obtain the distribution of water vapor density and temperature within the frost layer as

$$\rho_{\rm v}(x) = \rho_{\rm v,sat}(T_{\rm w})\cosh\left(\sqrt{\frac{\zeta}{D_{\rm eff}}}x\right) \tag{8}$$

$$T_{\rm f} = -\frac{D_{\rm eff}\Delta h_{\rm ig}\rho_{\rm v,s}(T_{\rm w})}{R_{\rm v}T_{\rm w}k_{\rm f}}\left[1-\cosh\left(\sqrt{\frac{\zeta}{D_{\rm eff}}}x\right)\right] + \frac{q''_{\rm tot}}{k_{\rm f}}x + T_{\rm w} \tag{9}$$

$$\zeta = D_{\rm eff}\left\{\frac{1}{\delta_f}\cosh^{-1}\left[\frac{\rho_{\rm v,s}(T_{\rm fs})}{\rho_{\rm v,s}(T_{\rm w})}\right]\right\}^2 \tag{10}$$

The densification and growth of the frost layer over time can be obtained from the following

$$\frac{{\rm d}\bar{\rho}_f}{{\rm d}t} = \frac{m''_\rho}{\delta_{\rm f}} \tag{11}$$

$$\frac{{\rm d}\delta_f}{{\rm d}t} = \frac{m''_{\rm v} - m''_\rho}{\bar{\rho}_{\rm f}} \tag{12}$$

Initial condition for frost thickness is taken as 1 μm. Initial condition for frost density is considered as 30 kg/m^3. The surface temperature of the fin-and-tube heat exchanger is assumed to be uniform throughout and is specified as an input.

3 Results and Discussion

Experiments were conducted in a purpose-built setup, the schematic of which is shown in Fig. 2. Three environmental sensors are placed at the inlet as well as the outlet of the evaporator to measure the psychometric condition of the air. Flow rate of the air through the duct is measured using an orifice flow meter. T-type thermocouples are placed on the surface of the evaporator at equal spacing along the length for temperature measurement.

The inlet conditions of the air are maintained at −7.5 °C and 70% relative humidity. Velocity of the air at the inlet is maintained at 0.70 m/s. The evaporated is connected to the refrigeration system, and the surface temperature is maintained at –15 °C. The model developed in the earlier section is used to predict the heat and mass transfer across the heat exchanger by providing the conditions mentioned earlier as an input to the model. Table 2 shows the comparison between the experimentally measured and results obtained from simulation for few selected parameters. It can be observed that a good agreement has been obtained between the same.

Figure 3 shows the comparison between the simulated and experimentally observed frost mass over a period of one hour. It can be observed that is a very good agreement with experimental data. Rate of frost growth is observed to be constant from the linear nature of the graph for the period of observation. The deviation at a

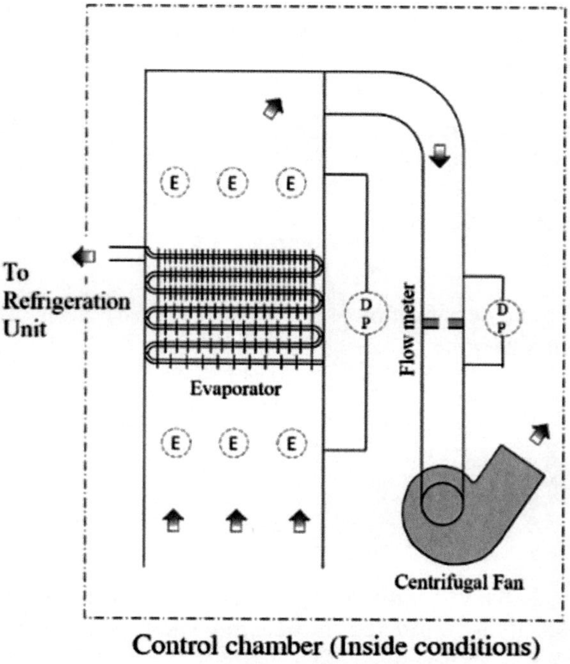

Fig. 2 Schematic of the experimental setup

Table 2 Comparison between simulation and experimental results

Description	Experimental value	Simulation result
Cooling capacity (W)	104	110
Air outlet temperature (°C)	−11.2	−11.8
Total frost mass @ 1 h (g)	18	18.6

later time is attributed to the slight change in the inlet temperature of the air in the experiments which was held constant in the simulations performed.

Figure 4 shows the rate of heat transfer across each tube row obtained from the numerical simulation. It can be seen that although tube row #1 has less heat transfer area due large fin pitch (see Table 1), the heat transferred is maximum across it.

This is because, the difference in temperature between the heat exchanger surface and air is maximum for first row, and then, it subsequently reduces due to the air being cooled after each tube row.

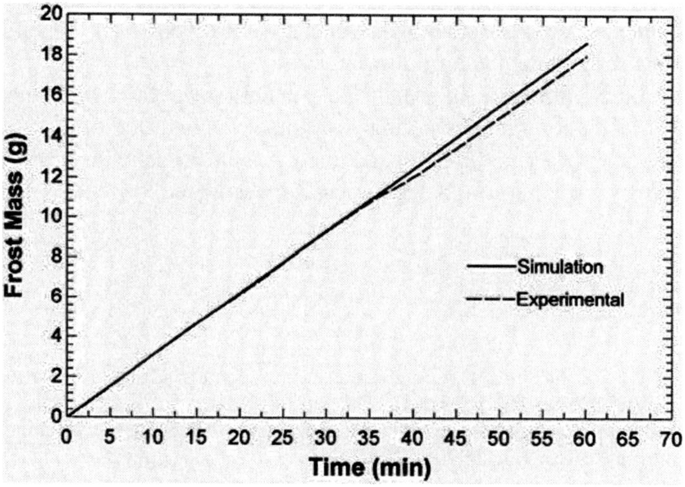

Fig. 3 Comparison between experimental and simulated frost growth

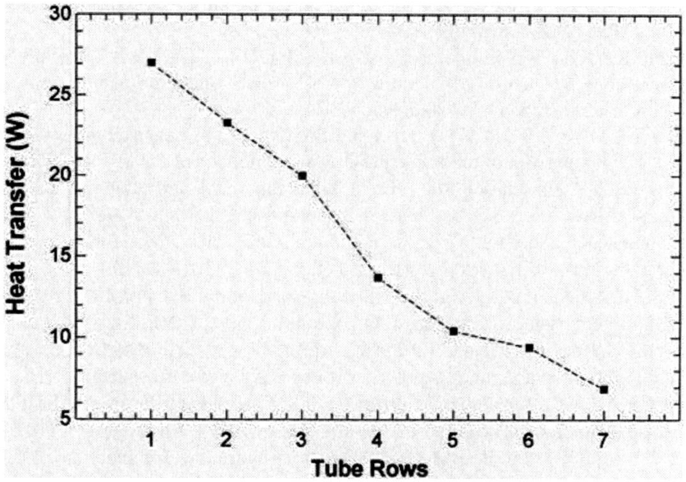

Fig. 4 Heat transfer rate across each tube row of the heat exchanger

4 Conclusion

Formation of frost on a typical fin-and-tube heat exchanger used in a domestic refrigerator is modeled, and the results obtained a comparison are made with the experimental observations. A good agreement is observed between the experimental and numerical simulation results. The rate of frost formation is observed to be uniform during the period of the study. Maximum heat transfer is observed in the starting rows of the heat exchanger which is ascribed to the larger temperature difference between the air and wall surface. The rate of heat transfer is observed to subsequently reduce across the tube rows along the air flow direction.

There are other directions in which the present work could be extended. The effect of frost formation rate on varying psychometric condition of the inlet air can be studied. Effect of varying fin pitch and the distribution of the fins across the tube rows on the frost formation can be numerically investigated.

References

1. Iragorry J, Tao YX, Jia S (2004) A critical review of properties and models for frost formation analysis. HVAC&R Res 10(4):393–420
2. Hayashi Y, Aoki A, Adachi S, Hori K (1977) Study of frost properties correlating with frost formation types. J Heat Transf 99(2):239–245
3. Sami SM, Duong T (1989) Mass and heat transfer during frost growth. ASHRAE Trans 95(3218):158–165
4. Tao YX, Besant RW, Rezkallah KS (1993) A mathematical model for predicting the densification and growth of frost on a flat plate. Int J Heat Mass Transf 36(2):353–363
5. Lee KS, Kim WS, Lee TH (1997) A one-dimensional model for frost formation on a cold flat surface. Int J Heat Mass Transf 40(18):4359–4365
6. Na B, Webb RL (2004) New model for frost growth rate. Int J Heat Mass Transf 47(5):925–936
7. Cheng CH, Cheng YC (2001) Predictions of frost growth on a cold plate in atmospheric air. Int Commun Heat Mass Transfer 28(7):953–962
8. Kondepudi SN, O'Neal DL (1993) Performance of finned-tube heat exchangers under frosting conditions: II. Comparison of experimental data with model. Int J Refrig 16(3):181–184
9. Seker D, Karatas H, Egrican N (2004) Frost formation on fin-and-tube heat exchangers. Part I—modeling of frost formation on fin-and-tube heat exchangers. Int J Refrig 27(4):367–374
10. Padhmanabhan SK, Fisher DE, Cremaschi L, Moallem E (2011) Modeling non-uniform frost growth on a fin-and-tube heat exchanger. Int J Refrig 34(8):2018–2030
11. Ribeiro RS, Hermes CJ (2014) Algebraic modeling and thermodynamic design of fan-supplied tube-fin evaporators running under frosting conditions. Appl Therm Eng 70(1):552–559
12. Da Silva DL, Hermes CJ, Melo C (2011) First-principles modeling of frost accumulation on fan-supplied tube-fin evaporators. Appl Therm Eng 31(14–15):2616–2621
13. Barbosa JR Jr, Melo C, Hermes CJ, Waltrich PJ (2009) A study of the air-side heat transfer and pressure drop characteristics of tube-fin 'no-frost'evaporators. Appl Energy 86(9):1484–1491
14. Karki S, Haapala KR, Fronk BM (2018) Thermal-hydraulic optimization of fan-supplied tube-fin evaporators for frosting conditions aiming at minimum energy consumption. In: International refrigeration and air conditioning conference, Purdue University
15. Zhang L, Jiang Y, Dong J, Yao Y, Deng S (2019) An experimental study on the effects of frosting conditions on frost distribution and growth on finned tube heat exchangers. Int J Heat Mass Transf 128:748–761

Numerical Analysis of Glauber Salt-Based Solar Energy Systems for Heating Cooling and Air Conditioning

Hiranmoy Samanta, Rohit Maity, Mrinal Ghosh, and Pradip Kumar Talapatra

Abstract The sun rays are concentrated through the parabolic disc mainly to the building roof, and the outside walls of the buildings exposed to the solar [1, 2] radiation are insulated as well as act as a heat storage unit with the use of PCMs and PCM mixed with metal slurry [3] mainly aluminum, steel, etc as thermal conductivity enhancer [4, 5] or heat transfer enhancer [6]. The paper mainly focuses on the total cycle of the PCM i.e. melting and solidification of the phase change materials. The numerical analysis for the melting cycle incorporates with complex transport phenomena. The mass, momentum and energy conservations equations are formed and numerically validated for the method from the experimental data. The Glauber salt-based heat storage devices contain high energy density and repeatability of usage over a period of time with chemical stability of the PCM. The experimental setup is made as well as numerical model validated through the heat exchanger unit by monitoring and collecting the continuous temperature over a prolonged period of time by using computerized data acquisition system. The rectangular fins [7] attached to the base plate of the aluminum are used as a thermal energy enhancer. The enthalpy update scheme-based 2D numerical model is used. TDMA solver is incorporated for solution. The paper indicates that the time for solidification of the system is studied with the melting pattern, temperature distribution and the stream function plots.

Keywords Phase change material · Solar heating · Glauber salt ($NaSO_4$, $10H_2O$) · Enthalpy update scheme · TDMA

H. Samanta (✉) · R. Maity · M. Ghosh · P. K. Talapatra
Department of Mechanical Engineering, Gargi Memorial Institute of Technology, Balarampur, Baruipur, Kolkata 700144, West Bengal, India
e-mail: hiru.samanta@gmail.com

R. Maity
e-mail: rohit.maity264@gmail.com

M. Ghosh
e-mail: Mrinalghosh2324@gmail.com

P. K. Talapatra
e-mail: pktpatra@rediffmail.com

M. Ramgopal et al. (eds.), *Advances in Air Conditioning and Refrigeration*, Lecture Notes in Mechanical Engineering, https://doi.org/10.1007/978-981-15-6360-7_22

Nomenclature

C	Morphological constant
c_p	Specific heat (J/kg K)
f_j	Liquid fraction of PCM
g	Gravitational acceleration (m/s^2)
h	Heat transfer coeff. (W/m^2 K)
ΔH	Enthalpy content (J/kg)
k	Thermal conductivity (W/m K)
P	Pressure (Pa)
t	Time (s)
T	Temperature (°C)
u_i	Velocity in *i*th direction (m/s)

Greek symbol

ρ	Density (kg/m^3)
μ	Viscosity (Pa s)
λ	Relaxation factor

Subscripts

a	Atmospheric
l	Liquid
melt	Melting
r	Room
ref	Reference

1 Introduction

The use of PCMs is widely used for energy storage due to their potential of containing high latent heat during melting and solidification. In the previous few decades, encapsulated PCM and PCM boards are widely used. The mixing of different PCMs is also useful, and high efficiencies are shown in thermal storage. PCMs provide opportunities for greater energy storage in many applications for residential buildings which include, but are not limited to, solar water heating, space heating/cooling [8] and waste heat recovery. The present global scenario of energy focuses mainly on the non-conventional methods of energy storage and utilization. The solar system is the most convenient green energy sources. The phase change materials (PCMs)-based thermal energy storage models are best in use due to high energy storage and other parameter during melting and solidification process. The performance analysis of numerous researchers has intensively studied phase change materials (PCMs) in thermal energy storage and building cooling because of their high thermal energy densities per unit volume mass and their availability in different fields of engineering with wide temperature ranges. This work comprises the melting of Glauber salt as well as the total cycle time observation over a period of time with the total storage of thermal energy in the form of heat, and the cooling slope is generated toward the building.

2 Literature Review

Doan et al. [9] have examined Green Star certification uptake and its relationship with Building Information Modeling. They have mentioned six themes, and Green Star is highlighted in New Zealand. The research showed a variety of suggestions to encourage Green Star development, with more extensive education playing a critical role, combined with BIM. Fong et al. [2] have introduced stratum ventilation (SV), a new indoor air distribution strategy. They have compared the conventional mixing ventilation with the stratum ventilation. They evaluated that the energy performance solar air conditioning can also be facilitated by high-temperature cooling like potential of Fernández et al. [10] did the study on optimum thermal solar power plant capacity with the help of different PCMs and the development of different solutions to improve the heat transfer Agyenim et al. [7] have experimentally validated a comparison of heat transfer enhancement in a medium-temperature thermal energy storage heat exchanger using fins for a concentric configuration. Jian-you [11] has numerically and experimentally validated the heat transfer in triplex tube with phase change material for thermal energy storage dominated by heat conduction. Ravikumar and Srinivasan [12] have found that optimal insulation consideration is needed for the buildings in the withering course region. PCM nearly having uniform roof bottom surface temperature is maintained for better result. Velraj [13] has experimentally and numerically modeled of inward solidification on a finned vertical tube for a latent heat storage unit. The internal longitudinal fin configuration has been done with v-shaped enclosure for vertical positioning. Zhang and Faghri [14] have analyzed thermal energy storage system with conjugate turbulent forced convection analytically. Watanabe and Kanzawa [15] have worked on exergy-based storage optimization with the use of a latent heat system with different PCMs having different melting points by numerical simulations and also from simple equation.

3 Experimental Setup

The experimental setup consists of control unit of temperature of HTF with flow control; 5 mm transparent plexiglass sheet is used to make the rectangular container attached with aluminum plate fitted with fin. The experimental setup considers cooling of a room (3.5 m × 3.5 m × 3.5 m) using the most efficient performing PCM. The present work focused on cooling of a room using PCM and enhancement of heat generation by using fin, and hence, TCE [3] is considered. Generally, the roof of the house is made of concrete. The Glauber salt (PCM) is used as a thermal storage device. The rectangular fin is considered as heat conductivity enhancer, and the aluminum slurry is used as thermal conductivity enhancer. The natural convection is the only way to carry away the heat from the roof. The PCM has a natural property to absorb heat if placed in atmosphere higher than its melting temperature and solidifies when the temperature is lower than the melting point by rejecting heat. The PCM absorbs the solar radiation in the daytime and releases the absorbed heat in atmosphere when temperature is lower than the melting point temperature. A 2D (x, y) rectangular cavity geometry is considered as per the problem definition. Anhydrous Glauber salt is selected as the PCM such that it absorbs heat during charging in the daytime and discharges by releasing heat at a lower temperature available at night. Aluminum base fitted with fins are considered for more efficient cooling and heating as well as to increase the thermal conductivity of the system. The temperature data of Bhubaneswar is presented in Table 1. The thermo-physical properties of the concrete, the PCM and the TCE used in the simulation are given in Table 2 (Fig. 1).

Table 1 Temperature data of Bhubaneswar (*source* Weather Forecast Department Bhubaneswar)

	Jan	Feb	Mar	Apr	May	June	July	Aug	Sept	Oct	Nov	Dec
Avg. temperature(°C)	22.1	24.9	28.4	30.8	31.7	30.6	28.6	28.3	28.6	27.7	24.7	21.9
Min. temperature (°C)	15.7	18.7	22.3	25.1	26.5	26.2	25.4	25.2	25.1	23.4	19.3	15.6
Max. temperature (°C)	28.6	31.2	34.6	36.6	37	35.1	31.8	31.5	32.1	32	30.1	28.3
Avg. temperature (°F)	71.8	76.8	83.1	87.4	89.1	87.1	83.5	82.9	83.5	81.9	76.5	71.4
Min. temperature (°F)	60.3	65.7	72.1	77.2	79.7	79.2	77.7	77.4	77.2	74.1	66.7	60.1
Max. temperature (°F)	83.5	88.2	94.3	97.9	98.6	95.2	89.2	88.7	89.8	89.6	86.2	82.9
Precipitation/rainfall (mm)	12	25	26	26	62	190	353	344	257	168	38	4

Table 2 Thermo-physical properties and system data (*Source* Ravikumar and Srinivasan [12])

Material name	Density (kg/m^3)	Thermal conductivity (W/m-K)	Specific heat (J/kg-K)	Melting point (K)	Melting point (°C)	Latent heat (kJ/kg)
Glauber salt solid (powdered form)	1485	0.544	1420	305.4	32.4	254
Anhydrous Glauber salt (Na_2SO_4, $10H_2O$)	1485	0.544	2100	–	–	–
Concrete (M20)	2300	1.28	1130	–	–	–
Aluminum pure at 300 k/30 °C	2790	237	903	903	660	398

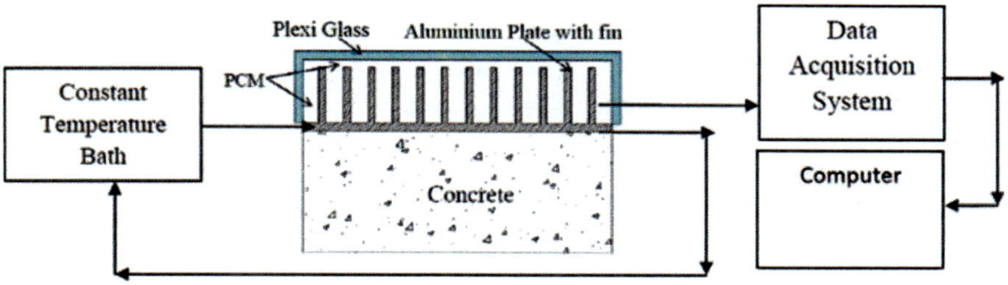

Fig. 1 Schematic diagram of experimental setup

4 Mathematical Modeling

The present work is focused on building cooling and heating in the hot summer regions. PCM-based cooling and heating methods are used, and the fins are attached for efficient heat generation, and the TCE [3] is enhancing the conductivity. The regions on the world map which are reported with a high temperature rise in summer as well as in post-summer days, the PCM used as thermal storage as well as building cooling where daytime is more than 10 hours follows with a small night time. The temperature is favorable for heat rejection at night. The charging and discharging of PCM in other words melting and solidification processes are modeled. The mathematical model is formulated by using fluid flow phenomenon and complex heat transfer criteria. These transport phenomena in the PCM are mathematically presented by a set of mass, momentum and energy conservation equations considering single-phase properties. The concrete material remains solid and fixed in the domain. The aluminum used in the modeling has constant thermo-physical properties.

4.1 Single-Phase Mass Conservation Equation

$$\frac{\partial(\rho u_j)}{\partial x_j} = 0 \tag{1}$$

4.2 Single-Phase Momentum Conservation

The single-phase momentum conservation equation for the *j*th direction is given as

$$\frac{\partial(\rho u_j)}{\partial t} + \frac{\partial(\rho u_j u_k)}{\partial x_k} = -\frac{\partial p}{\partial x_j} + \nabla.(\mu \nabla u_j) + S_j + \rho g \beta(T - T_{\text{ref}}) \tag{2}$$

In the model, the domain is treated as a porous medium during phase change of the PCM. The source term S_j [16] in the Eq. (2) is used for the purpose, which offers additional frictional resistance toward the fluid flow and given as

$$S_j = \text{Au}_j \tag{2a}$$

where *A* is defined as

$$A = -\frac{C(1 - f_{\text{L}})^2}{f_{\text{L}}^3 + b} \tag{2b}$$

In Eqs. (2a, 2b), f_{L} is the liquid fraction or the amount of the PCM melted from solid. The *C* is a morphological constant whose value is sufficiently large (~1.6×10^6). The term '*b*' is a computational constant introduced to avoid division by zero. $f_{\text{L}} = 1$ is considered as the liquid phase, and its solid to liquid transformation takes place transiently where f_{L} varies smoothly toward 0.

4.3 Energy Conservation

The single-phase energy equation can be given as:

$$\frac{\partial(\rho c_{\text{p}} T)}{\partial t} + \frac{\partial(\rho c_{\text{p}} u_j T)}{\partial x_j} = \nabla.(k \nabla T) + S_{\text{h}} \tag{3}$$

S_{h} is called source term. Source term indicates the absorption or rejection of the latent heat during melting phase change by the PCM considering the constant temperature. The corresponding latent heat source term is calculated and formulated as

$$S_h = -\frac{\partial}{\partial t}(\rho \Delta H) \quad (3a)$$

where

$$\Delta H = 0 \quad T_{sur} < T_{melt} \quad (3b)$$

$$\Delta H = L \quad T_{sur} > T_{melt} \quad (3c)$$

where $T = T_{sur,}$ surroundings or atmospheric temperature

The enthalpy update scheme is incorporated where the enthalpy value (ΔH) of the PCM is updated in every iteration by considering the enthalpy of the nth and ($n + 1$)th grid value (*Source* Brent et al. [16]) as:

$$[\Delta H_P]^{n+1} = [\Delta H_P]^n + \frac{a_P}{a_P^0}\lambda\left[\{H_P\}^n - cF^{-1}\{\Delta H_P\}^n\right] \quad (3d)$$

where ΔH_P is the latent heat stored or rejected at Pth node point of the unit computational cell, and n is the iteration number. Hence,

$a_P^0 = \rho \Delta V/\Delta t$, a_P is the coefficient of T_p in the discretized energy equation (*Source* Patankar [17]), λ is the relaxation factor, c is the specific heat and F^{-1} is the inverse latent heat function. The Anhydrous Glauber salt which is used as the PCM is considered as a pure substance; hence, F^{-1} is equal to T_{melt}. The liquid fraction of the PCM keeps on changing with rise and fall in temperature as well as with time, the different melting fractions are available at different zones where fraction of liquid at any cell is calculated by updating latent heat content (ΔH) from the surrounding cells,

$$f_L = \frac{\Delta H}{L_a} \quad (3e)$$

4.4 *Boundary Conditions Considered at the Faces of the Numerical Domain*

The mass, momentum and energy equation for the single phase are simulated effectively with appropriate temperature and fluid flow boundary conditions. The boundary conditions are considered as:

At the Top Face

The top surface of the PCM is exposed to sunlight where both solar heat in the form of radiation and convection are collected by the PCM. The top surface is represented by an equivalent surface temperature given as

$$T_{\text{ts}} = T_{\text{a}} + \frac{\alpha_{\text{a}} q_{\text{s}}}{h_{\text{a}}} \tag{4a}$$

$$\frac{\partial u_1}{\partial x_1} = 0 \text{ and; } u_2 = 0 \tag{4b}$$

At the Bottom Face

Convection boundary condition is applied at the roof bottom because heat convection takes place from the radiation heat absorbed by the roof.

$$-k\frac{\partial T}{\partial x} = h_{\text{r}}(T_{\text{rs}} - T_{\text{r}}) \tag{5a}$$

$$u_j = 0 \tag{5b}$$

At the Left Face

$$-k\frac{\partial T}{\partial x} = h_r(T_{rs} - T_r) \tag{6}$$

At the Right Faces

At the right face temperature, the fluid boundary conditions are kept constant

$$\frac{\partial T}{\partial x_1} = 0 \tag{7a}$$

$$\frac{\partial u_k}{\partial x_k} = 0 \tag{7b}$$

5 Numerical Modeling [18]

The finite volume method (FVM) is incorporated by using power law scheme to discretize the governing equations (*Source* Patankar [17]). The SIMPLER algorithm and tri-diagonal matrix algorithm (TDMA) solver are used to solve the 2D problem statement. After successful grid independency test suitable no of grids in *Y*- and *Z*-direction are selected as 102 × 82 uniform grid. The convergence criteria also considered for the equation solution of the equation and found that when $|(\phi - \phi_{\text{old}})/\phi_{\text{max}}| < 10^{-5}$ where ϕ stands for the solved variables at a grid point at the current iteration level. ϕ_{old} value is calculated from the previous iteration level. ϕ_{max} is the maximum value of the variable at the current iteration level at specified time in the total computational domain. The transient consideration is based upon the time

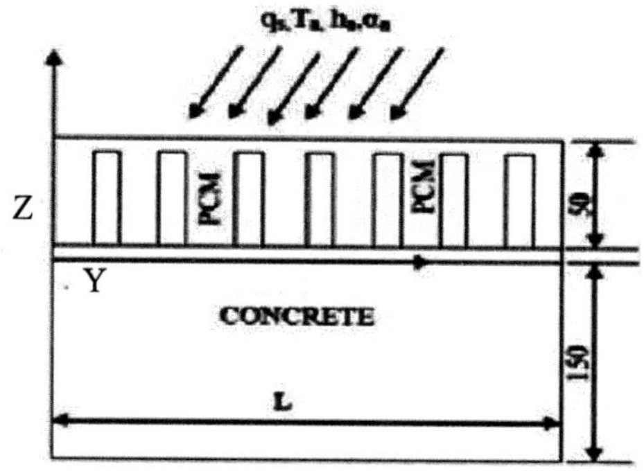

Fig. 2 A schematic of the computational domain

step of 0.1 s which represents a better convergence. Figure 2 represents the schematic diagram of computational domain.

5.1 Calculation of Average Room Temperature

Due to the total amount of sun rays falling on the system, it has found that the room temperature continuously changes from time to time. The room temperature depends on the heat transfer rate of both radiation and convection heat transfer to the room during the day. This is found as:

$$\frac{\mathrm{d}(\rho_a L C_p T_r)}{\mathrm{d}t} = q_r \tag{8a}$$

$$q_r = h_r(T_{rs} - T_r) \tag{8b}$$

6 Results and Discussion

The experimental setup accurately works on mainly three stages: melting, solidification and semisolid form.

a. Figure 3 represents the good agreement between the experimental and numerical data.
b. Figure 4 represents isotherm plot of the numerical [18] domain. Initially, the Glauber salt temperature is lower than the phase change temperature in the morning. As the temperature rises and meets the melting point temperature the PCM starts melting, the melt front proceeds with the melting of the PCM and proceeds further. The phase change temperature and the heat are stored as the

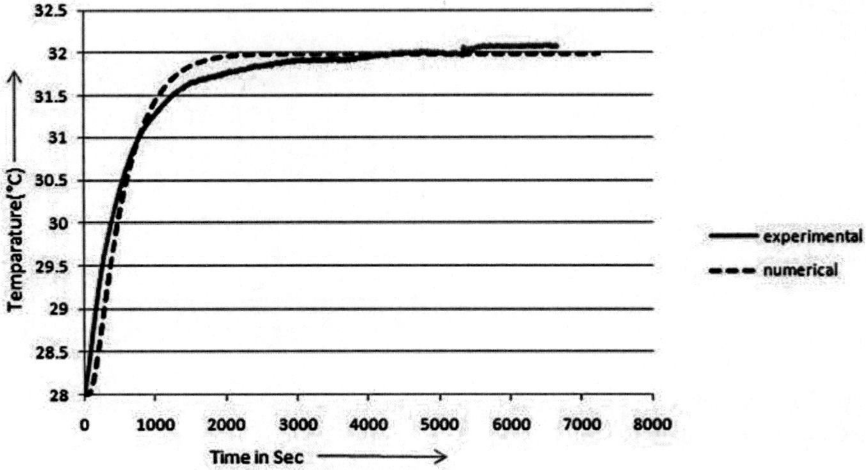

Fig. 3 Validation numerical data with experimental

Fig. 4 Isotherms plot of the domain

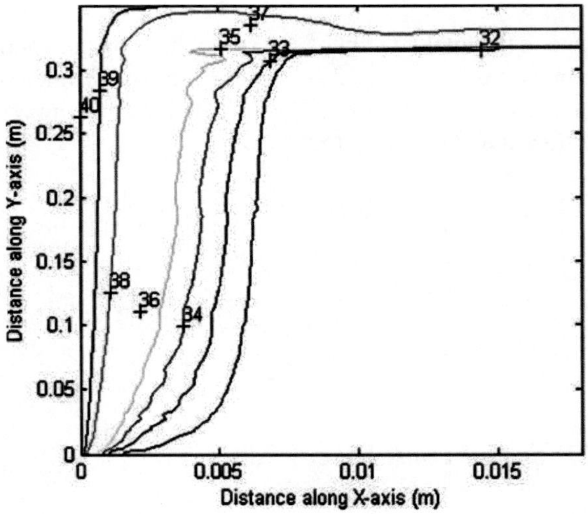

phase change latent heat. The heat generated passes through left the wall and progressed towards PCM zone.

c. Figure 5 shows the stream function plot in the domain. The stream function is visible in the liquid region, and the solid part of the PCM remains unchanged. As the melt front proceeds, the stream function changes with time. This stage will be ongoing until the PCM is melted, and the corresponding melting fraction increases from zero (0) to one (1).
d. Figure 6 shows the transient behaviour [8] of PCM where heat line, and heat flux shows the amount of heat flow in the region. Kimura and Bejan have explained

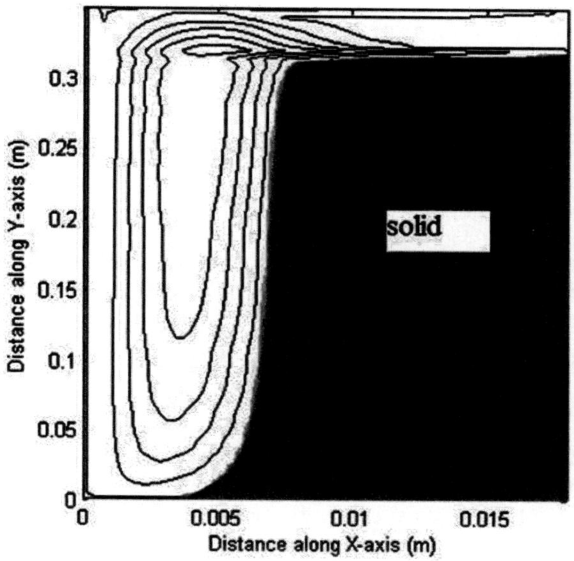

Fig. 5 Stream function generation the domain

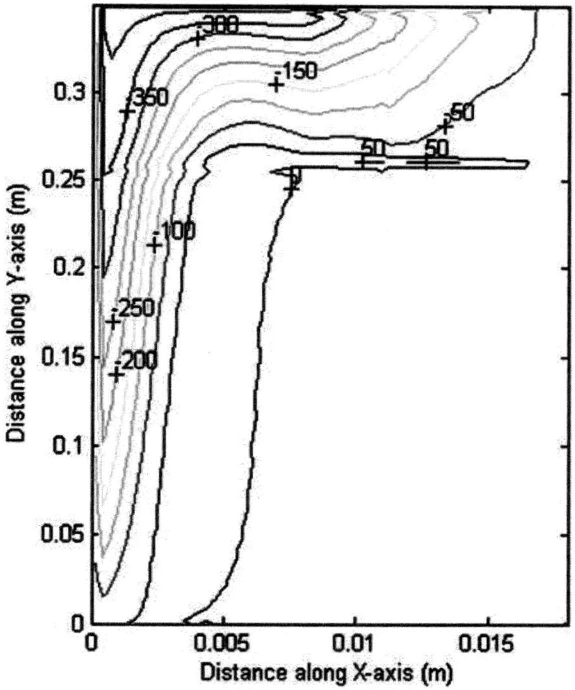

Fig. 6 Heat line and heat flux vector visualization

the border heat line and the heat flux in 2D region (x, y or r, z) where the heat line represents the constant heat flux line [19].

7 Conclusion

The numerical results are validated with the experimental data, and it shows a good agreement and correlation. The present works reflect only the melting phenomenon transiently. Temporal variation with time during melting at different locations of the PCM region has been studied. Finite Volume Method is incorporated to solve the problem by using the simplex solver and TDMA. The simulation predicts the melting and solidification behaviors of the phase change material with the temperature variation due to the available sunlight, and the radiation heat escapes in the form of convection. Correction factor is considered. Recycling of the PCM is possible over a period of time, but charging time is more compared to discharging time. Aluminum is used as thermal conductivity enhancer (TCE) for complete recycling of the PCM. The optimum as well as extra amount of the PCM is considered for more effective results. This research also represents the temporal variation, melt fraction and charging time. A threshold amount of heat and time is needed for the operation.

References

1. Peippo K, Kauranen P, Lund PD (1991) A multi component PCM wall optimized for passive solar heating. Energy Build 17:259–270
2. Fong KF, Lee CK, Lin Z (2019) Investigation on effect of indoor air distribution strategy on solar air-conditioning systems. Renew Energy 131:413–421, Elsevier, February 2019
3. Mukherjee S, Simlandi S, Barman N (2012) Studies on thermal behavior for cooling of a building using phase change material. Int J Earth Sci Eng ISSN 0974–5904: 695–701
4. Nayak KC, Saha SK, Srinivasan K, Dutta P (2006) A numerical model for heat sinks with phase change materials and thermal conductivity enhancers. IJHMT 49:1833–1844
5. Fan L, Khodadadi JM (2011) Thermal conductivity enhancement of phase change materials for thermal energy storage: a review. Renew Sustain Energy Rev 15:24–46
6. Velraj R, Seeniraj RV, Hafner B, Faber C, Schwarzer K (1999) Heat transfer enhancement in a latent heat storage system. Solar Energy 65:171–180
7. Agyenim F, Eames P, Smyth M (2009) A comparison of heat transfer enhancement in a medium temperature thermal energy storage heat exchanger using fins. Solar Energy 83(9):1509–1520
8. Kandasamy R, Wang X, Mujumdar AS (2008) Transient cooling of electronics using phase change material (PCM)-based heat sinks. Appl Therm Eng 28:1047–1057
9. Doan DT, Ghaffarianhoseini A, Naismith N, Ghaffarianhoseini A, Zhang T, Tookey J (2019) Examining Green Star certification uptake and its relationship with Building Information Modelling (BIM) adoption in New Zealand. J Environ Manage 250:109508
10. Fernández I, Renedo CJ, Pérez S, Carcedo J, Mañana M (2010) Advances in phase change materials for thermal solar power plants quality. International Conference on Renewable Energies and Power Quality (ICREPQ'11) Las Palmas de Gran Canaria (Spain), 13th to 15th April, 2010

11. Jian-you L (2008) Numerical and experimental investigation for heat transfer in triplex concentric tube with phase change material for thermal energy storage. Solar Energy 82(11):977–985
12. Ravikumar M, Srinivasan PSS (2008) Phase change material as a thermal energy storage material for cooling of building. J Theor Appl Inf Technol 503–511
13. Velraj R (1997) Heat transfer enhancement studies on a latent heat thermal storage systems for some solar applications. Ph.D. thesis, Anna University, Chennai, India
14. Zhang Y, Faghri A (1995) Analysis of thermal energy storage system with conjugate turbulence forced convection. J Thermophys Heat Transfer 9(4):722–726
15. Watanabe T, Kanzawa A (1995) Second law optimization of a latent heat storage system with PCMS having different melting points. Heat Recovery Syst CHP 15(7):641–653
16. Brent AD, Voller VR, Reid KJ (1988) Enthalpy-porosity technique for modeling convection-diffusion phase change: application to the melting of a pure metal. Numer Heat Transf 13:297–318
17. Patankar SV (1980) Numerical Heat Transfer and fluid flow. Hemisphere Publications, New York
18. Alawadhi EM, Amon CH (2002) Thermal analysis of a PCM thermal control Unit for portable electronics devices: experimental and numerical studies. Eight Intersociety conference on Thermal and Thermo-mechanical Phenomena in Electronic System, San Diego, California, pp 466–475
19. Kimura S, Bejan A (1983) The "Heatline" visualization of convective heat transfer. J Heat Transf 105(4):916–919. https://doi.org/10.1115/1.3245684

A Review on Energy-Efficient Building

Hiranmoy Samanta, Rohit Maity, Saheli Laha, and Pradip Kumar Talapatra

Abstract The change in climate and the depletion of the conventional energy storages pushes the thoughts of energy consumption as well as the energy efficient building and equipments in a large scale. The present work focuses on building energy assessment, optimization, recent advancement in energy-efficient building, etc. The different parameters affecting the building energy consumptions and how to minimize the use of energy by different methods are discussed. The case studies are needed for convincing both government agencies (responsible for ECBC [1] implementation) and builders and building design teams about the advantages of energy-efficient buildings. Monitoring of energy performance of buildings is a challenge due to non-installation or non-functioning energy information system (EIS) in majority of the buildings. The wide application of PCM-based material of different forms like encapsulation [2], slab and other forms like mixing with mortar and the wall and roof design applications are considered here.

Keywords Phase change material (PCM) · Energy consumption · ECBC

1 Introduction

The total energy produced throughout the world is mostly used for the HVAC [1] due the energy consumption by different types of building as per the application of the

H. Samanta (✉) · R. Maity · S. Laha · P. K. Talapatra
Department of Mechanical Engineering, Gargi Memorial Institute of Technology, Balarampur, Baruipur, Kolkata 700144, West Bengal, India
e-mail: hiru.samanta@gmail.com

R. Maity
e-mail: rohit.maity264@gmail.com

S. Laha
e-mail: Sahelilaha3@gmail.com

P. K. Talapatra
e-mail: pktpatra@gmail.com

M. Ramgopal et al. (eds.), *Advances in Air Conditioning and Refrigeration*, Lecture Notes in Mechanical Engineering, https://doi.org/10.1007/978-981-15-6360-7_23

building. The different inside building criteria keep on changing based on commercial, residential, hospital building and many other applications. The different government agencies, ECBC calculation and others are employed for energy consumption and minimization of the energy. The building location, climate condition, orientation shape, the area exposed to direct sunlight and the other factors are being considered for an energy-efficient building. Monitoring of energy performance of buildings is a challenge due to non-installation or non- functioning energy information system (EIS) in majority of the buildings. The use of PCM for cooling and heating the building is an optimized option with the aid of thermal energy storage in terms of encapsulation. The microcapsules [3] were characterized by the geometrical shapes, phase transition temperatures, mean particle sizes, chemical stability and their heating cooling cycling. The use of proper PCM, PCM mixture thermal conductivity enhancer put an remarkable effect of energy saving of a building.

2 Literature Review

Kheiri [4] reviewed on optimization methods applied in energy-efficient building geometry and envelope design. With geometric configurations and the building envelope, parameters can considerably influence the building energy performance. Madad et al. [5] reviewed Phase Change Materials for Building Applications and focused on New Perspectives. A detailed study of different PCM-, PCM-based LTEs and the computational methods are focused in this study. Gurav et al. [6] surveyed the design of a energy-efficient building. Bland et al. [7] focused on several disadvantages of PCM use in the building cooling application were identified and discussed on different parameters like super cooling, low thermal conductivity, phase segregation, fire safety and cost. Different issues caused by super cooling and phase segregation lead to thermal cycling degradation, limiting the useful lifecycle of the material, safety and cost.

Bayata and Temiz [8] developed a model software for energy efficiency optimization in the building design in turkey. Cui et al. [9] have Reviewed Phase Change Materials integrated in Building Walls for Energy Saving. Phase change material-based thermal energy storage system for building wall roof has been discussed in this work. Yang et al. [10] reviewed thermal comfort and building energy consumption implications. Kuznik et al. [11] investigated and reviewed on phase change materials integrated in building walls. Kosny et al. [12] studied the cost analysis of simple phase change material-enhanced building envelopes in southern US climates focusing on different parameters affecting the energy consumptions of building and different phase change material costs and the effect and energy savings cost. Geetha and Velraj [13] reviewed the passive cooling methods for energy-efficient buildings with and without thermal energy storage. Pacheco et al. [14] reviewed different aspects of energy efficient design of building. Aksoy and Inalli [15] investigated the impacts of some building passive design parameters like building shape and orientation on heating demand for a cold region. Mingfang [16] studied the influence of building length, depth and width parameters on the solar radiation received by a parallelepiped-shaped building. Florides et al. [17] quantified the effect of the shape factor on energy requirements for the heating and cooling of a building.

3 Building Energy Assessment

In a major impetus to institutionalize energy efficiency in the world, the building energy assessment considered different types of standards, bureau, codes and methods like Bureau of Energy Efficiency (BEE), Ministry of Power (MoP), energy efficiency measures (EEM), energy auditors or energy service companies (ESCO)

The methods are as follows:

a. **Asset Rating (AR)**: Assessment rating is mainly based on annual standard energy performance to assess the annual energy use of a building under standardized predefined conditions.
b. **Operational Rating (OR)**: Operational rating represents actual energy performance without any predefined values but actual assessment by the weighted sum of all energy used by the building, as measured by metres or other means. The both methods are based on the calculation, and the measurement-based method can be applied to existing buildings. Nowadays, to improve the building energy efficiency, many countries have recently developed AR methods for assessing the energy performance of existing buildings.

3.1 Factors Affecting the Energy Use of a Building

There are different parameters which influence the energy use of a building mainly considered as:

a. **External factors** (Geometrical location, building orientation, sunrays radiation, etc.)
b. **Internal factors** (Type of building, shape, type of building application, machineries, people)

3.2 Influence of Building Shape and Geometrical Location on the Energy Optimization of Buildings

Different geometrical locations and the building shapes influence the total solar energy (conductive, convective and irradiative) the building receives as well as its total energy consumption. The thermal radiation consumes by a building can increase energy requirements for cooling to up to 25%. During the design of the building, it is very important to consider the minimum value possible for the ratio between its outer surface and the total constructed volume attending towards the ideal case of a hemispherical shape. The different shapes and construction parameters and issues of the building have been researched and researchers have begun to study the performance of parallelepiped-shaped buildings and to vary the shape factor in order to find the best model. The researchers presented that in most of the cases the hexagonal or octagonal foundation plan, a curved or oval foundation plan or building without any specific geometric shape for obtaining the optimal dimensions design considered with for the specific geometric context.

The different parameters influence the regulation of heating and cooling requirements are the following:

(i) Compactness index
(ii) Geometrical shape factor
(iii) Weather or climate of a specific zones
(iv) Ascendancy of building shape on the life cycle of the building.

3.2.1 Compactness Index

The *compactness index* can be represented as the ratio of the volume to the outer surface of the building facade. High volumetric/surface ratio remarks a very compact building by condition that surface exposed to possible heat losses or gains is as small as possible. The relative compactness ratio presented for comparison between buildings (Figs. 1 and 2) [14].

$$R_c = \frac{\left(\frac{V}{A_{ext}}\right) \text{building}}{\left(\frac{V}{A_{ext}}\right) \text{reference}} \tag{1}$$

Fig. 1 Buildings with different compactness indices: (i) Building with a compactness index of 3.45 (Left); (ii) Building with a compactness index of 5 (Right)

Fig. 2 Shape factor of different shaped building

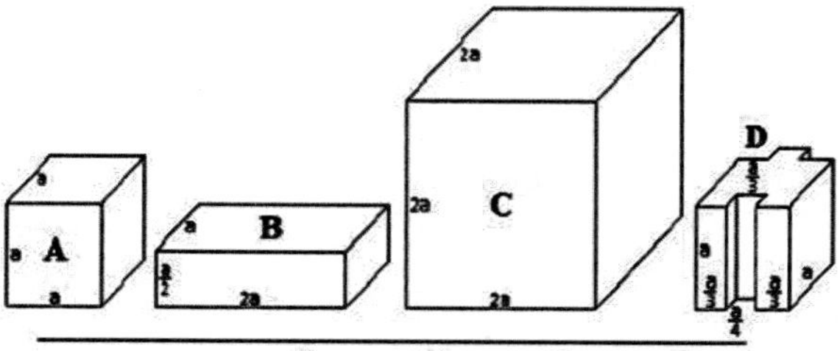

	A	B	C	D
Volume	a^3	a^3	$8a^3$	a^3
Thermal envelpe	$6a^2$	$7a^2$	$24a^2$	$7a^2$
Shape factor	6/a	7/a	3/a	7/a

3.2.2 Shape Factor

As per the civil engineering concept, there is always a proper relation between the length and the other parameters of the building constructed. Ratio of building length to building depth is represented as shape factor. Not only the building orientation and geometrical position, shape factor presents the percentage of the facade exposed at each cardinal point. The combined optimization of shape and orientation benefit the design; hence, the possible benefit lead to heat energy savings around 36%. It has been found that longest wall of the building facade facing the southern direction have the most appropriate position for a normal rectangular house [18]. A building model with a 1/2 shape factor (comparatively less outer wall surface exposed to sun with a southern orientation) requires almost 8.2% more energy for heating as well as cooling the building. This percentage of energy consumption increases considerably by (26.7%) in case of more roof insulation as the heat gained from solar radiation is prevented by the roof cladding. Mingfang studied and presented the different parameters which affect the collection of solar radiation based on building length, depth and width on a parallelepiped-shaped building. The volume was kept constant where the other parameters change, and Eq. (2) was applied.

$$\frac{Q}{Q_0} = \frac{(q_s + q_N)^3\sqrt{\lambda\beta} + (q_E + q_W) \times \sqrt[3]{\lambda^{-2}\beta + q_H} \times q_H \times \sqrt[3]{\lambda\beta^{-2}}}{q_S + q_N + q_E + q_W + q_H} \quad (2)$$

Q/Q_0 is the relative solar radiation received by the external surface of the buildings: q_S, q_N, q_Z, q_W, q_H are the solar radiation on the south-wall, north-wall, east-wall, west-wall and roof per unit area in daytime; λ is the length/depth ratio of the building; and β is the height/depth ratio of the building.

3.2.3 Climate and Shape Optimization

In very cold weather regions, as the building area increases, the more heat escapes through the building envelope, so in the cold climates zones more solar radiation in the form of heating collected and accumulated by increasing the building surface area that receiving the active as well as passive solar radiation. By increasing the shape factor, i.e. more external building surface for the same volume, lower compactness index directly proportional to the increase in the energy required for heating. In hot climate condition, fixed type of building performance cannot be determined as the relation of climate and the shape is not direct correction factors are present to determine the correlation.

3.2.4 Life-Cycle Cost Dependency on the Shape of the Building

Life span of a building counts from architectural conception to model design followed by construction time maintenance and till the date of demolition. The cycle includes the preliminary stage of design, construction period over months or years, occupation period, use, and maintenance, and dismantling. This process measures and evaluates the different tangible and intangible flow of manpower, material and energy in the system with the time flow towards the process. The optimal shape and design of building shape is performed and achieved through several years of a building's construction and service life, which is very useful data for calculations. The different optimal shapes like curve-shaped building, polygonal-shaped building are the fruit of the several year research. Wang et al. [19] investigated the effect of building shape on energy uses and the energy demand of the building. They calculated the life-cycle cost (LCC) and life-cycle environmental impact (LCEI) with Eqs. (3) and (4) (Fig. 3).

$$\mathrm{LCC} = \mathrm{IC}(X) + \mathrm{OC}(X) \tag{3}$$

$$\mathrm{LCEI}(X) = \mathrm{EE}(X) + \mathrm{OE}(X) \tag{4}$$

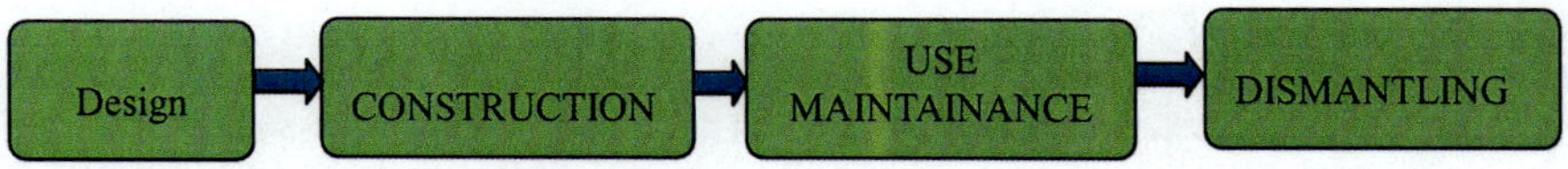

Fig. 3 Building life cycle

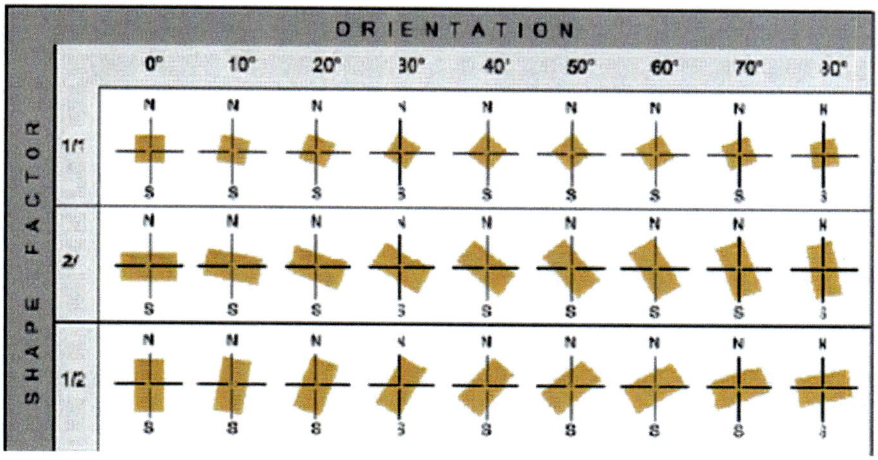

Fig. 4 Heating energy saving, depending on shape factor and orientation of the building. *Source* Aksoy and Inalli [15]

3.2.5 Building Orientation [14]

As far we have come across different parameters mainly based on the compactness factor, shape factor where active and passive solar heating affect the energy consumption of building. In the building design passive heating other than direct solar heating, the building orientation is the most important. Many researchers put their focus on passive heating and cooling of building. The quantity of direct solar radiation received on the building facade depends on the azimuth in the wall which is very important based on orientation angle of the building.

The optimal building orientation has the positive followings:

- It is a low-cost measure at initial stage
- Provide a decrease in energy demand
- More sophisticated passive systems are less used
- Increases the performance of other complex passive techniques enhanced
- The more quantity of daylight reduces the energy demand for artificial light; hence, internal heating load of the building decreases (Fig. 4).

3.2.6 Building Envelope and the Energy Demand

The external and internal heat loading of the building mostly depends on the building envelope. The building envelope (foundation, roof, walls, doors and windows) and the operation period of the heating system are the factors that have the greatest impact

Table 1 Energy consumption of an office unit at three different orientations

Energy consumption of an office unit at three different orientations						
Energy consumption at three orientations (kWh/Year)						
	South	%	East	%	West	%
Heating	186	0	231	24	219	18
Cooling	281	0	286	2	369	31
Total	467	0	517	11	588	26
T_{max} (°C)	26.4	266			27	

on the total energy consumption of the building. The envelope determines interior climate conditions based on the outer climate condition, hence the additional energy demand for heating and cooling. Actions on the elements and the proper selection of the envelope can have a positive impact on certain energy requirements and have a negative effect on others. Consequently, it is necessary to evaluate the performance of the building as a whole. The energy consumption of an office unit at three different orientations data is shown in Table 1.

The equation formed through the data collected for maximum heat loss in a reference building Eq. (6).

$$Q = U_0(t_i - t_{eo})(1 - X) + U_g\left(t_i - t_{eg}\right)X \tag{6}$$

3.2.7 Shading on Buildings

Shading is mainly used to minimize the direct solar radiation as well as the passive radiation entering to a building. Shading is used to not only control the maximum quantity of solar radiation in the hot climates but to increase the surface area. This kind of arrangements of shading provides most effective results when provided at building facade cavities since these are the elements that transmit the highest level of radiation to the inside of the building.

Shading Coefficient

$$\mathrm{SC} = \frac{\text{Solar Heat gain Factor of fenestration}}{\text{Solar heat gain factor of reference glass}} \tag{7}$$

$$\mathrm{SEC} = \frac{Q_\mathrm{S} \times (1 - \mathrm{SC} + \mathrm{QW} \times \mathrm{SC})}{Q_\mathrm{S} + Q_\mathrm{W}} \tag{8}$$

4 Passive Systems

Passive systems are nothing but the other methods to cool the room by means of circulation and heat removal from the room. In ancient times, the methods are employed in to remove the heat by moisture as well as removing the moisture from the room. The systems are low in cost as well as the efficiency is less. In the developing countries, the wide availability of electric power removes the obstacles of heating and cooling. Passive cooling system is eco-friendly and uses less amount of electricity compared to modern AC and other units. Nowadays, passive cooling is again in research focus with more technical developments.

4.1 Passive Cooling

Natural processes like evaporation, free convection, forced convection, radiation and circulation of air are incurred for passive.In this process, heat is directly rejected to the atmosphere or into the ground beneath buildings (i.e., conduction and convection) [20]. The classification of passive cooling systems was studied by Givoni [21], and different systems were also summarized (Fig. 5).

5 Glazing

Glazing is the process to ensure the amount of energy entering to the building as well as the amount of energy leaving the building. In hot climate, the glazing is done to enter less amount of heat to the building as well as the lighting up the area by natural means which is an effective way to save energy. Whereas in the cold regions, glazing is used to accumulate the heat to the building to heat up the building naturally, and the energy savings has been done through this process.

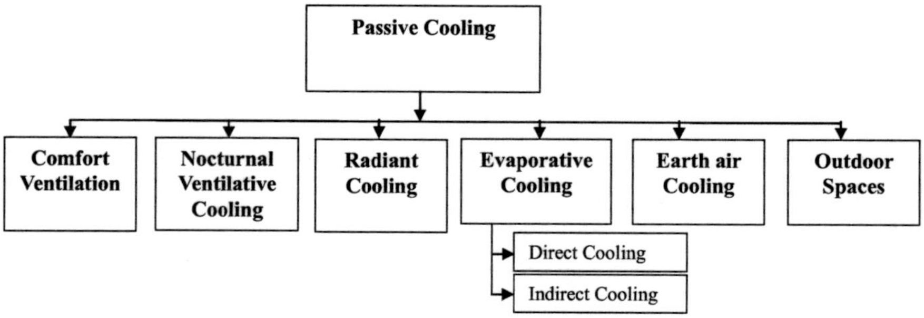

Fig. 5 Classification of passive cooling systems. *Source* Givoni [21]

5.1 Glazing Types [22]

Glazing that provides energy savings in terms of eat energy and the light can be classified in the following types

- **Heat-Absorbent Glass**: this types of glass or glass-like material allow solar radiation in heat energy to enter in the building (i.e. increasing its temperature) and circulate the heat throughout the room by convection and radiation to reduce the direct radiation through the glass. This is mainly used in cold climatic situation
- **Heat-Reflecting Glass**: this type of material uses layers or the reflecting coating or films, and sometimes, more than one layer has been used that blocks the entry of solar radiation into the building, effective for hot climate zone.
- **Low Radiation Glass**: this type of material coating or film reduces the heat transfer coefficient. It can also facilitate energy saving in winter.

6 Optimization

There are different methods to find the energy used by the domestic, official and commercial buildings. But a generalized optimization is needed to predict the energy consumptions for the building from the basic design assumptions lead to assumptions the post-assumptions to minimize the energy consumption. Where the old buildings are mainly based on architectural point of view and region and positioning constrained, new buildings are more optimized with the design and other considerations with the energy saving and efficient. Nowadays, the different designs and heat load and ECBC software are available to optimize the energy consumption and the energy uses reduction.

7 Recent Advancement in Energy Efficient Building

Recent advances in equipment design have yielded remarkable efficiency improvements, and there is considerable potential for further gains. The latest design of the double glass windows, PCM boards, roof insulation, rooftop gardening, etc., improved a lot in the field and contributed a huge amount of energy gain.

- Cool roofs [23] coated with materials containing specialized pigments reflect sunlight and absorb less heat than standard roofs. Expect these types of roof systems to get even "cooler" due to new fluorescent pigments developed by Lawrence Berkeley National Laboratory and PPG Industries that can reflect nearly four times the amount of sunlight of standard pigments.
- PCM mixed with building material and mortar.
- Low energy consumption electronic and electrical equipments.
- Low energy consumption lighting and ceiling end lighting.

8 Conclusion

The review on energy-efficient building considers the different criteria considering both internal and external factors and internal factors. The different bureau and the agencies are involved in energy assumption and building design and implementation as well as the demolition also. The energy calculation of the different types of building not only based on thermodynamic operation but the various other designed aspects to save energy.

- The sustainable design applications of buildings reduces the energy consumption and the energy demand for heating and cooling load as well as the total cost of the building [24].
- The data collected through the several years based on design, shape, orientation of the building are used for formulation.
- Different factors affect the total energy consumption like building orientation, shape, compactness.
- The radiation heat transfer of sun can be minimized by insulation, glazing, protective layer, etc.
- The economical and environmental conditions are not to meet the most for the energy consumption by a building [24].
- Conventional methods are not valid for the urban areas where the other heating factors are affecting the total outside heat load, and hence, the inside temperature conditions also change.

References

1. ASHRAE Guideline 14 (2002) Measurement of energy and demand savings. American Society of Heating, Refrigeration, and Air Conditioning Engineers Inc
2. Alkan C, Sarı A, Karaipekli A, Uzun O (2009) Preparation, characterization, and thermal properties of microencapsulated phase change material for thermal energy storage. Sol Energy Mater Sol Cells 93:143–147
3. Ozonur Y, Mazman M, Paksoy H, Evliya H (2006) Microencapsulation of coco fatty acid mixture for thermal energy storage with phase change material. Int J Energy Res 30:741–749
4. Kheiri F (2018) A review on optimization methods applied in energy-efficient building geometry and envelope design. Renew Sustain Energy Resour 92
5. Madad A, Mouhib T, Mouhsen A (2018) Phase change materials for building applications: a thorough review and new perspectives
6. Gurav P, Mayekar N, Kamble M, Pinage S, Bhosale AD (2018) Design of energy efficient building-survey. Int Res J Eng Technol (IRJET) 05(04). e-ISSN: 2395–0056
7. Bland A, Khzouz M, Statheros T, Gkanas EI (2017) PCMs for residential building applications: a short review focused on disadvantages and proposals for future development
8. Bayata O, Temiz I (2017) Developing a model and software for energy efficiency optimization in the building design process: a case study in Turkey. Turkish J Electr Eng Comput Sci
9. Cui Y, Xie J, Liu J, Pan S(2015) Review of phase change materials integrated in building walls for energy saving. In: 9th international symposium on heating, ventilation and air conditioning

(ISHVAC) and the 3rd international conference on building energy and environment (COBEE). Elsevier
10. Yang L, Yan H, Lam J (2014) Thermal comfort and building energy consumption implications—a review. Appl Energy 115:164–173
11. Kuznik F, David D, Johannes K, Roux J-J (2011) A review on phase change materials integrated in building walls. Renew Sustain Energy Rev, Elsevier 15(1):379–391
12. Kosny J, Shukla N, Fallahi A (2013) Cost analysis of simple phase change material-enhanced building envelopes in Southern U.S. Climates. Energy Eff Renew Energy
13. Geetha NB, Velraj R (2012) Passive cooling methods for energy efficient buildings with and without thermal energy storage—a review. Energy Sci Res 29(2):913–994
14. Pacheco R, Ordonez J, Martinez J (2012) Nano and biotech based materials for energy building efficiency. Renew Sustain Energy Rev 16:3559–3573
15. Aksoy UT, Inalli M (2006) Impacts of some building passive design parameters on heating demand for a cold region. Build Environ 41:1742–1754
16. Mingfang T (2002) Solar control for buildings. Build Environ 37:659–664
17. Florides GA, Tassou SA, Kalogirou SA, Wrobel LC (2002) Measures used to lower building energy consumption and their cost effectiveness. Appl Energy 73:299–328
18. Shi X, Memon S, Tang W, Cui H, Xing F (2014) Experimental assessment of position of macro encapsulated phase change material in concrete walls on indoor temperatures and humidity levels. Energy Build 71:80–87
19. Wang W, Rivard H, Zmeureanu R (2006) Floor shape optimization for green building design. Adv Eng Inf 20:363–378
20. Sanjay, Chand P (2004) Passive cooling techniques of buildings: past and present—a review. ARISER 4:37–46
21. Givoni B. (1994) Passive and low energy cooling of buildings. Wiley, Hoboken
22. Shaviv E (1981) The influence of the orientation of the main solar glazing on the total energy consumption of a building. Sol Energy 26:453–454
23. Saman W, Bruno F, Halawa E (2005) Thermal performance of PCM thermal storage unit for a roof integrated solar heating system. Solar Energy 78:341–349
24. Konuklu Y, Ostry M, Paksoy HO, Charvat P (2015) Review on using microencapsulated phase change materials (PCM) in building applications. Energy Build 106:134–155

Energy and Exergy Analysis of Vapour Absorption Cooling System Driven by Exhaust Heat of IC Engine

S. S. Bhatti, S. K. Tyagi, and Abhishek Verma

Abstract Traditionally, vapour compression system used in vehicle air conditioning causes reduction in engine power and increases fuel consumption 15–20%. In internal combustion engine, only 30–35% of fuel energy is converted into useful work and about 50% energy of fuel is released as heat to the atmosphere. The exhaust from IC engine has waste heat in the temperature range of 150–400 °C. This exhaust heat of IC engine can be employed for the vapour absorption system which requires low-grade heat source. In this article, the energy and exergy analysis of vapour absorption cooling system powered by waste heat of IC engine exhaust is presented. The effect of absorber and generator temperatures on the COP and exergetic efficiency of vapour absorption cooling system is determined. The exergy destruction in components and performance parameters like COP, exergy efficiency are also computed. The aim of present article is thermodynamic evaluation of the utilization of exhaust heat of IC engine for vapour absorption cooling system heat by both first law and second law approach.

Keywords Exergy · Internal combustion engines · Vapour absorption cooling system

1 Introduction

Emissions from internal combustion engines are one of the major sources of environmental degradation by global warming specially CO_2 emissions. Waste heat recovery from thermal energy systems is prospective technique to reduce CO_2 emissions and energy conservation. In IC engines, nearly one-third part of fuels energy is wasted through exhaust gases, which can be utilized for absorption refrigeration system.

S. S. Bhatti (✉) · S. K. Tyagi · A. Verma
Centre for Energy Studies, Indian Institute of Technology Delhi, Hauz Khas, New Delhi 110016, India
e-mail: s.bhatti21877@gmail.com

S. S. Bhatti
Mechanical Engg. Dept., Pusa Institute of Technology, DTTE Govt. of NCT Delhi, New Delhi, India

M. Ramgopal et al. (eds.), *Advances in Air Conditioning and Refrigeration*, Lecture Notes in Mechanical Engineering, https://doi.org/10.1007/978-981-15-6360-7_24

Thus, utilization of exhaust heat for absorption cooling not only increases the overall system efficiency, but also minimizes CO_2 emissions. The exhaust gases from internal combustion engine available at temperature range of 150–400 °C depend upon the speed of engine, and this heat which is low grade can be employed as a heating source for generator of vapour absorption system to heat lithium bromide solution. Manzela et al. 2010 [1] performed experiments on utilization of internal combustion engine exhaust gas for vapour absorption cooling systems and present the effects of absorption cooling system on performance parameters of engine. Izquierdo et al. [2] also applied this waste heat recovery technique; there is reduction in CO_2 generation, and also, there is economical savings by saving fuel, which otherwise used for driving conventional air conditioning mechanical compressor. Mostafavi et al. [3] carried out thermodynamic analysis of combined diesel engine and absorption refrigeration unit-supercharged engine. Mostafavi et al. [4] also studied combined diesel engine and absorption refrigeration system run by exhaust gas and computed cooling capacities. Agnew et al. [5] performed analysis of combined diesel absorption cycle as a combined power and cooling cycle for four different configurations in high ambient temperature conditions. Talbi et al. [6] examine the performance improvement by theoretical simulation of combined diesel engine and absorption refrigeration system. Anand et al. [7] used biogas to drive vapour absorption system to reduce greenhouse gases to save climate.

In the previous investigations, on the utilization of exhaust heat for the vapour absorption cooling systems are performed by energy analysis which is based on the first law of thermodynamics. The first law approach is not succeeded to find out actual energetic degradation or losses in thermal energy systems; on the other side, the second law of thermodynamic-based exergy analysis provides in-depth evaluation of processes or operations by pinpointing and quantifying the influence of irreversibilities in thermal energy system. In this study, vapour absorption system driven by exhaust heat of IC engine is analysed on energy and exergy-based analysis. In previous studies, the researchers Kaushik et al. [8]; Gomri [9]; Kaynakli and Kilic [10]; Grossman and Zaltash [11] have applied environmental temperature concept for evaluating irreversiblities in vapour absorption systems.

2 Description of Vapour Absorption Cooling System Driven by Exhaust Heat of IC Engine

Figure 1 depicts the schematic of vapour absorption cooling system driven by exhaust heat of IC engine. The exhaust heat coming out from IC engine is utilized to heat the lithium bromide solution in generator of single-effect vapour absorption system. Vapour absorption cooling system consists of different parts or components as shown in Fig. 1. The LiBr–water solution in the absorber having low LiBr concentration is pumped to the high-pressure generator. The exhaust heat from IC engine is supplied to generator for refrigerant (water) vapour generation, while the high concentration

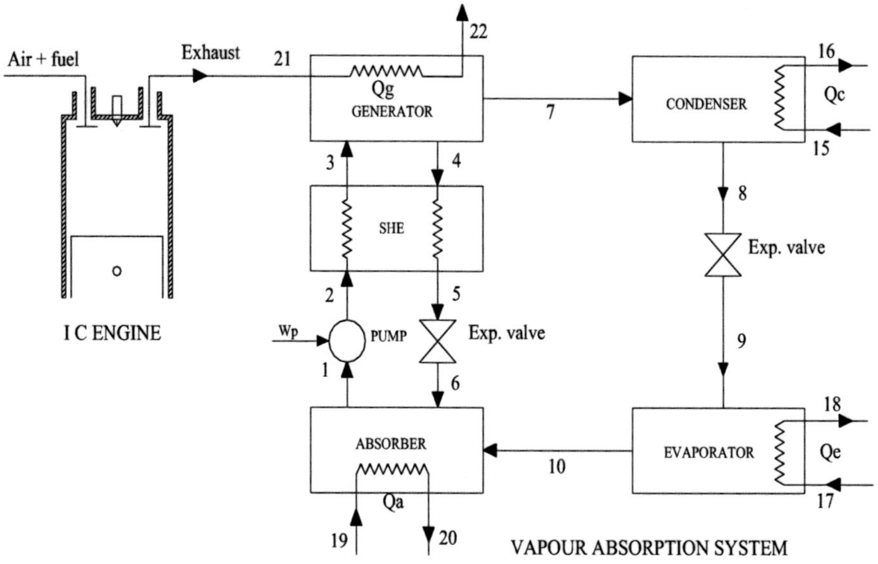

Fig. 1 Schematic of vapour absorption cooling system powered by exhaust heat of IC engine

LiBr solution is returned to the absorber through the solution heat exchanger. The solution heat exchanger improves the COP of system by input heat reduction. The refrigerant vapour from generator is condensed in condenser at constant pressure. The condensed refrigerant is expanded through the expansion valve to the evaporator pressure and temperature. The refrigerant absorbs the heat from the cooling space in the evaporator at constant pressure and becomes saturated. The saturated vapour from evaporator is further absorbed by LiBr–water solution in absorber.

Assumptions in the modelling of vapour absorption cooling system run with the aid of exhaust heat of IC engine:

1. The system is in steady-state condition.
2. Isenthalpic process is considered in expansion valves.
3. Pressure and heat losses in different parts are neglected.
4. Constant effectiveness of heat exchanger is considered.
5. For calculations, the values for reference temperature and pressure of 25 °C and 1 bar, respectively, are considered.

3 Thermodynamic Modelling of Vapour Absorption Cooling System Powered by Exhaust Heat of IC Engine

Thermodynamic modelling consists of the following equations for respective components of system:

$$\text{Mass balance} \sum \dot{m} = 0; \sum x\dot{m} = 0 \tag{1}$$

$$\text{Energy balance} \sum \dot{Q} + \sum \dot{W} + \sum \dot{m}h = 0 \tag{2}$$

$$\text{Entropy generation } \dot{S}_{\text{gen}} = \sum \dot{m}s_{\text{out}} - \sum \dot{m}s_{\text{in}} - \sum\left(\frac{\dot{Q}}{T}\right) \geq 0 \tag{3}$$

By using the above equations and concepts of the second law-based exergy analysis for all components, the governing equations for different components of vapour absorption cycle driven by exhaust heat of IC engine are shown in the Table 1. The state points are according to Fig. 1.

The total irreversible loss, by Gouy–Stodola equation [12], is

$$E\text{ loss} = T_o\dot{S}_{\text{gen}} \tag{4}$$

The COP of vapour absorption cooling system run by exhaust heat is evaluated by the following relation

$$\text{COP} = \frac{\dot{Q}_e}{\dot{Q}_g + \dot{W}_P} \tag{5}$$

Table 1 Equations used for IC engine exhaust-driven vapour absorption cooling system analysis

S. No.	Component	Energy analysis equations	Exergy analysis equations
1	Absorber	$\dot{m}_{10} + \dot{m}_{6.} = \dot{m}_1$ $\dot{m}_{10}h_{10} + \dot{m}_{6.}h_6 = \dot{m}_1h_1 + Q_a$	$\dot{S}_{\text{gen,a}} = \dot{m}_1s_1 - \dot{m}_{\text{ref.}}s_{10} - \dot{m}_6s_6 + (\dot{m}_{\text{ef}}c_{\text{p}})_{\text{a}} \ln \frac{T_{20}}{T_{19}}$
2	Pump	$\dot{W}_{\text{pump}} = \dot{m}_1(h_2 - h_1)$	$\dot{S}_{\text{gen,p}} = \dot{m}_1(s_2 - s_1)$
3	Solution heat exchanger	$\dot{Q}_{\text{HX}} = \dot{m}_1(h_3 - h_2) = \dot{m}_6(h_4 - h_5)$	$\dot{S}_{\text{gen,a}} = (\dot{m}_1c_{\text{p}})_1 \ln \frac{T_3}{T_2} + (\dot{m}_4c_p)_4 \ln \frac{T_5}{T_4}$
4	Generator	$\dot{Q}_{\text{g}} = \dot{m}_7h_7 + \dot{m}_4h_4 - \dot{m}_3h_3$ $= \dot{m}_{\text{exh,g}}C_{\text{p,exh}}(T_{21} - T_{22})$ $= \dot{m}_{\text{exh,g}}(h_{21} - h_{22})$	$\dot{S}_{\text{gen,gen}} = \dot{m}_3s_3 - \dot{m}_4s_4 - \dot{m}_{\text{ref.}}s_7 + (\dot{m}_{\text{exh g}}c_{\text{p exh}})_{\text{g}} \ln \frac{T_{22}}{T_{21}}$
5	Pressure reducing valve	$h_5 = h_6$	$\dot{S}_{\text{gen,pr v}} = \dot{m}_5(s_6 - s_5)$ $T_{6\text{r}} = T_6$
6	Condenser	$\dot{Q}_{\text{cond}} = \dot{m}_8(h_8 - h_7) = \dot{m}_{\text{ef,c}}(h_{16} - h_{15})$	$\dot{S}_{\text{gen,cond}} = \dot{m}_{\text{ref.}}(s_8 - s_7) + (\dot{m}_{\text{ef,c}}c_{\text{p,ef,c}})_{\text{cond}} \ln \frac{T_{16}}{T_{15}}$
7	Exp. valve	$h_8 = h_9$	$\dot{S}_{\text{gen,ev2}} = \dot{m}_{\text{ref.}}(s_9 - s_8)$
8	Evaporator	$\dot{Q}_{\text{e}} = \dot{m}_{10}(h_{10} - h_9)$ $= (\dot{m}_{\text{ef}}c_{\text{p}})_{\text{e}}(h_{\text{out,e}} - h_{\text{in,e}})$	$\dot{S}_{\text{gen,e}} = \dot{m}_{\text{ref.}}(s_{10} - s_9) + (\dot{m}_{\text{ef}}c_{\text{p}})_{\text{e}} \ln \frac{T_{18}}{T_{17}}$

The second law-based exergetic efficiency of vapour absorption cooling system run by exhaust heat is given by

$$\eta_{ex}=1-\frac{E_{LOSS}}{\dot{Q}_g\left(1-\frac{T_0}{T_g}\right)+\dot{W}_P} \tag{6}$$

4 Results

The following Table 2 shows the state points for the vapour absorption cooling system powered by exhaust waste heat of IC engine.

4.1 Energy and Exergy Analysis Results

Table 3 shows energy analysis and destruction of exergy results in different components for the vapour absorption cooling system powered by exhaust heat of the IC engine. The exhaust heat of exhaust gases of SI engine supplied to generator of an exhaust gas temperature varies between 100 and 300 °C which is used in this analysis [13]. At high rpm or high engine speeds, the temperature of exhaust gas increases which gives more cooling capacity.

It is observed that COP of the vapour absorption system driven by exhaust gases is 0.7615 when exhaust heat input to the generator is 3093 kW and cooling effect produced 2355 kW. The exergy degradation or destruction in system different components is shown in Table 3 and depicted in Fig. 2.

Table 2 State points for the vapour absorption cooling system driven by exhaust heat of IC engine

State points	*T* (°C)	*P* (kPa)	*h* (kJ/kg)	*s* (kJ/kg K)
1	37.79	1.016	91.46	0.2287
2	37.79	6.558	91.47	0.2287
3	66.15	6.558	149.3	0.4068
4	87.8	6.558	221.2	0.4791
5	52.8	6.558	156.2	0.2895
6	52.8	1.016	156.2	0.2896
7	87.8	6.558	2664	8.58
8	37.8	6.558	158.3	0.5428
9	7.2	1.016	158.3	0.5661
10	7.2	1.016	2514	8.968

Table 3 Energy analysis and exergy destruction results for the IC engine exhaust heat-driven vapour absorption cooling system

Component	Model validation [8]	Energy analysis (present work)	Exergy destruction (kW)
Absorber	$Q_a = 2945$ kW	$Q_a = 2943$ kW	130.5
Condenser	$Q_c = 2505$ kW	$Q_c = 2506$ kW	52.95
Evaporator	$Q_e = 2355$ kW	$Q_e = 2355$ kW	48.20
Generator	$Q_g = 3095$ kW	$Q_g = 3093$ kW	225.7
Ref. Exp. Valve	–	–	7.033
Sol. Exp. Valve	–	–	0.2171
Sol. heat exchanger	$Q_{shx} =$ 518 kW	$Q_{shx} =$ 522.5 kW	21.37
Pump	$W_p =$ 0.0314 kW	$W_p =$ 0.03256 kW	0.02831

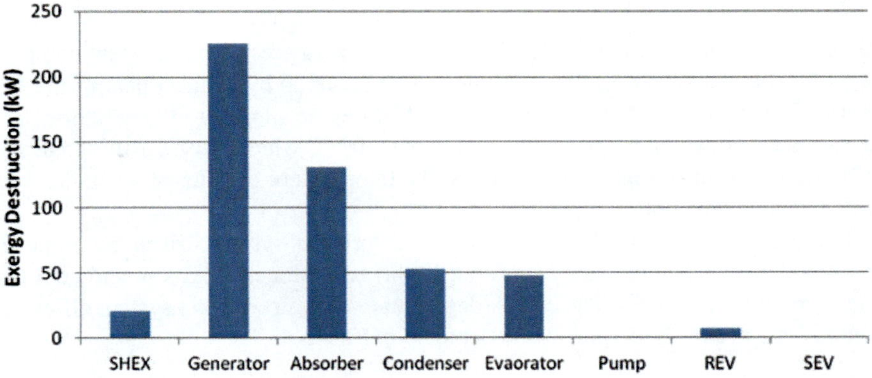

Fig. 2 Exergy destruction results for the exhaust heat-driven vapour absorption cooling system

4.2 Influence of Generator Temperature

Figure 3 depicts the effect of generator temperature on the coefficient of performance along with exergetic efficiency of vapour absorption cycle driven by IC engine exhaust. The COP is shown in y-scale on left side, and exegetic efficiency is shown in y-scale on right-hand side. It shown that at high generator temperatures, COP becomes constant and exergetic efficiency decreases. The figure shows that there is rise in COP in the initial stage with rise in temperature of generator. The COP of the vapour absorption cooling system varies between 0.5 and 0.75, and by further rise in temperature, no gain in COP is achieved. The exergetic efficiency of absorption systems increases considerably initially and attains maxima and then declines continuously as the temperature of generator increases.

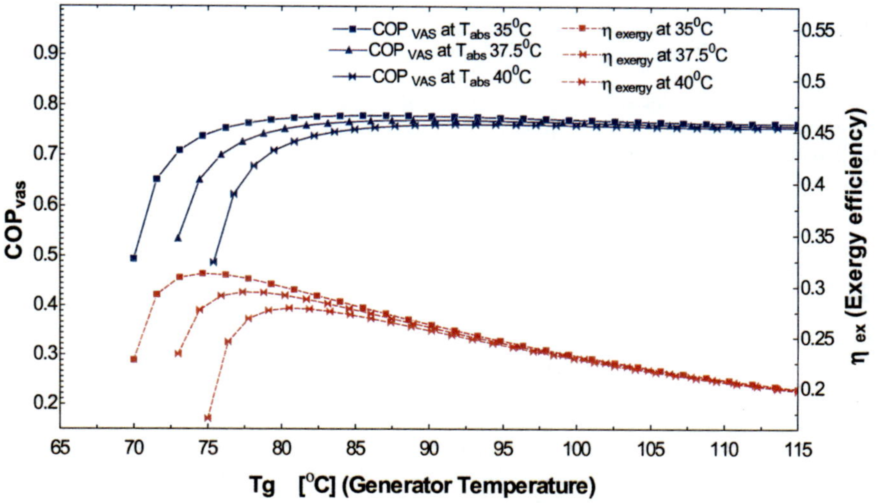

Fig. 3 Influence of generator temperature (T_g) on COP and exergy efficiency of vapour absorption cooling system

5 Conclusions

In this work, energy and exergy analysis of vapour absorption cooling system powered by exhaust heat of IC engine has been presented. The results show quantitative difference in exergy destruction and efficiency of vapour absorption cycle driven by IC engine exhaust by applying the second law approach. The conclusions drawn from the present study are that maximum destruction of exergy found in generator, second in absorber, third in condenser, fourth in evaporator, fifth in solution heat exchanger and then in expansion valve. Pump and solution expansion valve have negligible exergy destruction. The COP and exergetic efficiency rise with generator temperature, up to certain values for different absorber temperatures. The optimum conditions are in the range of 75–80 °C of the generator temperature for exhaust heat-operated single-effect vapour absorption cooling system. Integration of vapour absorption with IC engine saves energy for cooling systems and also decreases in greenhouse gases.

Acknowledgements The authors are grateful to Prof. S.C. Kaushik (CES, IIT Delhi) for their important inputs.

References

1. Manzela AA, Hanriot SM, Gómez LC, Sodré JR (2010) Using engine exhaust gas as energy source for an absorption refigeration system. Appl Energy 87:1141–1148
2. Izquierdo M, Vega MD, Lecuona A, Rodriguez P (2000) Entropy generated and exergy destroyed in lithium bromide thermal compressors driven by the exhaust gases of an engine. Int J Energy Res 24:1123–1140
3. Mostafavi M, Agnew B (1996) Thermodynamic analysis of combined diesel engine and absorption refigeration unit-supercharged engine. Appl Therm Eng 16(6):509–514
4. Mostafavi M, Agnew B (1997) Thermodynamic analysis of combined diesel engine and absorption refrigeration unit-naturally aspirated diesel engine. Appl Therm Eng 17(5):471–478
5. Agnew B, Talbi M, Mostafavi M (1999) Combined power and cooling, an analysis of the combined Diesel-absorption cycle. Appl Therm Eng 19(10):1097–1105
6. Talbi M, Agnew B (2002) Energy recovery from diesel engine exhaust gases for performance enhancement and air conditioning. Appl Therm Eng 22:693–702
7. Anand S, Gupta A, Tyagi SK (2014) Critical analysis of a biogas powered absorption system for climate change mitigation. Clean Technol Environ Policy 16(3):569–578
8. Kaushik SC, Arora A (2009) Eenrgy and exergy analysis of single effect and series flow double effect ater-lithium bromide absorption refrigeration systems. Int J Refrig 32(6):1247–1258
9. Gomri R, Hakmi R (2008) Second law analysis of double effect vapour absorption cooler system. Energy Convers Manag 49:3343–3348
10. Kaynakli O, Kilic M (2007) Second law based thermodynamic analysis of water-lithium bromide absorption refrigeration system. Energy 3:1505–1512
11. Grossman G, Zaltash A (2001) ABSIM-modular simulation of advanced absorption systems. Int J Refig 24:531–543
12. Bejan A (1996) Entropy generation minimization. CRC Press, New York
13. Bhatti SS, Verma S, Tyagi SK (2019) Energy and exergy based performance evaluation of variable compression ratio spark ignition engine based on experimental work. Therm Sci Eng Progr 332–339

Parametric Estimation of Wall Temperature in a Parabolic Trough Solar Collector Using Supercritical CO_2 as Heat Transfer Fluid for Process Heat Production

Ravindra Vutukuru and Maddali Ramgopal

Abstract In this paper, studies on a commercial parabolic trough solar collector (PTSC) using supercritical CO_2 (s-CO_2) as the heat transfer fluid (HTF) are presented. When compared to other HTFs, supercritical CO_2 can operate at higher temperatures which is useful for power plants. The collector module used for this study is LS-2 module. The effect of mass flow rate and inlet temperature of the HTF on receiver wall temperature, thermal loss, heat transfer coefficient and thermal efficiency is analysed. CO_2 is taken in supercritical state in order to facilitate the direct integration of the collector to high-temperature power cycles. It is found that higher mass flow rate results in lower wall temperature and higher heat transfer coefficient. At any particular mass flow rate, lower operating pressure results in higher wall temperature. Rise in HTF inlet temperature results in an increase in wall temperature and reduction in thermal efficiency of the collector. However, variation in operating pressure results in negligible change in wall temperature.

Keywords Supercritical CO_2 · Wall temperature · Heat transfer coefficient · PTSC

1 Introduction

Solar thermal systems have been gaining increasing prominence due to their environment-friendly and sustainable nature. Among various technologies, concentrated solar power (CSP) is a promising technology for high-temperature heat and power applications. The traditional HTFs for solar thermal systems are liquid fluids like Therminol VP1, Dowtherm Q and molten salts such as Hitec XL. However, these fluids have a temperature constraint. Liquid heat transfer fluids provide stable operation up to 400 °C, while molten salt can operate up to 550 °C. In addition, there

R. Vutukuru (✉)
School of Energy Science and Engineering, Indian Institute of Technology Kharagpur, Kharagpur 721302, WB, India
e-mail: vutukuru.ravindra@gmail.com

M. Ramgopal
Mechanical Engineering Department, Indian Institute of Technology Kharagpur, Kharagpur 721302, WB, India

M. Ramgopal et al. (eds.), *Advances in Air Conditioning and Refrigeration*, Lecture Notes in Mechanical Engineering, https://doi.org/10.1007/978-981-15-6360-7_25

are issues associated with Hitec XL such as the need for freeze protection and high material cost.

Recently, studies on usage of gaseous heat transfer fluids in CSP systems have been undertaken. Gases eliminate the temperature constraint in CSP systems. Another possibility with gases is their ability to directly integrate the solar collector with components like turbine and compressor, thereby eliminating the need for separate secondary heat exchange loop. However, developing gas-based systems which operate at high pressures presents operation and maintenance issues which need to be analysed. Ouagued et al. [1] studied a numerical model of parabolic trough collector under Algerian climate conditions. The receiver model was divided into number of segments and energy balance equations applied to each segment. The performance with seven different thermal oils was studied. Various studies on parabolic trough collectors (PTC) using different HTFs were presented by Bellos et al. [2–6].

Vutukuru et al. [7] studied the suitability of various HTFs for solar thermal systems at high temperatures. It was found that s-CO_2 is better in terms of inventory cost of the fluid which is important for setting up the solar field. Muñoz-Antón et al. [8, 9] presented theoretical and experimental studies on CO_2 as a thermodynamic process fluid and as HTF in solar thermal receivers.

In this paper, estimation of wall temperature using s-CO_2 in a LS-2 module is presented. For this collector, the effect of s-CO_2 flow rate and inlet temperature of fluid on wall temperature is analysed. The effect on wall temperature at various operating pressures for s-CO_2 is also studied. The parametric studies of wall temperature are useful in determining the thermal losses in PTC, which influence the thermal efficiency.

2 Thermal Modelling of the Collector

Figure 1 shows the schematic diagram of a LS-2 collector module. LS-2 collector has a length (L) and width (W) of 7.8 m and 5 m, respectively. The concentration ratio (C) is 22.74, and the focal distance of the module (F) is 1.71 m. The receiver inner (D_{reci}) and outer (D_{reco}) diameters are 0.066 m and 0.070 m, respectively. The corresponding inner (A_{reci}) and outer (A_{reco}) surface areas of the receiver are 1.617 m^2 and 1.715 m^2, respectively. The dimensions of the cover inner (D_{coveri}) and outer (D_{covero}) diameters are 0.109 m and 0.115 m, respectively. The corresponding inner (A_{coveri}) and outer (A_{covero}) surface areas of the cover are 2.671 m^2 and 2.818 m^2, respectively. Aperture area (A_{ap}) of the LS-2 module is 39 m^2.

Available solar energy (Q_{s}) is estimated as

$$Q_{\text{s}} = A_{\text{ap}}.G_{\text{b}} \tag{1}$$

Energy balance equation of the receiver is presented below.

$$Q_{\text{s}}.\eta_{\text{opt}}(\theta) = Q_{\text{u}} + Q_{\text{loss}} \tag{2}$$

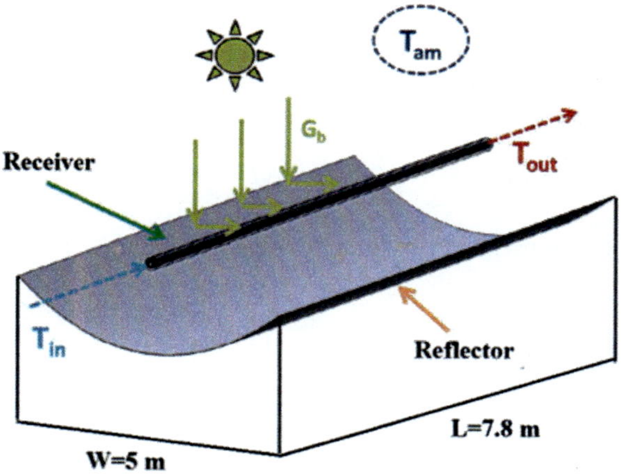

Fig. 1 LS-2 collector module

Optical efficiency (η_{optical}) of the collector is expressed as shown below [10].

$$\eta_{\text{optical}} = \rho.\gamma.\tau.\alpha.K(\theta) \tag{3}$$

where ρ is the reflectance of solar concentrator, γ is the intercept factor, τ is the cover transmittance and α is the absorber absorbance. The corresponding values for LS-2 collector module are 0.83, 0.99, 0.95 and 0.96, respectively. Incident angle modifier (K) depends on the incident angle θ. For our study, incident angle is 0°, and corresponding incident angle modifier value is 1. Maximum optical efficiency of the collector is taken as 75% [10].

Useful energy gain (Q_u) can be calculated as

$$Q_u = m.C_p.(T_{\text{outlet}} - T_{\text{inlet}}) \tag{4}$$

where m is mass flow rate, T_{inlet} is the inlet temperature, T_{outlet} is the outlet temperature and C_p is the mean specific heat capacity

Useful energy gain can also be calculated by using the following equation.

$$Q_u = h_f.A_{\text{reci}}.(T_{\text{rec}} - T_{\text{fm}}) \tag{5}$$

T_{rec} is the mean temperature of the receiver.

T_{rec} is calculated by taking into account the thermal loss associated with the collector. Equation (6) considers radiation thermal loss, whereas Eq. (7) takes into consideration the radiation and convection losses. [2].

$$Q_{\text{loss}} = \frac{A_{\text{reco}}.\sigma.\left(T_{\text{rec}}^4 - T_{\text{cover}}^4\right)}{\frac{1}{\varepsilon_{\text{rec}}} + \frac{1-\varepsilon_{\text{rec}}}{\varepsilon_{\text{rec}}}\left(\frac{A_{\text{reco}}}{A_{\text{coveri}}}\right)} \tag{6}$$

Cover temperature is denoted by T_{cover} and σ is the Stefan–Boltzmann constant. The cover emittance 'ε_{cover}' and receiver emittance 'ε_{rec}' are assumed as 0.9 and 0.2, respectively [10].

$$Q_{loss}=A_{covero}.h_{air}.(T_{cover}-T_{amb})+A_{covero}.\sigma.\varepsilon_{cover}.\left(T_{cover}^{4}-T_{sky}^{4}\right) \quad (7)$$

Temperature of sky (T_{sky}) can be calculated as [4].

$$T_{sky}=T_{amb}-8 \quad (8)$$

Convection heat transfer coefficient (h_{air}) between cover and ambient is estimated as [4].

$$h_{air}=\frac{4.V_{air}^{0.58}}{D_{covero}^{0.42}} \quad (9)$$

where V_{air} is the air velocity.

Thermal efficiency (η_{th}) of the PTC is calculated as [4].

$$\eta_{th}=\frac{Q_u}{Q_s} \quad (10)$$

Nusselt number is calculated from the Dittus–Boelter equation which is found to be applicable to s-CO_2 also [4].

$$\mathrm{Nu}=0.023.(\mathrm{Re})^{0.8}.(\mathrm{Pr})^{0.4} \quad (11)$$

Heat transfer coefficient (h_f) is calculated as [4].

$$h_f=\frac{\mathrm{Nu}.k_f}{D_{reci}} \quad (12)$$

where k_f is thermal conductivity of the fluid.

In the above equation, the thermo-physical properties are obtained from Engineering Equation Solver (EES software) at the mean fluid temperature (T_{fm}) given by [10]:

$$T_{fm}=\frac{T_{inlet}+T_{outlet}}{2} \quad (13)$$

In our study, T_{rec} is taken as the wall temperature (T_{wall}). The heat transfer between absorber pipe and fluid is dominated by the convective heat transfer mechanism. This is justified because the thickness of the absorber pipe is negligible compared to inner diameter and thermal conductivity of copper material ($K_{pipe}=400$ W/mK). Hence, the thermal resistance ($R_{1-2\,(conv)}$) is expressed as

$$R_{1,2(\text{conv})} = \frac{1}{\Pi D_{\text{reci}}\,\text{Lh}_{\text{f}}} \tag{14}$$

Wall temperature can be calculated as

$$Q_{\text{u}} = \frac{T_{\text{wall}} - T_{\text{fm}}}{R_{1,2(\text{conv})}} \tag{15}$$

3 Results and Discussion

The ambient temperature (T_{amb}) is taken as 32 °C, and the default value of direct beam irradiance (G_b) is assumed as 900 W/m^2. For the parametric analysis, the irradiance value is varied from 600 to 1000 W/m^2. S-CO_2 mass flow rate is assumed as 0.5 kg and varied from 0.5 to 1.5 kg for the parametric study. Default value of collector inlet temperature is taken as 126.85 °C. For the parametric analysis, the s-CO_2 inlet temperature is varied from 100 to 200 °C. Operating pressure is fixed at 200 bar.

3.1 *Effect of s-CO_2 mass Flow Rate*

The effect of s-CO_2 mass flow rate on T_{wall} and Q_{loss} at various G_b values keeping operating pressure fixed at 200 bar is shown in Fig. 2. The wall temperature and thermal loss decrease with increase in s-CO_2 flow rate. Reduction in thermal loss

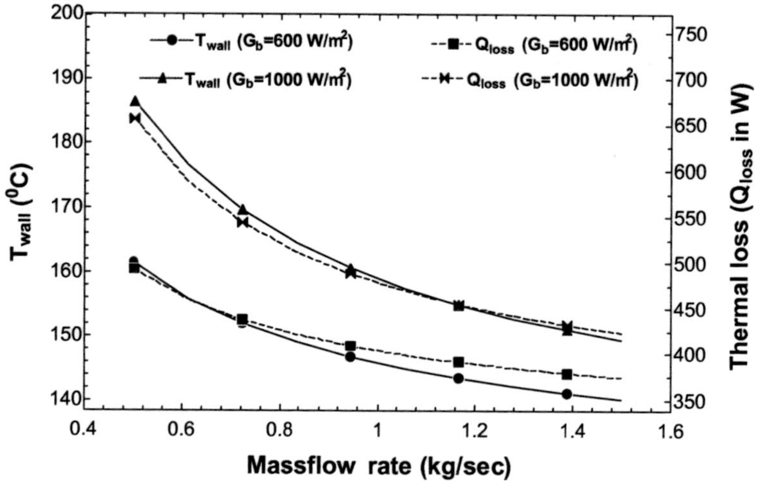

Fig. 2 Effect of mass flow rate of s-CO_2 on wall temperature and thermal loss

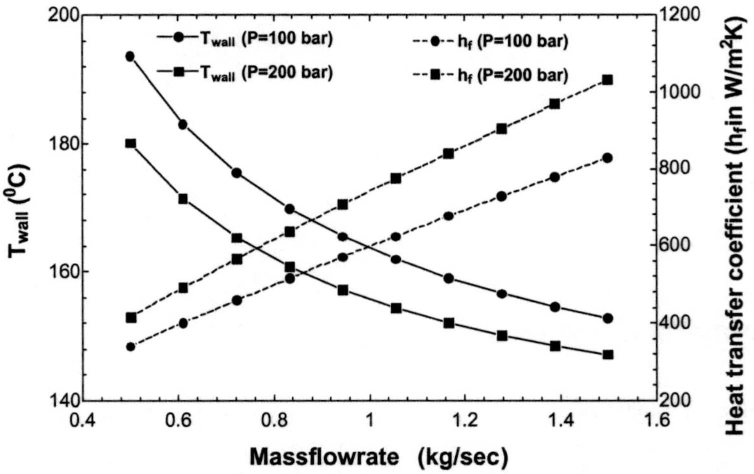

Fig. 3 Effect of mass flow rate of s-CO_2 on wall temperature and heat transfer coefficient at various operating pressures

improves the thermal efficiency. An increase in s-CO_2 mass flow rate results in rise in heat transfer coefficient due to increase in Reynolds number. The convective thermal resistance between the HTF and the absorber pipe decreases due to rise in heat transfer coefficient. The mean fluid temperature reduces with increase in s-CO_2 flow rate. Due to these factors, T_{wall} decreases with rise in s-CO_2 flow rate. The effect of s-CO_2 flow rate on T_{wall} and h_{f} for s-CO_2 at various operating pressures keeping G_{b} fixed at 900 W/m^2 is presented in Fig. 3. The plots are shown at pressure values of 100 and 200 bar favourable for CO_2-based power plant applications.

3.2 *Effect of s-CO_2 Fluid Inlet Temperature*

The effect of T_{inlet} on T_{wall} and η_{th} at various solar irradiance values keeping operating pressure fixed at 200 bar is presented in Fig. 4. The plots are made at the default values of the ambient temperature (32 °C) and mass flow rate (0.5 kg/s). T_{wall} rises with increase in fluid inlet temperature. The thermal loss increases causing a reduction in useful heat gain. As a result, the thermal efficiency reduces. The convective resistance between the absorber pipe and the HTF decreases with rise in fluid inlet temperature due to reduction in heat transfer coefficient of the fluid. Figure 5 shows the influence of T_{inlet} on T_{wall} for s-CO_2 at two different values of operating pressure. Rise in operating pressure results in marginal change in wall temperature.

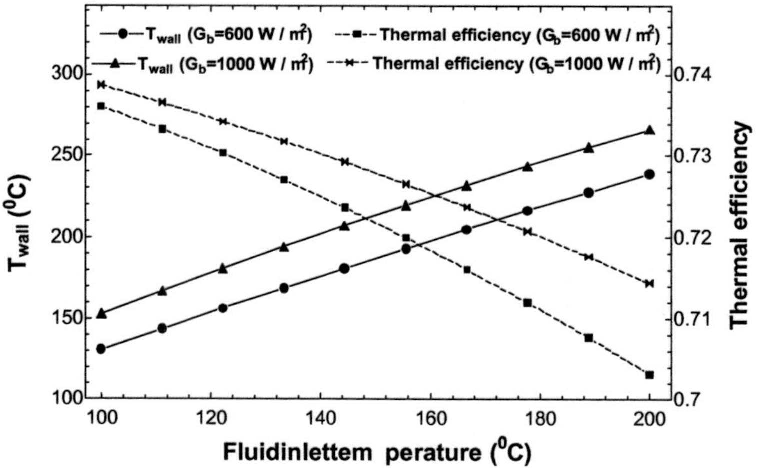

Fig. 4 Effect of s-CO_2 fluid inlet temperature on wall temperature and thermal efficiency

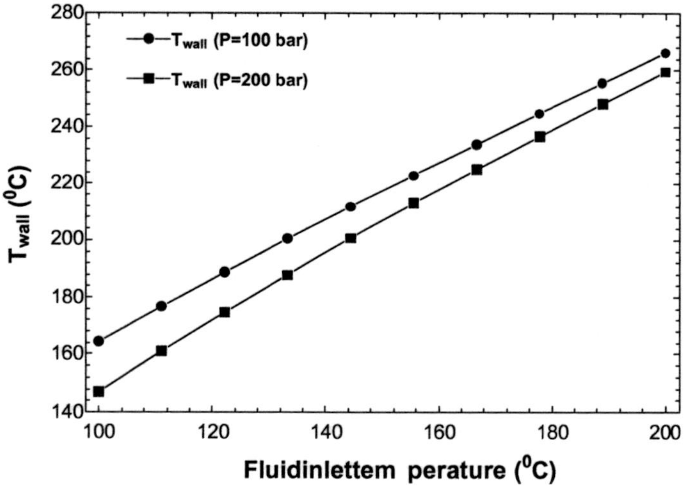

Fig. 5 Effect of s-CO_2 fluid inlet temperature on wall temperature at various operating pressures

4 Conclusions

The effect of mass flow rate, inlet temperature and fluid operating pressure on wall temperature, thermal loss, heat transfer coefficient and thermal efficiency of absorber tube of PTSC was investigated with s-CO_2 as the HTF. Higher mass flow rate results in lower wall temperatures and better heat transfer coefficients. Secondly, an increase in HTF inlet temperature reduces the thermal efficiency due to higher wall temperature.

In additions-CO_2 can be used for power generation, co- and trigeneration systems. So, parabolic trough solar collectors can be directly integrated in such systems.

References

1. Ouagued M, Khellaf A, Loukarfi L (2013) Estimation of the temperature, heat gain and heat loss by solar parabolic trough collector under Algerian climate using different thermal Oils. Energy Convers Manag 75:191–201
2. Bellos E, Tzivanidis C, Antonopoulos KA, Daniil I (2016) The use of gas working fluids in parabolic trough collectors—an energetic and exergetic analysis. Appl Therm Eng 109:1–14
3. Bellos E, Tzivanidis C, Antonopoulos KA (2017) A detailed working fluid investigation for solar parabolic trough collectors. Appl Therm Eng 114:374–386
4. Bellos E, Tzivanidis C (2017) Parametric investigation of supercritical carbon dioxide utilization in parabolic trough collectors. Appl Therm Eng 127:736–747
5. Bellos E, Tzivanidis C, Daniil I, Antonopoulos KA (2017) The impact of internal longitudinal fins in parabolic trough collectors operating with gases. Energy Convers Manag 135:35–54
6. Bellos E, Tzivanidis C, Daniil I (2017) Energetic and exergetic investigation of a parabolic trough collector with internal fins operating with carbon dioxide. Int J Energy Environ Eng 8:109–122
7. RavindraVutukuru A (2019) Saikiran Pegallapati, Ramgopal Maddali, suitability of various heat transfer fluids for high temperature solar thermal systems. Appl Therm Eng 159:113973
8. Muñoz-Antón J, Rubbia C, Rovira A, Martínez-Val JM (2015) Performance study of solar power plants with CO_2 as working fluid. A promising design window, Energy Convers Manag 92:36–46
9. Muñoz-Anton J, Biencinto M, Zarza E, Díez LE (2014) Theoretical basis and experimental facility for parabolic trough collectors at high temperature using gas as heat transfer fluid. Appl Energy 135:373–381
10. Bellos E, Tzivanidis C (2018) Analytical expression of parabolic trough solar collector performance. Designs 2:9

Thermodynamic Analysis and Performance of Various Binary and Ternary Mixtures to Replace R410A

Adithya Kumar and Shaik Saboor

Abstract For residential air conditioning applications, R410A has been used extensively. The global warming associated with the refrigerant is very high. This has led to further complications in the environment which was identified much later. Initiatives have been directed toward solving this dilemma. This research primarily focuses on creating binary and ternary refrigerant replacements for R410A. The conformability to the pressure–temperature characteristics, the increase in performance and the reduction in global warming potential was analyzed using computational methods. A MATLAB code was generated for the purpose of the calculation of the thermodynamic properties. The code was verified using REFPROP software. The accuracy of the code was very high. Four binary refrigerants and four ternary refrigerants were considered as replacements, and the results were obtained using the MATLAB code. One particular ternary refrigerant RT4 is resulted in a 34.938% reduction in the global warming potential while giving an increase of 4.335% in the performance of the refrigeration system.

Keywords R-410A replacement · Global warming potential · Coefficient of performance · Binary and ternary mixtures

1 Introduction

The demand for the facilities of refrigeration and air conditioning has risen exponentially through the years. From being a luxury service limited to use by a few elites in the society, air conditioning and refrigeration have penetrated to various strata in the society. The advent of new technology that provides these services with higher efficiency and lower cost is credited for the drastic infiltration of the service. R410a is a refrigerant that inflicts zero harm to the ozone layer and provides better performance as compared to R22. R410a is superior to R22 even upon comparing the prices of both refrigerants, R410a providing cheaper rates. Among the available refrigerants,

A. Kumar · S. Saboor (✉)
School of Mechanical Engineering, Vellore Institute of Technology, Vellore 632014, Tamil Nadu, India
e-mail: saboor.nitk@gmail.com

M. Ramgopal et al. (eds.), *Advances in Air Conditioning and Refrigeration*, Lecture Notes in Mechanical Engineering, https://doi.org/10.1007/978-981-15-6360-7_26

R410a is the most suitable replacement for R22 refrigerant. After the implementation of the Montreal Protocol [1], was the predicament of ozone depletion caused by chlorofluorocarbons was actively targeted. Further, it was long after the implementation of the Montreal Protocol that the excessively enormous global warming potential values of hydrofluorocarbons were identified. The motive for research has transformed into the identification of new refrigerant blends or the discovery of new refrigerants which would reduce the GWP of the HFCs while maintaining an ozone depletion potential value of zero. In residential applications, R410A is the refrigerant that is predominantly used. It has a GWP of 2088, which indicates that the global warming capability of R410A is 2088 times more potent as compared to carbon dioxide.

2 Refrigerant Selection

The objective of the research is to develop a refrigerant mixture whose GWP is lower than that of R410A yet does not sacrifice on the coefficient of performance. A viable refrigerant was HFO1234yf which is mildly flammable which was blended with R410A—whose flammability index is A1—to create a new refrigerant mixture that may replace R410A. The GWP of HFO1234yf is 4, which will lead to a drastic reduction in the GWP of the mixture. The ODP of the mixture still remains zero, which implies that the new refrigerant mixture would not lead to any impact on the ozone layer. By not constricting the research to the confines of binary mixtures, ternary mixtures can also be evaluated. Ternary mixtures of R410A, HFO1234yf (R1234yf) and R290 in different compositions were evaluated. R290 has a higher flammability limit as it is a hydrocarbon. But, smaller quantities of hydrocarbons such as R290 can help the blend mix with the mineral oils used as lubricants and also help improve the performance of the refrigeration system [2]. Both binary and ternary mixtures will lead to refrigerants with zeotropic characteristics. The normal boiling point of HFO1234yf, R410A and R290 are 243.71K [3], 224.65K and 231.15K [4], respectively. The difference in temperature between R410A and R1234yf is greater than 20°, thus R1234yf cannot constitute a larger amount in the defined mixture. R290 cannot be added in equal proportion despite its acceptable normal boiling point associated with its higher flammability (Tables 1 and 2).

The ratios in the tables are in terms of mass fractions and the refrigerants will be referred to, as mentioned in the table, henceforth.

Table 1 Binary mixtures

R410A	R290	Name
95	5	RB1
90	10	RB2
85	15	RB3
80	20	RB4

Table 2 Ternary mixtures

R410A	R1234yf	R290	Name
80	10	10	RT1
75	15	10	RT2
70	20	10	RT3
65	25	10	RT4

3 Numerical Methodology

Several equations of state have been proposed over the years, by various renowned scientists with the purpose of relating the state variables of pressure, temperature and volume using a single relation [5, 6]. This research uses the Martin-Hou equation of state [7] to find the properties of the state variables of the refrigerants. The Martin-Hou equation relates the pressure, temperature and volume as shown in Eqs. (1)–(5). The coefficients t_1, t_2, t_3, t_4 and t_5 are in turn functions of the critical pressures, critical temperature and critical volume of the refrigerant considered.

$$p = \sum_{i=1}^{5} \frac{t_i}{(v - b)^i} \tag{1}$$

$$t_2 = A_2 + B_2 T + C_2 e^{-5.475\frac{T}{T_c}} \tag{2}$$

$$t_3 = A_3 + B_3 T + C_3 e^{-5.475\frac{T}{T_c}} \tag{3}$$

$$t_1 = RT \tag{4}$$

$$t_4 = A_4 \tag{5}$$

$$t_5 = B_5 T \tag{6}$$

In the equations, T represents the temperature, T_C is the critical temperature and v is the specific volume. Ai, Bi and Ci are coefficients which are dependent on the critical values of the refrigerants. The coefficients B4, C4, A5 and C5 have been assumed zero, and the coefficient B3 is an anomaly as it depends on the slope of the saturated pressure vs. the saturated temperature graph at constant critical volume [8]. This further leads to a predicament as for the Martin-Hou equation to yield the results in the saturated properties that must be calculated prior. To solve this quandary, the Ambrose–Walton corresponding states equation (Eq. 7) was employed [9]. The coefficients were a function of the ratio of saturated temperature to critical temperature. Using this equation, a direct relationship between the saturated pressure and the saturated temperature was obtained. The required information to obtain B3 was acquired.

$$\text{In } P_{\text{vpr}} = f^{(0)}(T_r) + \omega f^{(1)}(T_r) + \omega^2 f^{(2)}(T_r) \tag{7}$$

In the above equation, P_{vpr} stands for vapor pressure which is the ratio of the saturated pressure to the critical pressure, T_r represents the ratio of saturated temperature to the critical temperature and ω represents the acentric factor of the refrigerant. All the coefficients are dependent on the critical values of the state variables. A MATLAB code was generated using the Martin-Hou equation and its accuracy was verified using REFPROP. The code was used to generate results for a pure refrigerant R134a, an imaginary binary mixture R134a-R152a in 50–50 proportion and a ternary mixture R134a-R152a-R290 in the ratio 44–52–4, respectively. The results of the code have been displayed in the graphs below.

As observed in the graphs above, the curve generated by the MATLAB code and the results generated by REFPROP are nearly congruent. This leads to the conclusion that the MATLAB code is accurate through the temperature range 243.15 and 353.15 K.

The state variables have been calculated using the methodology mentioned above. On observing the Ambrose–Walton corresponding states method, it can be deduced that the relation of pressure is directly proportional to temperature as the acentric factor is a constant for a particular refrigerant and the only variables in the equation are the saturated pressure and temperature. This will lead to the monotonically increasing curves depicted in Fig. 1, but the increase does not conform to a linear pattern. To perform thermodynamic analysis of the refrigerant, thermodynamic properties of the refrigerant such as liquid enthalpy, liquid entropy, vapor enthalpy and vapor entropy are quintessential. The following equations have been used to calculate the properties.

$$x_i = \frac{n_i}{\sum_{k=1}^{n} n_k} \tag{8}$$

$$p_c = \sum_{i=1}^{n} x_i p_{C_1} \tag{9}$$

$$v^L = \frac{\sum_{i=0}^{n} x_i v_i^L}{M} \tag{10}$$

$$s^L = \frac{\sum_{i=0}^{n} x_i s_i^L - R \sum_{i=0}^{n} x_i \text{ In } x_i}{M} \tag{11}$$

In the above equations, x_i and n_i represent the mole fraction and the mole. The subscript i is a reference to the refrigerant in consideration in binary and ternary mixtures. p_C represents the critical pressure of the mixture, and the molar mass, the critical volume and the critical temperature of the mixture can be found out similarly. v^L represents the specific volume in liquid state and M represents the molar mass of the mixture. The specific volume in its SI units is m^3/kg while x_i

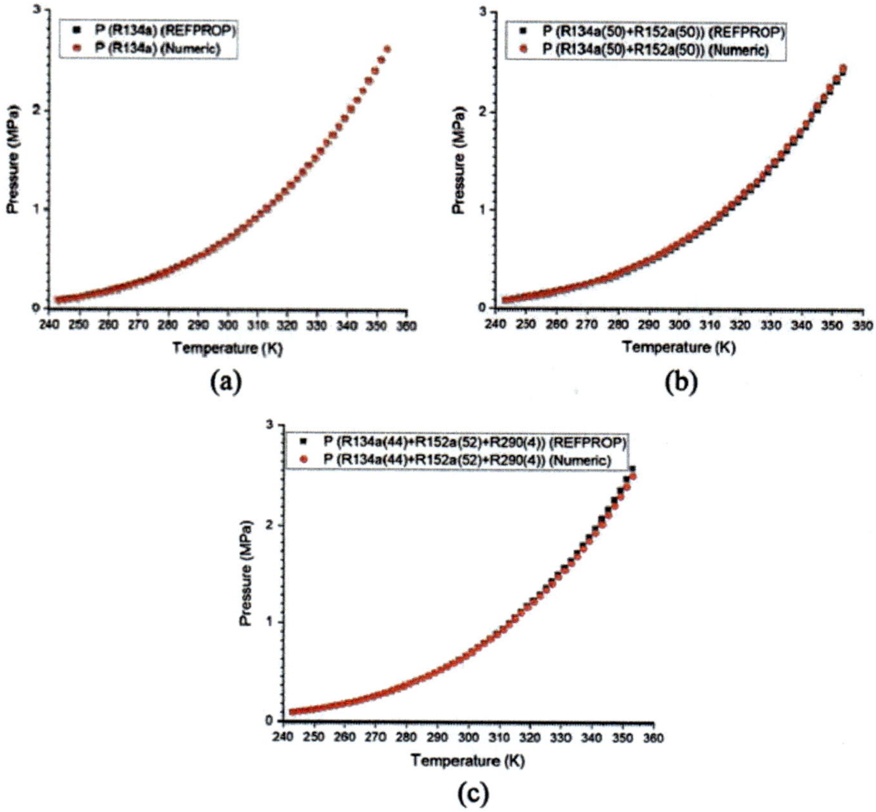

Fig. 1 Comparison of the curves generated by MATLAB code and REFPROP. **a** for unary refrigerant R134a, **b** for binary refrigerant R134a (50) + R152a (50), **c** for ternary refrigerant R134a (44) + R152a (52) + R290 (4)

is the mole fraction. Hence, the specific volume must be converted to similar units for dimensional accuracy. s^L is the liquid entropy in Eq. (11). In the equations, T_C represents the critical temperature, s_{fg} and h_{fg} represent the entropy and the enthalpy of the liquid–gas inter-phase, respectively.

$$h_{fg} = RT_C\left(7.08\tau^{0.354} + 10.95\omega\tau^{0.456}\right) \tag{12}$$

$$s_{fg} = \frac{h_{fg}}{T_{sat}} \tag{13}$$

$$GWP_{mix} = \sum_{i=1}^{n} (\%\,\text{of component})_i * GWP_{(i)} \tag{14}$$

$$\mathrm{COP} = \frac{\mathrm{RE}}{W_{\mathrm{in}}} \tag{15}$$

4 Results and Discussion

The pressure–temperature characteristics were plotted and compared with the pressure of R410A. The degree of superimposition of the binary and ternary curves on R410A pressure curve indicates the degree of change required in the refrigeration system.

From Fig. 2, it can be deduced that the ternary mixtures considered would be a better replacement for R410A in the temperature range considered as the curves are over-lapping better. For binary mixtures, the pressure at the terminals of the temperature range is accurate, but the discrepancy in the values that lie between have larger deviations. This implies that the changes that need to be incorporated in the refrigeration system would be higher. But the objective is to replace R410A without significant changes to the system. In ternary mixtures, as the content of R1234yf is increased the conformability of the curve also increases. But there is a major discrepancy at the higher end of the temperature range.

The COP was calculated considering the ARI conditions. The difference in the COP has been exhibited in Fig. 3a [10]. Among the refrigerants that have been considered, only RT1 has a drastic drop in COP as compared to R410A. The other refrigerants have a minimum increase of 3.086% in the performance of the system. The binary mixtures RB3 and RB4 have higher COP values even in comparison with the ternary mixtures. This indicates that the addition of R290 has a more positive effect on the COP as compared to R1234yf. Similarly, the GWP of the refrigerants has been calculated using the equations and depicted in Fig. 3b. There is a linear reduction

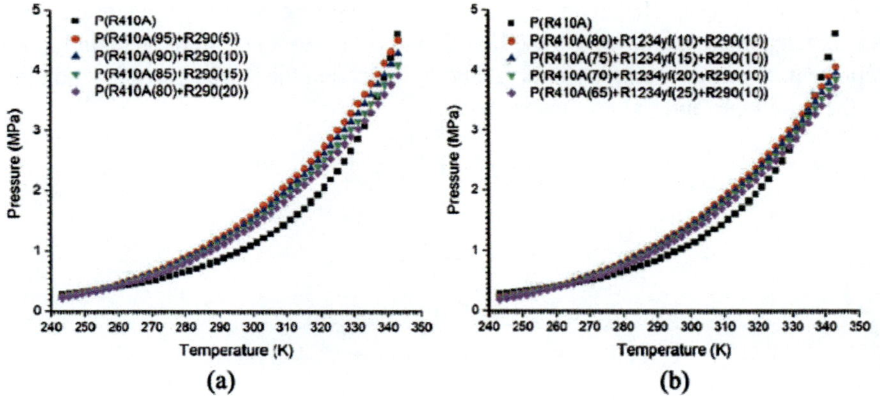

Fig. 2 Pressure versus temperature curves. **a** for binary mixtures and **b** for ternary mixtures

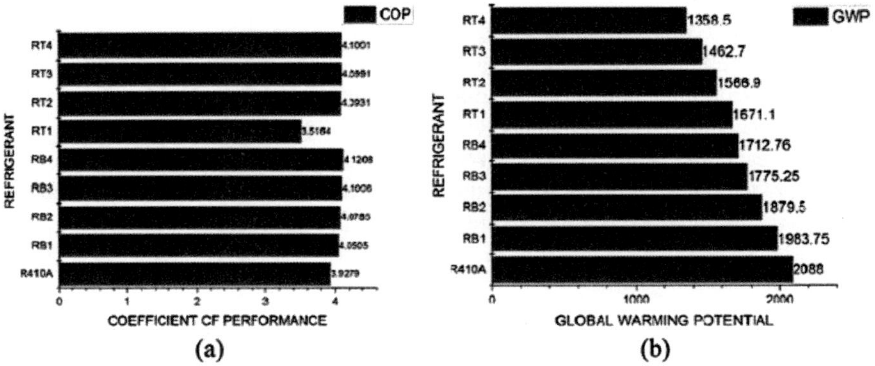

Fig. 3 **a** COP of refrigerants and **b** GWP of refrigerants

of the GWP from R410A to RT4. This unique observation is due to the similar GWP of R1234yf and R290 which are 4 and 3, respectively. In binary mixtures, only R290 was present which impacted the GWP. The quantity of R290 could not be high so the changes in the GWP were not very significant. In the ternary mixtures, higher proportion of the mixture could be substituted by a blend of R1234yf and R290 which resulted in a drastic drop of GWP. The lowest GWP obtained was 1358.5, which is still a very high value, but, in the comparison results are approximately a 35% decrease in the GWP. The ternary refrigerant RT4 had the best conformability among all the refrigerants indicating that the changes that it will pose on the refrigeration system would be minimal in comparison with the others.

5 Conclusion

- From the results, it is evident that as the quantities of R290 and R1234yf are increased, the conformability of the curve improves, which indicates that the ternary refrigerants would be the optimal choice for refrigerant replacement. Binary curves require a higher value of R290 which would adversely affect the flammability of the mixture.
- In Fig. 3, it can be interpreted that the COP values are higher for all refrigerants barring RT1. The highest increase in COP can be observed for RB4, with an increase of 4.855%, but this cannot be considered optimal due to the large amount of R290 in the mixture. In ternary mixtures, the highest COP is due to RT4, with an increase of 4.335%, owing to the large quantities of R1234yf and R290. The addition of R290 in binary and the addition of R290 and R1234yf in the ternary lead to better COP values.
- The flammability of the refrigerants has been increased due to the inclusion of R290. This can be confirmed because R290 is highly flammable and R1234yf is

also relatively not resistant to flame. This would imply that the increased concentration of these components in the mixture would increase the flammability. But, binary mixtures are more prone as compared to ternary mixtures as binary contains only R290 while ternary contains R1234yf which is relatively resilient.

- The lowest GWP is observed for RT4 with a GWP of 1358.5 which is a reduction by 34.938%. The conformability of the curve is also highest, and the COP increase is significant making RT4 the best replacement in this research.

References

1. Polonara F, Kuijpers JML, Peixoto RA (2017) Potential impacts of the Montreal Protocol Kigali Amendment to the choice of refrigerant alternatives. Int J Heat Technol 35(Special Issue 1):S1–S8. https://doi.org/10.18280/ijht.35sp0101
2. Jabaraj DB, Avinash P, Mohan Lal D, Renganarayan S (2005) Experimental investigation of HFC407C/HC290/HC600a mixture in a window air conditioner. Energy Conserv Manage 47:2570–2590. https://doi.org/10.1016/j.enconman
3. Brown JS, Polonara F, Di Nicola G, Fedele L, Bobbo S (2012) Vapor Pressure of Hydrofluoroolefins: critical review of experimental data and models. In: International refrigeration and air conditioning conference, Paper 1316. http://docs.lib.purdue.edu/iracc/1316
4. ASHRAE (2009) Handbook fundamentals (SI). Chapter 29. Refrigerants, pp 29.1–29.10
5. Poling BE, Prausnitz JM, O'Connell J. The properties of gases and liquids. McGraw-Hill, New York
6. Li S, Zhang YD, Li Y, Liao RQ (2015) Equilibrium calculation and technological parameters optimization of natural gas liquefaction process with mixed refrigerant. Int J Heat Technol 33(No.2):123–128. http://dx.doi.org/10.18280/ijht.330220
7. Martin JJ, Kapoor RM, De Nevers N (1959) An improved equation of state for gases, A.I.Ch.E J 5(2):159-160https://doi.org/10.1002/aic.690050207
8. Dong L, Zhang Y, Li S, Wei S, Zhang J, Qi Y (2012) An empirical equation to directly calculate parameter B4 of the Martin-Hou equation of state. Chem Eng Commun 199(5):577–586. https://doi.org/10.1080/00986445.2011.599899
9. An H, Yang W (2012) A new generalized correlation for accurate vapor pressure prediction. Chem Phys Lett 543:188–192. http://dx.doi.org/10.1016/j.cplett.2012.06.029
10. ASHRAE (2009) Handbook fundamentals (SI). Chapter 29, Refrigerants, pp 30.1–30.75

Numerical Investigation of Unsteady Thermal Characteristics of Lightweight Concrete for Energy-Efficient Buildings

A. Chelliah and S. Saboor

Abstract Building envelope in houses is accountable for the enormous heat gain. This numerical study focuses on the influence of lightweight aggregate in the concrete for the heat gain in the building due to the sun's radiation. Four lightweight aggregates, such as gravel, stalite, lytag, leca and argex were selected for inclusion in concrete and silica fume where it was added to the mortar. Transient thermal characteristics subjected to regular thermal excitation were solved using the admittance method through a computer simulation program. From the result, it is observed that argex concrete wall showed a better thermal performance with a higher time lag (6.218 h) and a lower decrement factor (0.5051). Mortar with silica fume showed a better thermal performance than a mortar with a higher time lag (0.60 h) and a lower decrement factor (0.991).

Keywords Structural lightweight concrete · Admittance method · Time lag · Decrement factor

Nomenclature

w	Admittance
U	Transmittance
k	Thermal conductivity
C_p	Specific heat
δ	Decrement factor
ξ	Thermal heat capacity
ρ	Density
τ	Time lag
φ	Temperature

A. Chelliah · S. Saboor (✉)
School of Mechanical Engineering, Vellore Institute of Technology, Vellore 631014, Tamil Nadu, India
e-mail: saboor.nitk@gmail.com

M. Ramgopal et al. (eds.), *Advances in Air Conditioning and Refrigeration*, Lecture Notes in Mechanical Engineering, https://doi.org/10.1007/978-981-15-6360-7_27

Acronyms

AC	Argex concrete
ACW	Argex concrete wall
CP	Cement plaster
LC	Lytag concrete
LCW	Lytag concrete wall
LEC	Leca concrete
LECW	Leca concrete wall
M	Mortar
NGC	Normal gravel concrete
NGCW	Normal gravel concrete wall
SC	Stalite concrete
SCW	Stalite concrete wall
SFCP	Silica fume cement plaster
SFM	Silica fume mortar

1 Introduction

Air conditioning and daylighting consume a major amount of energy for the building. Passive cooling, suitable building materials and its thickness are the ways to keep the building cool and reduce the air conditioning load. All over world-building construction showing sustainable growth which leads to the massive consumption of raw materials, water, fossil fuels that result in environmental problems and natural resource depletion. The buildings are responsible for around 40% of world total power consumption which accounts for 40% of greenhouse gas emission. The majority of the Indian building constructions are far from energy-efficient buildings [1]. The heat wave travels sinusoidally from outside environments of hot climatic conditions to the inside building environment. The heat wave has two parameters and they are wavelength and amplitude. Wavelength signifies the time to reach the inside building surface. The amplitude implies temperature magnitude and it depends upon solar radiation and convection between outside building surface and atmospheric air. The decrement factor is the ability of building materials to prevent the heat from the outside environment to the inside surface of the building. The time lag is the lagging of the peak temperature from the outside surface to the inside surface of the building. High time lag and low decrement factor are suitable for good building envelope [2]. Asan numerically investigated and reported that thermo-physical properties, insulation thickness and its position of the building materials play an important role in time lag and decrement factor [3, 4]. Ulgen reported that experimental findings were contrasted from analytically calculated values of time lag and decrement factor [5]. In this study, the admittance method was used to calculate the unsteady thermal characteristic of the structural lightweight concrete walls.

2 Literature Review

Normal gravel concrete uses gravel as aggregate which gives strength as well as weight to concrete. Structural lightweight aggregate concrete uses lightweight coarse aggregate (LECA) which provides low density and better thermal insulation. Stallite is the lightweight aggregate that is produced from gray striated rock which has parent materials as meta-argiliite. Lytag is the lightweight aggregate which is produced from pyro-processing of fly ash. It has a good absorption capacity and it gives the low density and good resistance to impact loading. Leca and argex are the lightweight aggregates which are produced by clay pellet exposed to high temperature in the rotary kiln so that outer surface melts and sintered lightweight, but argex gives better thermo-physical properties than leca. Demirboga studied the effect of silica fume as cement replacement (10, 20 and 30%) in the mortar which showed a reduction in the thermal conductivity as 17, 31, and 40% [6]. Xu and Chung studied the inclusion of 15% of silica fume in the mortar. They revealed that the silica fumes decrease the thermal conductivity by 6%, increase the specific heat by 7% and increase the density by 8% [7]. Wegian reported that the usage of lightweight coarse aggregate in concrete leads to a reduction in the dead weight, cost-saving, tensile and compressive strength [8]. Real et al. accessed and reported the heating and cooling energy saving (5 and 3%) with the use of structural lightweight aggregate concrete (SLWAC) than conventional concrete [9]. Yun et al. experimented with the addition of lightweight aggregate like stalite, argex and asanolite to concrete. Stalite gave better thermal insulating properties compared to other aggregates [10]. Liu et al. experimented with the lightweight aggregate of oil palm shell foamed geopolymer concrete (OPSFGC) and reported that reduction in the thermal conductivity from 22 to 48% than conventional concrete [11]. Oktay et al. experimentally investigated with addition lightweight aggregate to the concrete and reported that reduction in the compressive strength, density and thermal conductivity and also they reported the addition of silica fume to the concrete gave lower density, lower thermal conductivity and higher specific heat than conventional concrete [12]. Shafigh et al. reported that the concrete with low thermal conductivity and high specific heat give better insulation for energy conservation [13]. Asadi et al. reviewed that increasing 1% of porosity of the concrete contributes to a 0.6% reduction in thermal conductivity [14]. Saboor and Babu studied the effect of air thickness in the concrete and reported that 6.23% of increment in time lag for 0.02 m air gap thickness [15]. Balaji et al. numerically investigated the thermal conductivity, unsteady thermal characteristics of the building materials. They have compared the thermal performance of alternate and conventional building materials [16].

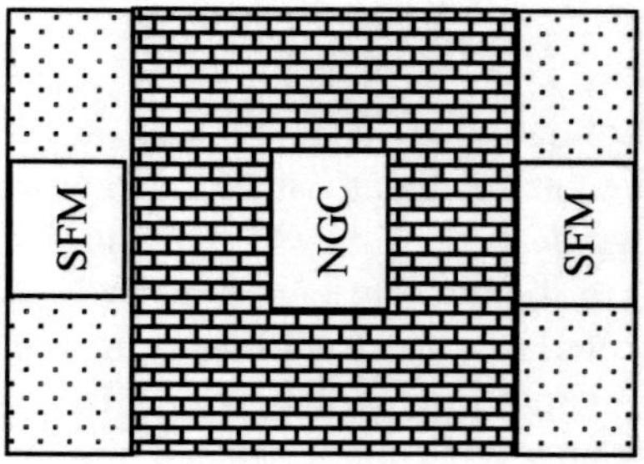

Fig. 1 Cross-section of concrete wall

Table 1 Thermo-physical properties of concrete and mortar

SI no	Building materials	Thermal conductivity (k) (W/m K)	Specific heat (C_p) (J/kg K)	Density (ρ) (kg/m^3)
1	NGC	1.98	741	2248
2	SC	1.21	932	1811
3	LC	1.14	951	1739
4	LEC	1.06	945	1659
5	AC	0.94	1000	1541
6	CP	0.58	642	2040
7	SFCP	0.54	705	2200

Note Thermo-physical properties of building materials are taken from Refs. [7, 9]

3 Materials and Methods

3.1 Configuration of Concrete Wall

The most generally normal gravel concrete is surrounded by mortar. For structural lightweight aggregate concrete is used for the thermal insulating purpose. Figure 1 represents a cross-section of the structural lightweight aggregate concrete wall. Table 1 presents the thermo-physical properties of the concrete and mortar. It is observed that mortar with silica fume and of structural lightweight aggregate concrete show better thermal insulating properties than other mortar and normal weight concrete. Table 2 shows the unsteady thermal characteristic of mortar, mortar with silica fume and structural lightweight aggregate concrete wall.

3.2 Methodology

This admittance method is used to solve transient thermal characteristics of the concrete wall and mortar like transmittance admittance, time lag, decrement factor and thermal heat capacity.

Table 2 Unsteady thermal characteristics of concrete and mortar

SI no.	Code	Wall configuration	U (W/m^2 K)	μ	$\varphi(h)$	W (W/m^2 K)	ξ (kJ/m^2 K)
1	M	0.03CP	4.51	0.994	0.49	4.55	11.61
2	SFM	0.03SFCP	4.43	0.991	0.60	4.49	13.87
3	NGCW	0.03SFM + 0.15NGC + 0.03SFM	2.8	0.529	5.38	4.63	116.57
4	SCW	0.03SFM + 0.15SC + 0.03SFM	2.47	0.496	6.05	4.5	110.58
5	LCW	0.03SFM + 0.15LC + 0.03SFM	2.42	0.498	6.09	4.47	109.43
6	LECW	0.03SFM + 0.15LEC + 0.03SFM	2.37	0.51	6.06	4.42	107.6
7	ACW	0.03SFM + 0.15AC + 0.03SFM	2.27	0.505	6.22	4.38	105.58

$$\frac{\partial^2 \varphi}{\partial X^2} = \frac{1}{\alpha}\frac{\partial \varphi}{\partial \tau} \tag{1}$$

$$T = [A\sinh(\alpha x + i\alpha x) + B\cosh(\alpha x + i\alpha x)]\exp(i2\pi\tau/p) \tag{2}$$

where $\alpha = \sqrt{\rho c_\text{p}/kp}$, x—finite thickness of the slab, p—period.

$$\begin{bmatrix} \varphi_\text{i} \\ q_\text{i} \end{bmatrix} = \begin{bmatrix} \cos h(k+ik) & (\sin h(k+ik))/y \\ y\sin h(k+ik) & \cos h(k+ik) \end{bmatrix} \begin{bmatrix} \varphi_O \\ q_O \end{bmatrix} \tag{3}$$

where cyclic thickness $k = \alpha x$, φ—cyclic temperature, q—cyclic heat flux.

Characteristic admittance of slab (y) = $\sqrt{j}\,2\pi k\rho c_\text{p}/p$

$$\begin{bmatrix} c_1 & c_2 \\ c_3 & c_1 \end{bmatrix} = \begin{bmatrix} a+ib & (c+id)/y \\ y(-d+ic) & a+ib \end{bmatrix} \tag{4}$$

where $a = \cosh k\cos k$, $b = \sinh k\sin k$, $c = (\cos hk\sin k + \sinh k\cos k)\,(1/\sqrt{2})$, $d = (\cos hk\sin k - \sin hk\cos k)\,(1/\sqrt{2})$.

Matrix for internal and external surface resistance is given by

$$H_\text{si} = \begin{bmatrix} 1 & 1/h_i \\ 0 & 1 \end{bmatrix} \quad H_\text{so} = \begin{bmatrix} 1 & 1/h_o \\ 0 & 1 \end{bmatrix} \tag{5}$$

Transmission matrix for multilayer walls with convection resistance is given by

$$\begin{bmatrix} \varphi_i \\ q_i \end{bmatrix} = \begin{bmatrix} 1 & 1/h_i \\ 0 & 1 \end{bmatrix} \begin{bmatrix} e_1 & e_2 \\ e_3 & e_1 \end{bmatrix} \begin{bmatrix} f_1 & f_2 \\ f_3 & f_1 \end{bmatrix} \begin{bmatrix} 1 & 1/h_o \\ 0 & 1 \end{bmatrix} \begin{bmatrix} \varphi_e \\ q_e \end{bmatrix} \tag{6}$$

where *e* and *f* represent individual building materials.

$$\begin{bmatrix} \varphi_i \\ q_i \end{bmatrix} = \begin{bmatrix} g_1 & g_2 \\ g_3 & g_1 \end{bmatrix} \begin{bmatrix} \varphi_e \\ q_e \end{bmatrix} \tag{7}$$

Thermal transmittance (U) is the addition of thermal resistance of the concrete and mortar with silica fume.

$$U = \frac{1}{R_i + \frac{l_1}{\kappa_1} + \frac{l_2}{\kappa_2} + R_o}. \tag{8}$$

Thermal admittance (W) is the amount of energy reaching the inner surface of the building materials per unit temperature swing.

$$w_i = \frac{g_1}{g_2}$$

$$W = |w_i| \tag{9}$$

Decrement factor (δ) is the ability of concrete control the magnitude of temperature from outside to inside surface of the building materials.

$$\delta = 1/\mathrm{ug}_2 \tag{10}$$

Time lag (τ) is the time difference between maximum outside temperature and an inside temperature of the building materials.

$$\tau = \frac{12}{\pi} \arctan\left[\frac{\mathrm{im}\left(-\frac{1}{\mathrm{ug}_2}\right)}{\mathrm{rea}\left(-\frac{1}{\mathrm{ug}_2}\right)}\right] \tag{11}$$

Thermal capacity (ξ) is the amount of energy stored in the building materials per unit temperature swing during the first cycle of the heat flow.

$$\xi = \frac{t}{2}\left[\frac{g_1 - 1}{g_2}\right] \tag{12}$$

Equations (8)–(12) were used to find unsteady thermal characteristics of concrete wall and mortar. The inner and outer surface heat transfer coefficients of the concrete wall are 25 and 7.7 W/m^2 K [17].

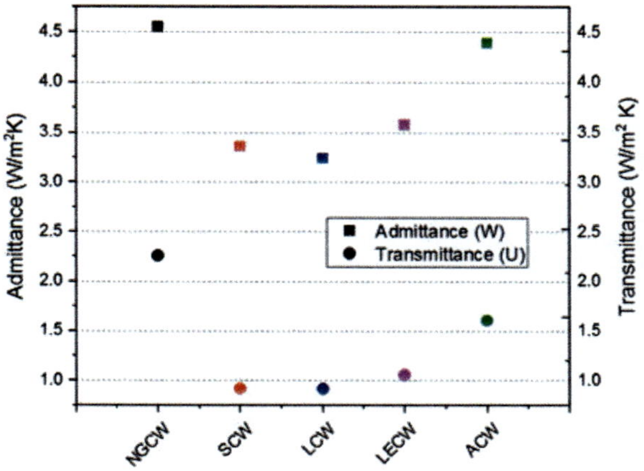

Fig. 2 Transmittance and admittance of concrete wall

4 Results and Discussion

4.1 Transmittance and Admittance of Concrete Wall

The admittance value defines the thermal storage of the wall. Transmittance value defines the thermal resistance of the wall. High admittance value and low transmittance give good thermal comfort [16, 18, 19]. Table 2 shows mortar with silica fume that gives lower transmittance (4.43 W/m^2 K) than mortar. Mortar gives higher admittance (4.54 W/m^2 K) than mortar with silica fume and Fig. 2 shows the transmittance and admittance of the concrete wall. It is observed that among all other configuration, argex concrete wall (ACW) (2.27 W/m^2 K) shows less transmittance than normal gravel concrete wall (NGCW), stalite concrete wall (SCW), lytag concrete wall (LCW) and leca concrete wall (LECW). It is observed that among all other configurations, NGCW (4.63 W/m^{2}K) gives higher admittance than SCW, LCW, LECW and ACW.

4.2 Decrement Factor of Concrete Wall

From Table 2, it shows mortar with silica fume that gives lower decrement factor than mortar. Figure 3 shows the decrement factor of the concrete wall. It is observed that among all other configurations, SCW (0.4955) shows lower decrement factor than NGCW, LCW, LECW and ACW.

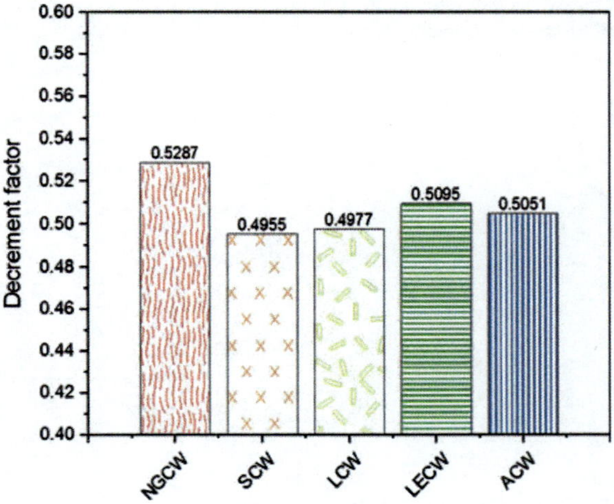

Fig. 3 Decrement factor of the concrete walls

4.3 *Decrement Factor of Concrete Over Various Wall Thicknesses*

Figure 4 shows the decrement factor of concrete walls over various wall thicknesses. It is observed that as the thickness of the homogenous concrete walls increases the decrement factor values reduce.

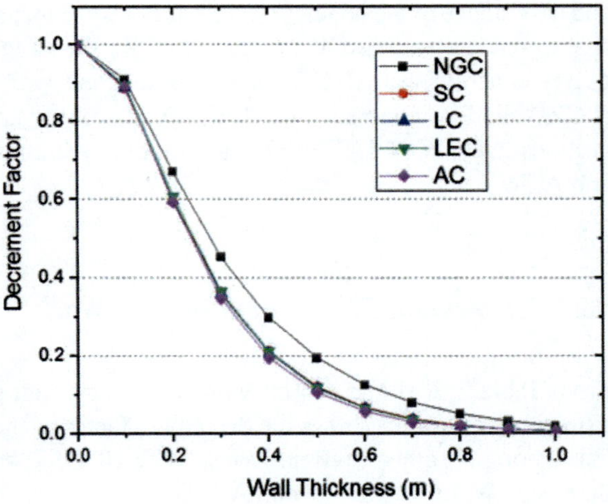

Fig. 4 Decrement factor of concrete over various wall thicknesses

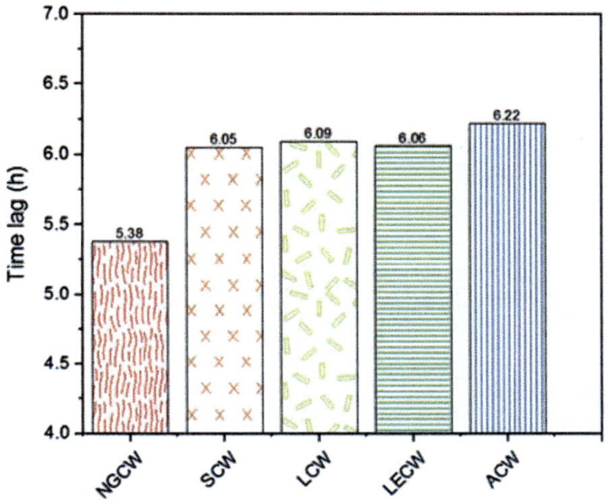

Fig. 5 Time lag of concrete walls

4.4 Time Lag of Concrete Wall

Table 2 shows mortar with silica fume (0.6 h) gives higher time lag than mortar. Figure 5 shows the time lag of the concrete wall. It is observed that among all other configurations, ACW (6.22 h) shows a higher time lag than NGCW, SCW, LCW and LECW.

4.5 Time Lag of Concrete Over Various Wall Thicknesses

Figure 6 shows the time lag of homogenous concrete over various wall thicknesses. All the homogenous concrete walls show an increase in the time lag as the increase in the thickness of homogenous concrete.

4.6 Thermal Capacity of Homogenous Concrete Over the Various Thicknesses

Table 2 shows mortar with silica fume that gives high thermal capacity (13.87 kJ/m^2 K) than mortar. Figure 7 shows the maximum thermal capacity of concrete. It is observed that the 0.3 m thickness is the optimum thickness to have the maximum thermal capacity for all the concrete walls studied. The NGCW (85.33 kJ/m^2 K) gives maximum thermal capacity than SCW, LCW, LECW and ACW. After 0.3 m thickness of the wall, there is a decrease in the thermal capacity of homogenous concrete as depicted in Fig. 7.

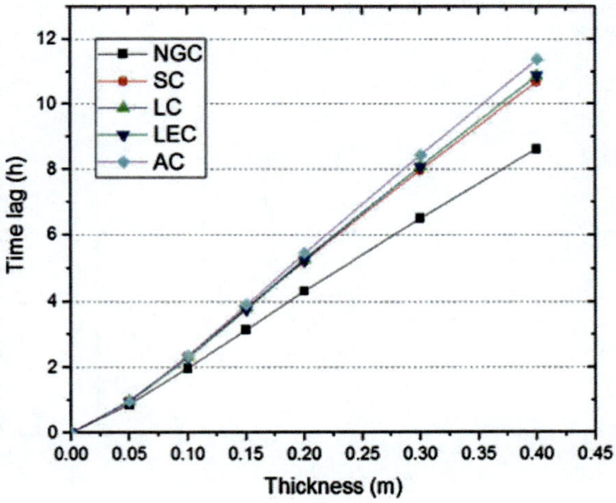

Fig. 6 Time lag of homogenous concrete walls

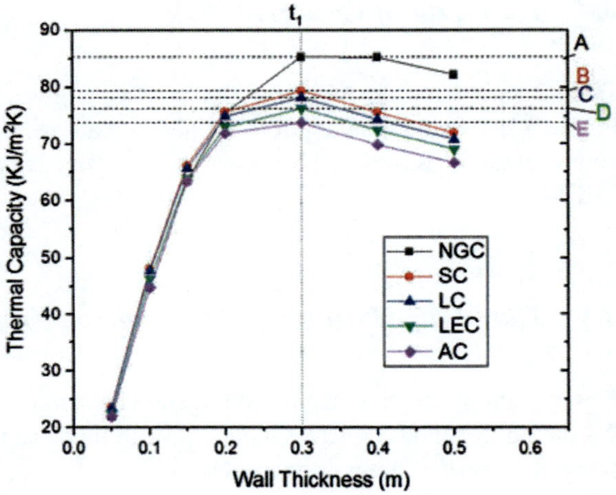

Fig. 7 Thermal capacity of homogenous concrete over various thicknesses

5 Conclusion

The admittance method was used to explore the unsteady thermal characteristics of various concrete walls.

- From Table 3, it is observed that the mortar with silica fume showed a lower transmittance, a lower decrement factor, a higher time lag and a higher thermal capacity than mortar.
- From Figs. 3 and 5, it is observed that the argex concrete wall (ACW) showed the lowest decrement factor and the highest time lag values than other concrete walls, and therefore ACW is observed to be energy efficient than other concrete walls studied.

- From Figs. 4 and 5, it is observed that the increase in the thickness of the homogenous concrete showed a reduction in the decrement factor and a rise in the time lag.

References

1. GRIHA (2010) Green rating for integrated habitat assessment GRIHA manual. 1:129
2. Duffin RJ (1984) A passive wall design to minimize building temperature swings. Sol Energy 33:337–342
3. Asan H (2000) Investigation of wall's optimum insulation position from maximum time lag and minimum decrement factor point of view. Energy Build 32:197–203. https://doi.org/10.1016/S0378-7788(00)00044-X
4. Asan H (2006) Numerical computation of time lags and decrement factors for different building materials. Build Environ 41:615–620. https://doi.org/10.1016/j.buildenv.2005.02.020
5. Ulgen K (2002) Experimental and theoretical investigation of effects of wall's thermophysical properties on time lag and decrement factor. Energy Build 34:273–278. https://doi.org/10.1016/S0378-7788(01)00087-1
6. Demirboğa R (2003) Influence of mineral admixtures on thermal conductivity and compressive strength of mortar. Energy Build 35:189–192. https://doi.org/10.1016/S0378-7788(02)00052-X
7. Xu Y, Chung DDL (2000) Effect of sand addition on the specific heat and thermal conductivity of cement. Cem Concr Res 30:59–61. https://doi.org/10.1016/S0008-8846(99)00206-9
8. Wegian FM (2012) Strength properties of lightweight concrete made with LECA grading. Aust J Civ Eng 10:11–22. https://doi.org/10.7158/C10-668.2012.10.1
9. Real S, Gomes MG, Moret Rodrigues A, Bogas JA (2016) Contribution of structural lightweight aggregate concrete to the reduction of thermal bridging effect in buildings. Constr Build Mater 121:460–470. https://doi.org/10.1016/j.conbuildmat.2016.06.018
10. Yun TS, Jeong YJ, Han TS, Youm KS (2013) Evaluation of thermal conductivity for thermally insulated concretes. Energy Build 61:125–132. https://doi.org/10.1016/j.enbuild.2013.01.043
11. Liu MYJ, Alengaram UJ, Jumaat MZ, Mo KH (2014) Evaluation of thermal conductivity, mechanical and transport properties of lightweight aggregate foamed geopolymer concrete. Energy Build 72:238–245. https://doi.org/10.1016/j.enbuild.2013.12.029
12. Oktay H, Yumrutaş R, Akpolat A (2015) Mechanical and thermophysical properties of lightweight aggregate concretes. Constr Build Mater 96:217–225. https://doi.org/10.1016/j.conbuildmat.2015.08.015
13. Shafigh P, Asadi I, Mahyuddin NB (2018) Concrete as a thermal mass material for building applications—a review. J Build Eng 19:14–25. https://doi.org/10.1016/j.jobe.2018.04.021
14. Asadi I, Shafigh P, Abu Hassan ZFB, Mahyuddin NB (2018) Thermal conductivity of concrete—a review. J Build Eng 20:81–93. https://doi.org/10.1016/j.jobe.2018.07.002
15. Saboor S, Ashok Babu TP (2015) Effect of air space thickness within the external walls on the dynamic thermal behaviour of building envelopes for energy efficient building construction. Elsevier B.V. https://doi.org/10.1016/j.egypro.2015.11.564
16. Balaji NC, Mani M, Venkatarama Reddy BV (2019) Dynamic thermal performance of conventional and alternative building wall envelopes. J Build Eng 21:373–395. https://doi.org/10.1016/j.jobe.2018.11.002
17. CIBSE (2006) CIBSE environmental design guide A. The Chartered Institution of Building Services Engineers, London. https://doi.org/10.4324/9781315671796

18. Shaik S, Talanki Puttaranga Setty AB (2016) Influence of ambient air relative humidity and temperature on thermal properties and unsteady thermal response characteristics of laterite wall houses. Build Environ 99:170–183. https://doi.org/10.1016/j.buildenv.2016.01.030
19. Saboor S, Ashok Babu TP (2016) Optimizing the position of insulating materials in flat roofs exposed to sunshine to gain minimum heat into buildings under periodic heat transfer conditions. Environ Sci Pollut Res 23(10):9334–9344. https://doi.org/10.1007/s11356-015-5316-7

Experimental Study of the Temperature Distribution Inside an Indirect Solar Dryer Chamber

Mohammed El Hadi Attia, Zied Driss, Mokhtar Ghodbane, Ahmed Kadhim Hussein, Sachindra Kumar Rout, and Dong Li

Abstract In this work, the temperature distribution in an indirect solar dryer chamber has been studied experimentally, where this chamber operates under the forced pregnancy influence. The experimental work aims to improve the temperature distribution in the solar dryer chamber, where the chamber is divided into two parts by an aluminum-coated polystyrene board, which has eight regularly distributed holes with a diameter of up to 2 cm. The obtained results confirmed that the panel improves the temperature distribution inside the solar dryer. The regularity of heat distribution in the upper part took a short period and this part is the largest. Despite its size compared to the lower part, the top part has the longest duration of heat distribution.

Keywords Solar dryer · Temperature distribution · Thermal insulation · Forced convection

M. E. H. Attia
Laboratory of LABTHOP, Faculty of Exact Sciences, University of El Oued, El Oued 39000, Algeria

Z. Driss
Laboratory of LASEM, ENIS, University of Sfax, BP 1173, Sfax 3038, Tunisia

M. Ghodbane
Department of Mechanical Engineering, University of Blida 1, Blida, Algeria

A. K. Hussein
College of Engineering, Mechanical Engineering Department, University of Babylon, Babylon City, Hilla, Iraq

S. K. Rout (✉)
C.V. Raman Global University, Bhubaneswar 752054, Odisha, India
e-mail: Sachindra106@gmail.com

D. Li
School of Architecture and Civil Engineering, Northeast Petroleum University, Fazhan Lu Street, Daqing 163318, China

M. Ramgopal et al. (eds.), *Advances in Air Conditioning and Refrigeration*, Lecture Notes in Mechanical Engineering, https://doi.org/10.1007/978-981-15-6360-7_28

1 Introduction

Solar energy is one of the clean sources and sustainable renewable energies in the world and has been known to use it in many areas that are indispensable in the lives of everyday communities, such as electricity production [1–3], refrigeration and air conditioning [4, 5], water heating [6–9], cooking [10, 11] and drying products [12]. Solar energy will be exploited through efficient thermal conversion systems called solar collectors, which are divided into two families, namely flat solar collectors and solar collectors with the linear or point concentrations effect [13]. Also, numerous scientific researches have confirmed that nanoparticles can be used to improve the thermal transfer coefficient of the based fluid [14–17].

As mentioned earlier, one of the most important axes of solar energy uses is the drying of various products, where the vegetable, fruit and marine products worldwide suffer from a lack of control over the conservation methods, transport and storage that help maintain the quality of the final product, because moisture causes the destruction of these products after harvest [18]. Also, the drying process is of great importance compared to classical preservation methods such as freezing and canning. In addition, many agricultural products consumed in large quantities are not always available during different seasons, and drying is a convenient and safe solution to this problem. It provides easy transportation and storage and maintains the nutritional benefit of the product (vitamins, proteins, etc.). However, the drying process usually requires a high and expensive heating energy input necessary for evaporation of the water to be removed from the plant or foodstuff to be preserved, but due to the fossil energy cost continues to increase or fluctuate, solar energy can provide us with the amount of heat needed to dry the agricultural product with ease [19].

Drying products through the sun is a very old preservation technique, it is enough to expose the product to be dried to the sun, and then lose its moisture. There are many discreet scientific researches that deal with solar drying, including the work of Kuan et al. who performed a numerical study of a solar dryer assisted by heat pump [20]. Experimentally, Lakshmi et al. conducted a study of the forced convection effect in the mixed mode on the solar dryer performance to dry stevia leaves [21]. In addition, Vijayan et al. conducted an experimental study on an indirect solar dryer to dry bitter gourd slices [22]. Also, Lamrani et al. analyzed the energy and environment of an indirect solar hybrid wood dryer using TRNSYS [23]. It was also found that Mustayen et al. conducted a review study on the performance of many solar dryers and the factors that affect their efficiency [24]. Besides, Fudholi et al. conducted a review work on the use of solar dryers in the agricultural and marine fields [25]. In addition, Fudholi and Sopian conducted a review study on solar drying using a flat air collector [26]. This is a group of discreet scientific works that touched the solar drying of various kinds.

Thus, the drying purpose is to vaporize the liquid content in the solid product in order to keep it at room temperature or to reduce its weight. Therefore, the drying purpose is to remove the liquid that accompanies a solid product under the action of heat. The departure of this liquid is carried out by evaporation, by vaporization or by sublimation. So, drying is a simultaneous transfer operation of matter and heat.

This work addresses the problem of thermal distribution in an indirect thermal solar dryer. Modifications made to improve heat distribution in the drying chamber through the experimental study of the proposed model. Two parts of the chamber were created with an aluminum-coated polystyrene panel with eight holes with 2 cm of diameter and regularly distributed. Therefore, we divided the drying room into the upper and lower part. Will these changes succeed in the regular distribution of heat in the dryer?

2 Experimental Setup

This work aims to manufacture a drying chamber using simple and inexpensive tools to study the temperature distribution in an indirect solar dryer (forced convection). The experimental solar dryer consists of several components, namely.

2.1 Drying Chambers

As shown in Fig. 1a, b, the inner volume of the drying tray is divided into two parts (upper and lower) to improve the thermal distribution in order to obtain a desiccant product with homogeneous properties, where a polystyrene sheet covered with aluminum has been used to divide the interior size of the drying chamber into two parts. The sheet has holes that allow the heat-saturated air stream to pass from the bottom to the top.

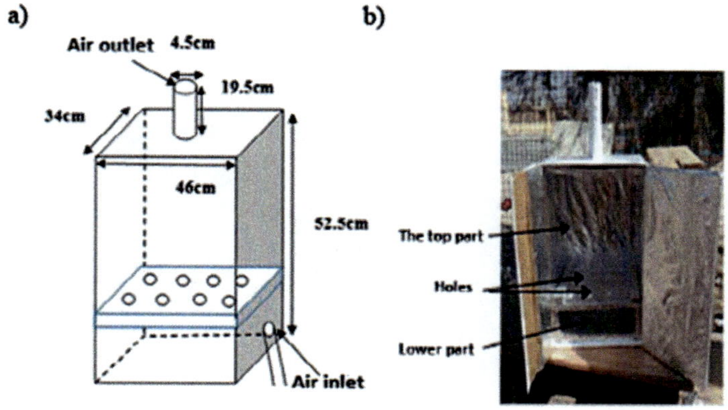

Fig. 1 Drying chamber of the studied system: **a** Illustration scheme for rooms, **b** the dryer rooms

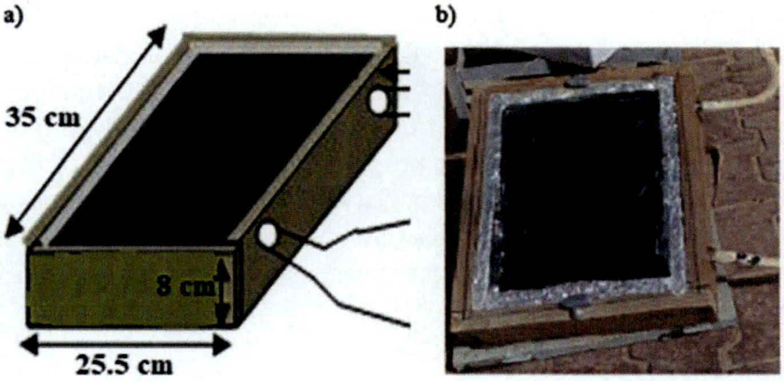

Fig. 2 Flat solar collector: **a** Illustration scheme, **b** Experimental solar collector

2.2 *Flat Solar Collector*

As presented in Fig. 2a, b, the solar collector used is a wooden box of 35 cm × 25.5 cm × 8 cm insulated from all sides with aluminum polystyrene. This box contains inside a special black painted metal plate. The metal plate is topped by a black copper tube of 3 mm thick, where the air propelled by the fan through the copper tube to be heated by the greenhouse effect inside the glass-covered box, since it is the global solar radiation that falls on the solar collector surface in order to move the greenhouse process in the presence of the cover glass [5, 26].

2.3 *Fan Supply Air*

The fan is used in the experiment as shown in Fig. 3 where it operates with a current of 0.13 (A) and a voltage of 9 (V). This fan draws air from the outside air and pushes it toward the inlet of the copper tube inside the box of the flat solar collector.

Fig. 3 Fan used experimentally

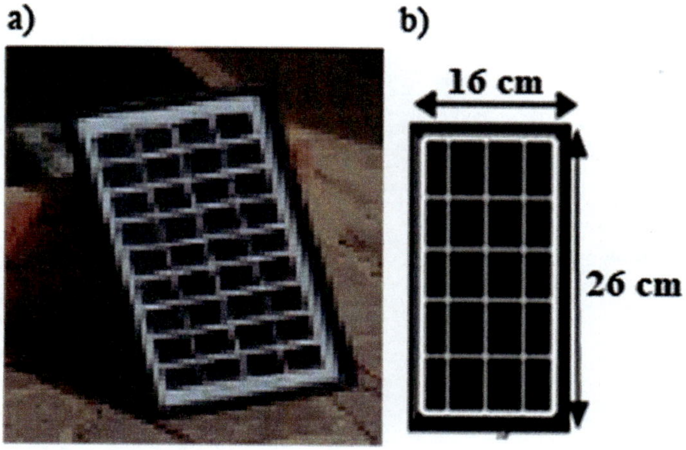

Fig. 4 Solar cell: **a** Experimental solar cell, **b** Illustration scheme for solar cell

The fan is electrically fed to a solar cell with dimensions of 16 cm × 26 cm as shown in Fig. 4a, b.

3 Experimental Study

The solar dryer starts by pulling the ambient air by the fan to the flat solar collector (forced convection) through a tube. Then, it is heated by converting the global solar radiation that reaches the solar collector surface to thermal energy absorbed by the air flowing inside the copper tube and becomes high temperature. The drying chamber is heated by hot air from the solar collector to the lower chamber of the drying chamber by a tube of 9 cm length and 1.5 cm diameter.

The hot air rises through the eight homogeneous holes in the panel from the lower chamber values to the upper chamber section containing the product to be dried. Since there is a temperature difference between the product and the heated air, heat exchange occurs between them, which leads to saturation of the air with humidity and comes out of the drying chamber, while the product is dried.

The experimental work was conducted in the El Oued area (latitude 33°22′06″ N, longitude 6°52′03″ E and altitude 76 m). The weather is hot and dry in summer in this region, while winter is very cold and dry.

As for the experimental work, it was conducted on April 29, 2018 in the Renewable Energies Laboratory at El Oued University. Table 1 shows the climatic data for experiment day.

The experimental installation has been prepared as shown in Fig. 5 at a solar collector inclination angle of 30°.

Table 1 Summary of weather data for April 29, 2018

Sunrise	05:45 am
Sunset	06:47 pm
Ambient air temperature	20–40 °C
Humidity	33%
Wind speed	13 km/h

Fig. 5 Experimental setup

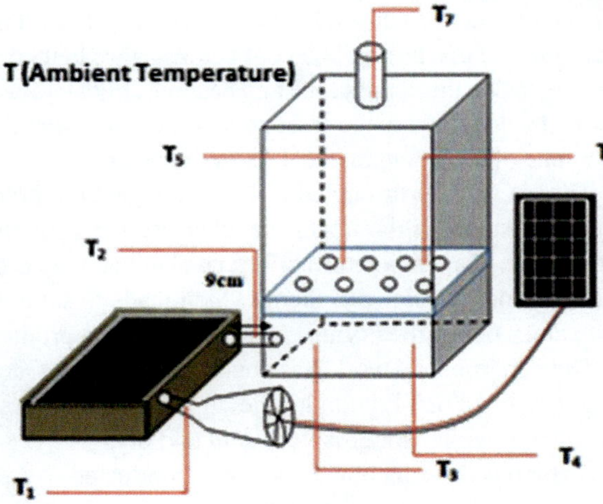

Fig. 6 Scheme of thermocouples installation inside the drying chamber

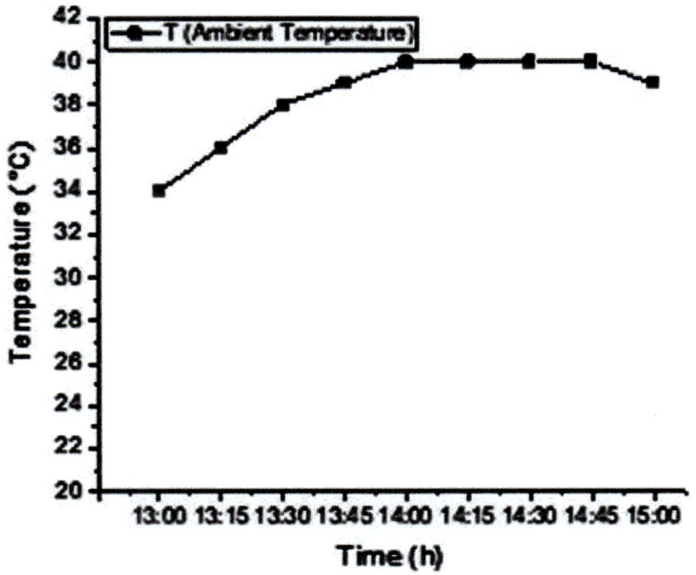

Fig. 7 Change in ambient air temperature

The thermocouples were installed inside the chamber as shown in Fig. 6, and the temperature was measured every 15 min.

4 Results and Discussion

4.1 Ambient Temperature

Figure 7 shows the change in the ambient air temperature from 1 to 3 pm, where the temperature at one was 34 °C and reached about 40 °C on the second, then dropped to 38 °C at 3 pm.

4.2 Solar Collector Temperature

Figure 8 illustrates the change in air temperature at the inlet and the outlet of the solar collector, where a wide difference is observed between the curves the air that enters and the curve of the air that comes out. This difference is due to the amount of heat that the air gains from the solar collector, since the solar collector absorbs a significant amount of solar energy. The global solar radiation is trapped between the

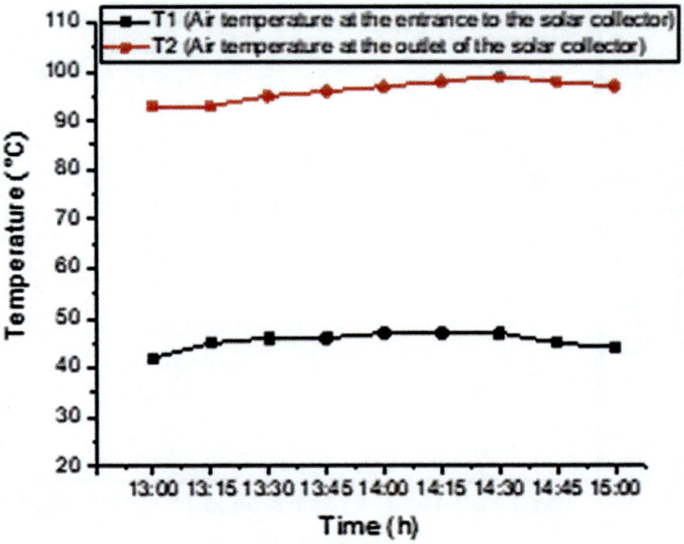

Fig. 8 Air temperature variation at the inlet and the outlet of the solar collector

glass plate that covers the flat solar collector and the copper tube, which is a passage for air propelled by the fan.

4.3 Drying Chamber Temperature

Figure 9 shows the temperature distribution in the bottom part of the dryer room, it is noticeable that there is a very small spacing at the beginning of the experiment between "T3 = 43.5 °C" and "T4 = 42.5 °C". After 75 min, it was observed that the values of T3 are equal to the values of T4, and this indicates the harmonious distribution of heat inside the bottom part of the drying room.

As shown in Fig. 10, the same observations have appeared at the top of the drying chamber, where it was found that the T5 is very close to the T6 in the first 45 min of the experiment. After this period, the two values become equal.

Figure 11 shows the air temperature in the outlet of the drying chamber. From Fig. 9, it has been observed that air temperature varies from 42 at 1 pm to 52 °C at 2:45 pm, and then decreases to 48 °C at 3 pm.

Figure 12 summarizes the temperature distribution within the solar dryer room.
Figure 12 summarizes the temperature distribution within the solar dryer room.

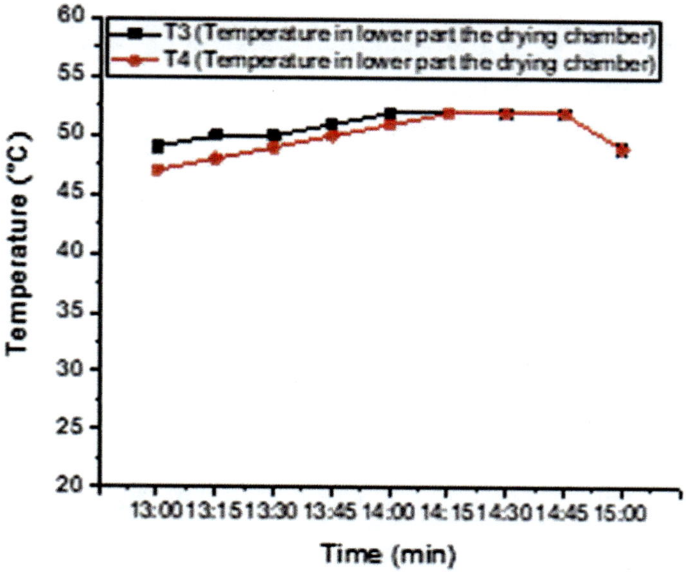

Fig. 9 Temperature distribution at the bottom part of the dryer room

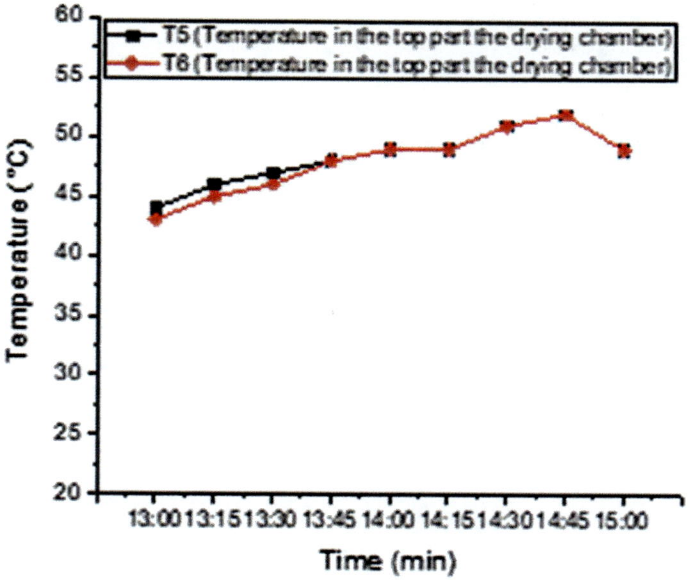

Fig. 10 Temperature distribution at the top part of the dryer room

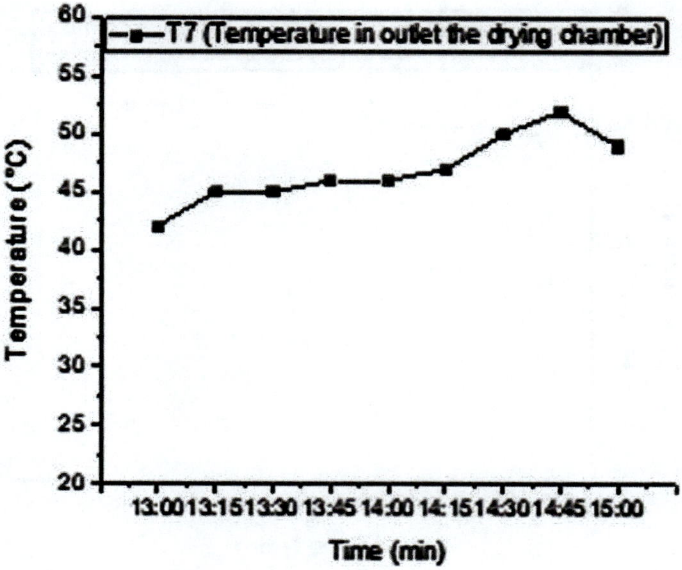

Fig. 11 Air temperature in the outlet of the solar drying chamber

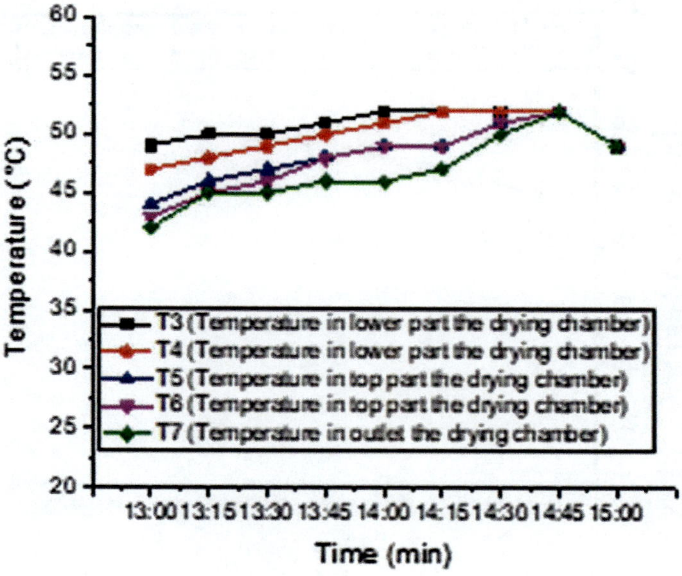

Fig. 12 Temperatures distribution in the solar drying chamber

5 Conclusions

In this experimental work, an indirect solar dryer containing an aluminum-coated polyester plate with eight holes has been studied to improve the distribution of hot air coming from a flat solar collector. The most important results that have been reached are:

- Ambient air temperature is an important factor to get good air temperature distribution;
- The solar collector provided hot air exceeding 90 °C at its outlet;
- The heat distribution at the bottom part of the drying chamber took a long time to homogenize (about 75 min) despite its smallness;
- In a short period, the uniformity of heat distribution is more rapid (about 45 min) in the upper part of the solar drying chamber despite its large size;
- The air temperature at the outlet is slightly lower compared to the rest of the chamber;

Therefore, the results lead us to improve the heat distribution within the solar drying chamber.

In the future work, it will be proposed to increase the number of holes distributed on the surface of the board with the possibility of changing its diameter, lowering or lifting the board for better results.

References

1. Ghodbane M, Boumeddane B, Said Z, Bellos E (2019) A numerical simulation of a linear Fresnel solar reflector directed to produce steam for the power plant. J Clean Prod 231:494–508
2. Chahmi A (2019) Study of photovoltaic systems with differences connecting configuration topologies for applications in renewable energy systems. Int J Energ 4(1):28–36
3. Boukelia TE (2017) Potential assessment of using dry cooling mode in two different solar thermal power plants. Int J Energ 2(2):18–24
4. Ghodbane M, Boumeddane B, Hussein AK (2019) Performance analysis of a solar-driven ejector air conditioning system under El-Oued climatic conditions, Algeria. J Therm Eng
5. Ghodbane M, Boumeddane B, Moummi N, Largot S, Berkane H (2016) Study and numerical simulation of solar system for air heating. J Fundam Appl Sci 8(1):41–60
6. Ghodbane M, Boumeddane B, Said N (2016) A linear Fresnel reflector as a solar system for heating water: theoretical and experimental study. Case Stud Therm Eng 8(C):176–186
7. Ghodbane M, Boumeddane B, Said N (2016) Design and experimental study of a solar system for heating water utilizing a linear Fresnel reflector. J Fundam Appl Sci 8(3):804–825
8. Said Z, Ghodbane M, Hachicha AA, Boumeddane B. Optical performance assessment of a small experimental prototype of linear Fresnel reflector. Case Stud Therm Eng
9. Ghodbane M, Boumeddane B (2017) A parabolic trough solar collector as a solar system for heating water: a study based on numerical simulation. Int J Energ (IJECA) 2(2):29–37
10. Noman M, Wasim A, Ali M, Jahanzaib M, Hussain S, Ali HMK, Ali HM (2019) An investigation of a solar cooker with parabolic trough concentrator. Case Stud Therm Eng 14:100436

11. Barba FJ, Gavahian M, Es I, Zhu Z, Chemat F, Lorenzo JM, Khaneghah AM (2019) Solar radiation as a prospective energy source for green and economic processes in the food industry: from waste biomass valorization to dehydration, cooking, and baking. J Clean Prod 220:1121–1130
12. Raman SVV, Iniyan S, Goic R (2012) A review of solar drying technologies. Renew Sustain Energy Rev 16(5):2652–2670
13. Kalogirou SA (2004) Solar thermal collectors and applications. Prog Energy Combust Sci 30(3):231–295
14. Hussein AK (2015) Applications of nanotechnology in renewable energies—a comprehensive overview and understanding. Renew Sustain Energy Rev 42:460–476
15. Hussein AK (2016) Applications of nanotechnology to improve the performance of solar collectors—recent advances and overview. Renew Sustain Energy Rev 62:767–792
16. Said Z, Abdelkareem MA, Rezk H, Nassef AM (2019) Fuzzy modeling and optimization for experimental thermophysical properties of water and ethylene glycol mixture for Al_2O_3 and TiO_2 based nanofluids. Powder Technol 353:345–358
17. Said Z, Assad MEH, Hachicha AA, Bellos E, Abdelkareem MA, Alazaizeh DZ, Yousef BA (2019) Enhancing the performance of automotive radiators using nanofluids. Renew Sustain Energy Rev 112:183–194
18. Kumar M, Sansaniwal SK, Khatak P (2016) Progress in solar dryers for drying various commodities. Renew Sustain Energy Rev 55:346–360
19. Atalay H (2019) Comparative assessment of solar and heat pump dryers with regards to exergy and exergoeconomic performance. Energy 189:116180
20. Kuan M, Shakir Y, Mohanraj M, Belyayev Y, Jayaraj S, Kaltayev A (2019) Numerical simulation of a heat pump assisted solar dryer for continental climates. Renew Energy 143:214–225
21. Lakshmi DVN, Muthukumar P, Layek A, Nayak PK (2019) Performance analyses of mixed mode forced convection solar dryer for drying of stevia leaves. Sol Energy 188:507–518
22. Vijayan S, Arjunan TV, Kumar A (2020) Exergo-environmental analysis of an indirect forced convection solar dryer for drying bitter gourd slices. Renew Energy 146:2210–2223
23. Lamrani B, Khouya A, Draoui A (2019) Energy and environmental analysis of an indirect hybrid solar dryer of wood using TRNSYS software. Sol Energy 183:132–145
24. Mustayen AGMB, Mekhilef S, Saidur R (2014) Performance study of different solar dryers: a review. Renew Sustain Energy Rev 34:463–470
25. Fudholi A, Sopian K, Ruslan MH, Alghoul MA, Sulaiman MY (2010) Review of solar dryers for agricultural and marine products. Renew Sustain Energy Rev 14(1):1–30
26. Fudholi A, Sopian K (2019) A review of solar air flat plate collector for drying application. Renew Sustain Energy Rev 102:333–345

Analytical Computation of Thermodynamic Performance of Various New Eco-friendly Alternative Refrigerants Applicable for Air Conditioners

Sharmas Vali Shaik, T. P. Ashok Babu, Debasish Mahapatra, Saboor Shaik, Kiran Kumar Gorantla, and V. Sai Siva Subramanyam

Abstract The objective of the present investigation is to do the theoretical thermodynamic analysis of various new eco-friendly R22 substitutes used in vapour compression refrigeration (VCR) cycle. In this work, nine mixture refrigerants were considered at different compositions. Thermodynamic properties of all the considered refrigerants were developed and the same properties were used in the performance analysis of alternative refrigerants. Standard VCR cycle was considered for the thermodynamic assessment of alternative refrigerants. The working conditions considered are expressed as $T_e = 7.2$ °C, $T_k = 54.4$ °C, $\Delta T_{sup} = 11.1$ °C and $\Delta T_{sub} = 8.3$ °C, respectively. Results revealed that the COP of mixture refrigerant MR20 (R600a/R134a/R1270 5/47.5/47.5 in mass%) was 2.02% higher than the COP of R22 and other nine investigated refrigerants. Discharge temperature of compressor obtained for MR20 was 11.79 °C lower compared to that of R22. Compressor power obtained for MR20 was 1.96% lower than that of R22. Volumetric refrigeration capacity obtained for MR20 was relatively closer to that of R22. GWP_{100} of MR20 (619) was lower compared to the GWP_{100} of R22 (1760). Overall, the performance of mixture refrigerant MR20 was better compared to all the considered R22 alternatives, and therefore it might be an appropriate candidate to replace R22 used in air conditioners.

Keywords Alternative refrigerants · COP · GWP · Discharge temperature · Volumetric capacity

---

S. V. Shaik (✉) · T. P. Ashok Babu · D. Mahapatra
Department of Mechanical Engineering, National Institute of Technology Karnataka, Surathkal, Mangalore 575025, India
e-mail: sharmasvali.nitk@gmail.com

S. Shaik
School of Mechanical Engineering, Vellore Institute of Technology, Vellore 632014, Tamil Nadu, India

K. K. Gorantla · V. Sai Siva Subramanyam
Department of Mechanical Engineering, Sasi Institute of Technology & Engineering, Tadepalligudem 534101, Andhra Pradesh, India

M. Ramgopal et al. (eds.), *Advances in Air Conditioning and Refrigeration*, Lecture Notes in Mechanical Engineering, https://doi.org/10.1007/978-981-15-6360-7_29

1 Introduction

Refrigerant R22 was generally used in air conditioners and in heat pumps due to its favourable characteristics [1]. As per Montreal Protocol hydrochlorofluorocarbon (HCFC) refrigerant R22 is going to phase out by the year 2030, since all the HCFC refrigerants contain ozone depleting substance called chlorine [2, 3]. In this context, many countries are spending much effort to develop their own R22 alternatives used in air conditioning and heat hump industries. From past several years, various performance studies were done on R22 alternatives. Refrigerants R410A and R407C were considered as foremost replacements for R22 [4]. Experimental study revealed that, the flammability nature of hydrocarbon refrigerants could be reduced by blending the hydrocarbons with hydrofluorocarbon (HFC) refrigerants [5]. Experimental investigation was performed on a heat pump with mixture refrigerant R290/R134a (45/55 by mass percentage) and R290 as R22 alternatives [6]. Results exhibited that, the above mixture performed marginally lower compared with R290 and R22, and also the above mixture was less flammable when compared to R290. Experimental study was done with R424A and R417A as R22 replacements [7]. Results revealed that R424A would be a suitable replacement to R22 when compared to R417A. Theoretical analysis of several refrigerant mixtures showed that R444B refrigerant would be a suitable retrofit candidate to R22 [8]. Experimental investigation was done on split air conditioner working with several HFCs such as R424A, R422D, R422A and R417A as R22 alternatives [9]. Results showed that all the considered HFCs performed lower when compared to R22. Experimental study with refrigerant R290 was conducted as substitute to R134a used in vending machine [10]. Results shown that R290 performed better compared to R134a. Theoretical thermodynamic analysis was done with R419A and R410A as substitutes to R22 [11]. Results showed that, the COP of R410A was very nearer to that of R22, whereas COP of R419A was 13.78% lower compared to R22. The main drawback of refrigerants such as R407C, R410A R417A, R419A, R422A, R422D and R424A was higher GWP compared to R22. To overcome this problem, natural refrigerants like hydrocarbons could be used. The main limitation of hydrocarbons was flammability in nature. Therefore, in this study, an attempt is made to establish the new R22 alternatives which will have zero ODP, low GWP and less flammability.

2 Materials and Methods

2.1 Establishment of Eco-friendly R22 Alternatives

In this work, nine mixture refrigerants consist of R134a, RE170, R1270, R600a, R32, R125 and R290 were developed at various compositions. Designation of various considered refrigerants and their properties are presented in Table 1.

Table 1 Properties of several alternative refrigerants

Refrigerants	Composition (in mass%)	MW (kg/kmol)	P_c (MPa)	T_c (K)	BP (°C)	T_{bub} (°C)	T_{dew} (°C)	T_{glide} (°C)	ODP	GWP (100 years)	ASHRAE Safety class
R22	Pure refrigerant	86.46	4.99	369.3	−40.81	–	–	–	0.055	1760	A1
MR10 (R600a/R134a/R290)	5/47.5/47.5	61.39	4.118	359.6	–	−45.28	−42.56	2.72	0	619	A2[a]
MR20 (R600a/R134a/R1270)	5/47.5/47.5	59.51	4.348	361.5	–	−45.96	−42.83	3.13	0	619	A2[a]
MR30 (RE170/R134a/R32)	5/77.5/17.5	83.02	4.766	372.5	–	−36.60	−29.90	6.7	0	1126	A2[a]
MR40 (RE170/R134a/R32)	10/67.5/22.5	76.27	5.058	374.7	–	−37.88	−30.28	7.6	0	1030	A2[a]
MR50 (RE170/R134a/R32)	15/72.5/12.5	78.34	4.885	380.4	–	−32.84	−27.23	5.61	0	1027	A2[a]
MR60 (RE170/R134a/R32)	25/54.5/20.5	67.98	5.352	383.2	–	−35.28	−28.33	6.95	0	848	A2[a]
R407C (R32/R125/R134a)	23/25/52	86.20	4.639	359.2	–	−43.63	– 36.63	7.0	0	1624	A1
R410A (R32/R125)	50/50	72.58	4.901	344.4	–	−51.44	−51.36	0.08	0	1924	A1
R419A (R125/R134a/RE170)	77/19/4	109.34	4.060	356.6	–	−42.64	−35.85	6.79	0	2688	A1

[a] Computed RF number of refrigerants

Temperature glide (T_{glide}) of refrigerant mixtures as presented in Table 1 is computed as the difference between dew point temperature and bubble point temperature of a given mixture at atmospheric pressure (0.101325 MPa). Therefore, $T_{glide} = (T_{dew} - T_{bub})$ at atmospheric pressure (0.101325 MPa).

From Table 1, it is evident that mixture refrigerants (MR10 to MR60) are classified into mildly flammable group (A2), since their refrigerant flammability number is below 30 kJ/g. Safety class of refrigerants (MR10–MR60) are not available in literature, and hence RF number correlation is used in this work and it is expressed as [12]:

$$\text{RF number} = \left[\left(\frac{\text{UFL}}{\text{LFL}}\right)^{0.5} - 1\right]\frac{Q}{M} \tag{1}$$

where M is molecular mass (kg/kmol); Q is combustion heat (kJ/mol); LFL is lower limit of flammability (by vol.%); UFL is upper limit of flammability (by vol.%).

ASHRAE safety class of several fluids were taken from related literature [12, 13]. Thermodynamic characteristics of refrigerants are required to do the performance computation of various developed R22 alternative refrigerants. Therefore, in this research work, Martin-Hou equation of state and several correlations were taken from literature [14–16]. The methodology for the establishment of properties of various considered refrigerants was derived from related literature. The equation of state and correlations used in this work are presented below [14–16].

Wagner vapour pressure equation is expressed as

$$\ln\left(\frac{P_{\text{sat}}}{P_{\text{c}}}\right) = \left(\frac{1}{1-x}\right)\left[Ax + Bx^{1.5} + Cx^{2.5} + Dx^{5}\right] \tag{2}$$

Liquid density correlation is expressed as

$$\rho_{\text{r}} = \frac{\rho}{\rho_{\text{c}}} = 1 + 0.85(1 - T_{\text{r}}) + (1.6916 + 0.984\omega)(1 - T_{\text{r}})^{1/3} \tag{3}$$

Martin-Hou equation of state is expressed as

$$P = \frac{RT}{V-b} + \frac{A_2 + B_2T + C_2\mathrm{e}^{\frac{-5.475T}{T_C}}}{(V-b)^2} + \frac{A_3 + B_3T + C_3\mathrm{e}^{\frac{-5.475T}{T_C}}}{(V-b)^3} + \frac{A_4}{(V-b)^4} + \frac{B_5T}{(V-b)^5} \tag{4}$$

Clausius–Clapeyron equation is given as

$$\frac{\mathrm{d}P_{\text{sat}}}{\mathrm{d}T} = \frac{h_{\text{fg}}}{T \times V_{\text{fg}}} \tag{5}$$

where $V_{\text{fg}} = V_{\text{g}} - V_f$.

Refrigerants computed thermodynamic properties have been compared with experimental properties available in the literature [17]. The deviation obtained for the computed properties of alternative refrigerants is within 2.15% compared with ASHRAE [17]. Hence, the methodology used can be considered as reliable and could be applied for establishing the properties of other fluids. For example, the deviations in computed thermodynamic characteristics of refrigerant R22 are presented in Table 2.

Table 2 Validation of R22 calculated properties with ASHRAE at AHRI conditions ($T_k = 54.4$ °C and $T_e = 7.2$ °C)

Thermodynamic characteristics	Properties of R22 from ASHRAE at 7.2 °C	Computed properties for R22 at 7.2 °C	Deviation (%)	Properties of R22 from ASHRAE at 54.4 °C	Computed properties for R22 at 54.4 °C	Deviation (%)
P_{sat} (MPa)	0.62843	0.63084	−0.38	2.15364	2.15621	−0.12
h_g (kJ/kg)	407.66	407.35	0.07	417.63	426.22	−2.05
h_f (kJ/kg)	208.69	208.61	0.03	269.67	274.03	−1.61
S_g (kJ/kg K)	1.7403	1.7396	0.04	1.6787	1.7018	−1.37
S_f (kJ/kg K)	1.03107	1.03072	0.03	1.2271	1.2460	−1.54
V_g (m^3/kg)	0.03757	0.03747	0.26	0.01033	0.01049	−1.54
ρ_f (kg/m^3)	1256.07	1252.79	0.26	1059.46	1056.95	0.23

3 Methodology

3.1 Thermodynamic Analysis of Air Conditioner with Alternative Refrigerants

The main purpose of thermodynamic analysis is to find suitable replacement to R22. Standard VCR cycle was most widely used while performing thermodynamic analysis of several refrigerants [8, 18–21]. Figure 1 shows the standard VCR cycle on pressure-enthalpy (P-h) chart. Several processes occur in the standard VCR cycle are superheating (1-1″), isentropic compression (1″-2″), constant pressure condensation (2″-3″), subcooling (3-3″), isenthalpic expansion (3″-4″) and constant pressure evaporation (4″-1″). In standard VCR cycle, influence of superheating and subcooling on the system performance was taken into account, and hence this work emphasis on thermodynamic analysis of several ecological R22 alternatives by considering the standard VCR cycle. In this cycle, pressure drops and heat losses occurred in various devices of the system are ignored for the simplicity of calculations.

The operating conditions of air conditioner are expressed as evaporator temperature of $T_e = 7.2$ °C, condenser temperature of $T_k = 54.4$ °C, degree of superheating (ΔT_{sup}) as 11.1 °C and degree of subcooling (ΔT_{sub}) as 8.3 °C, respectively [22, 23]. In this work, system capacity (Q_c) is considered as 1.5TR (5.275 kW or 211 kJ/min). Results obtained from this analysis for several developed R22 alternatives are presented in Table 3. The formulas used in thermodynamic analysis of refrigerants are specified below.

Refrigerant mass flow rate is calculated as

$$\dot{m} = Q_c/\mathrm{RE} \tag{6}$$

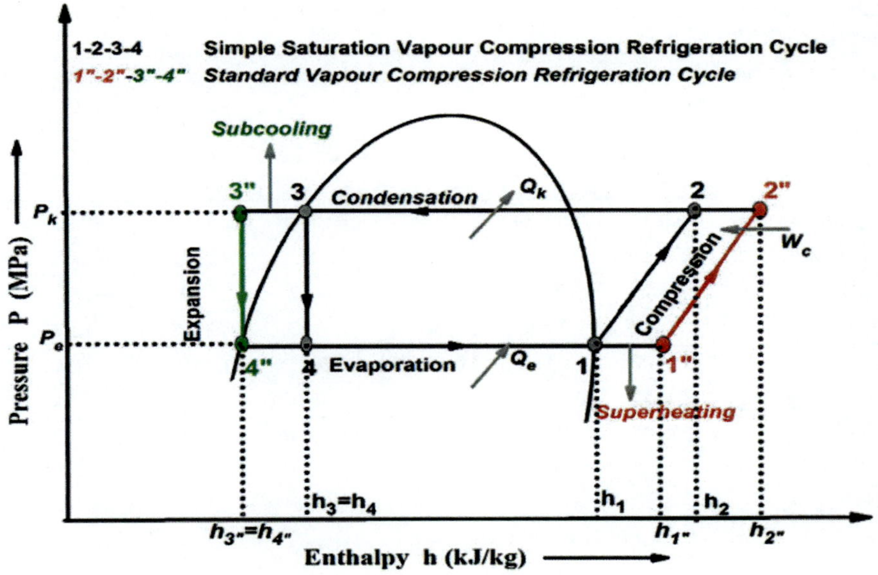

Fig. 1 Representation of standard VCR cycle on P-h chart

Table 3 Performance characteristics of various R22 alternative mixture refrigerants

Refrigerants	$\dot{m}$ (kg/min)	RE (kJ/kg)	W_c (kJ/kg)	COP	T_d (°C)	W_{cp} (kW)	VRC (kJ/m^3)
R22	2.1372	148.086	35.718	4.145	84.26	1.272	3730
MR10	1.4669	215.755	53.087	4.064	71.67	1.297	2820
MR20	1.3587	232.930	55.070	4.229	72.47	1.247	3376
MR30	1.8464	171.408	48.496	3.534	84.07	1.492	2896
MR40	1.6931	186.925	53.688	3.481	86.93	1.515	3006
MR50	1.6770	188.720	50.593	3.730	80.02	1.414	2751
MR60	1.4704	215.245	59.891	3.593	85.11	1.467	2827
R407C	2.1185	149.393	44.172	3.382	83.97	1.559	3334
R410A	1.9786	159.960	37.614	4.252	89.26	1.240	5349
R419A	3.1682	99.898	29.322	3.406	67.35	1.548	3611

Refrigeration effect is calculated as

$$\text{RE} = (h_{1''} - h_{4''}) \tag{7}$$

Isentropic work of compression is calculated as

$$W_c = h_{2''} - h_{1''} \tag{8}$$

Coefficient of performance is calculated as

$$COP = RE/W_c \tag{9}$$

Compressor power is calculated as

$$W_{cp} = \dot{m}(h_{2''} - h_{1''}) \tag{10}$$

Discharge temperature of compressor can be computed as

$$\begin{aligned} S_{2''} &= S_3 + C_{p2} \times \ln(T_{2''}/T_{3''}) = S_{1''} \\ S_{1'''} &= S_1 + C_{p1} \times \ln(T_{1''}/T_1) \end{aligned} \tag{11}$$

Volumetric refrigeration capacity is calculated as

$$VRC = \rho_{1''}(h_{1''} - h_{4''}) \tag{12}$$

4 Results and Discussion

4.1 Coefficient of Performance

The energy performance metric of an air conditioner is indicated by the coefficient of performance (COP). COP depends on both compressor work and refrigeration effect. From Fig. 2, it is noticed that, the COP of mixture refrigerants like MR20 and R410A are 2.02 and 2.58% higher than the R22 and other nine studied fluids. Hence, MR20 and R410A would be suitable replacements to R22 from stand point of COP. Similarly, COP of refrigerants (MR10, MR30, MR40, MR50, MR60, R407C and R419A) is lower in the range of 1.95–18.4% when compared with baseline candidate R22.

4.2 Discharge Temperature of Compressor

An important parameter which indicates the lifetime and stability of compressor is the discharge temperature of compressor. Refrigerant with high discharge temperature causes the burnt out of windings of the compressor motor, and thus it decreases the compressor stability and lifetime. With reference to Fig. 3, it is evident that, the discharge temperature of compressor for several blends like MR10, MR20 and R419A are lower in the range of 11.79–16.91 °C compared to refrigerant R22, and hence these fluids exhibit greater lifetime and stability of compressor motor.

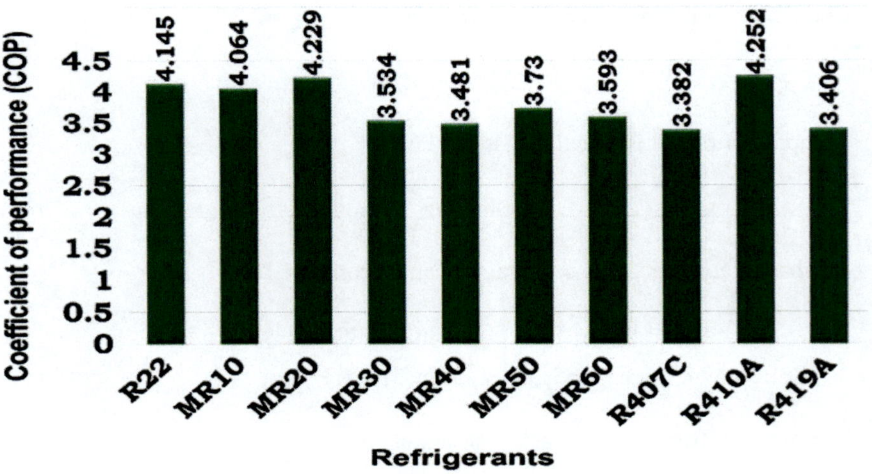

Fig. 2 COP of R22 alternative mixture refrigerants

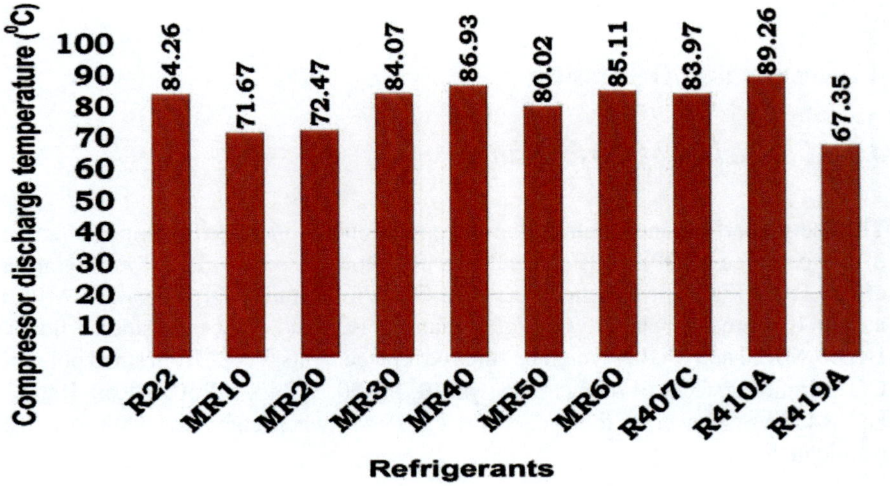

Fig. 3 Discharge temperature of R22 alternative mixture refrigerants

4.3 Compressor Power

Figure 4 depicts the compressor power of various R22 alternatives. From Fig. 4, it is evident that, the compressor power of MR20 and R410A is 1.96 and 2.51% lower than the R22 and other considered refrigerants. This is due to their better energy efficiency and low pressure ratio compared to R22. Therefore, MR20and R410A would be suitable replacements to R22 from the stand point of low compressor

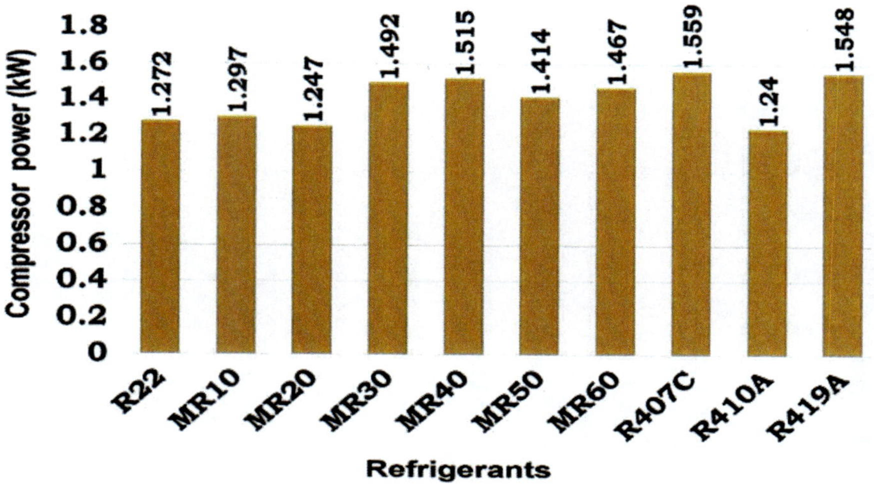

Fig. 4 Compressor power of R22 alternative mixture refrigerants

power. Similarly, compressor power of refrigerants (MR10, MR30, MR40, MR50, MR60, R407C and R419A) is higher in the range of 1.96–22.56% when compared with R22.

4.4 Volumetric Refrigeration Capacity

Volumetric refrigeration capacity (VRC) indicates the size of compressor needed to produce the desired cooling effect. It depends on vapour density at outlet of evaporator as well as on the refrigeration effect. With reference to Fig. 5, it is evident that the refrigerant R410A has highest volumetric capacity among all the considered refrigerants and it is 43.40% higher when compared with R22. Therefore, R410A needs smaller compressor and also it demands the redesign compressor completely which might be costly. Fluids exhibit closer volumetric capacity when compared to R22 that would be taken as suitable replacements to R22. Volumetric capacity of blends like MR20, R407C and R419A are reasonably lower in the range of 3.19–10.61% when compared with R22, and therefore these blends could be taken as suitable R22 substitutes from the view point of volumetric capacity.

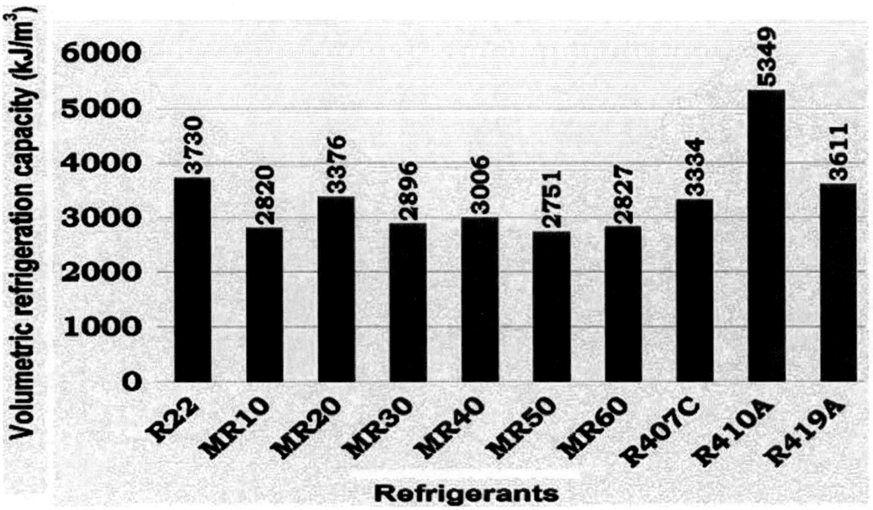

Fig. 5 Volumetric refrigeration capacity of R22 alternative mixture refrigerants

5 Conclusions

The present study deals with the analytical computation of thermodynamic performance of various eco-friendly alternative mixture refrigerants that can be applicable for air conditioners. The conclusions drawn from this study are presented below.

(i) COP of refrigerants such as MR20 and R410A were 2.02 and 2.58% higher than the R22 and other nine considered refrigerants. Compressor discharge temperature of refrigerants like MR10, MR20 and R419A were lower in the range of 11.79–16.91 °C when compared with baseline candidate R22, and hence these refrigerants show better durability and lifetime of compressor motor.

(ii) Compressor power of MR20 and R410A were 1.96 and 2.51% lower than that of R22.

(iii) Volumetric capacity of MR20, R407C and R419A were relatively closer to that of R22, and hence these refrigerants could be used in existing R22 compressor.

(iv) Global warming potential (GWP_{100}) of MR10 (619) and MR20 (619) was lower compared to that of R22 (1760) and the other nine investigated fluids. Mixture refrigerants (MR10 to MR60) were classified into mildly flammable group (A2). Hence, safety measures should be followed while using these refrigerants.

(v) Overall, mixture refrigerant MR20 (R600a/R134a/R12705%/47.5%/47.5% by mass) would be considered as sustainable R22 alternative that can be applicable for air conditioners from the stand point of COP, GWP, ODP, volumetric capacity, compressor discharge temperature and compressor power.

References

1. Mohanraj M, Jayaraj S, Muraleedharan C (2009) Environment friendly alternatives to halogenated refrigerants-A review. Int J Greenh Gas Control 3:108–119
2. United Nations Environmental Programme (1987) Montreal protocol on substances that deplete the ozone layer. Final Act New York, United Nations
3. Powell RL (2002) CFC Phase out; have we met the challenge. J Fluorine Chem 114:237–250
4. Calm JM, Domanski PA (2004) R-22 replacement status. ASHRAE J 46:29–39
5. Zhao Y, Bin L, Haibo Z (2004) Experimental study of the inert effect of R134a and R227ea on explosion limits of the flammable refrigerants. Exp Therm Fluid Sci 28:557–563
6. Kim MS, Mulroy WJ, Didion DA (1994) Performance evaluation of two azeotropic refrigerant mixtures of HFC-134a with R-290 (propane) and R-600a (isobutane). J Energy Resour Technol 116:148–154
7. Oruc V, Devecioglu AG (2015) Thermodynamic performance of air conditioners working with R417A and R424A as alternatives to R22. Int J Refrig. 55:120–128
8. Devecioğlu AG, Oruç V (2015) Characteristics of some new generation refrigerants with low GWP. Energy Procedia 75:1452–1457
9. Oruc V, Devecioglu AG, Berk U, Vural I (2016) Experimental comparison of the energy parameters of HFCs used as alternatives to HCFC-22 in split type air conditioners. Int J Refrig. 63:125–132
10. Alkhaledi KA, Means K (2018) A study on the use of propane (R-290) in vending machines as a substitute for R-134a to minimise the global warming potential. Int J Global Warm 14:131–141
11. Bolaji BO, Huan Z (2012) Computational analysis of the performance of ozone-friendly R22 alternative refrigerants in vapour compression air-conditioning systems. Environ Prot Eng 38:41–52
12. Kondo S, Takahashi A, Tokuhashi K, Sekiya A (2002) RF number as a new index for assessing combustion hazard of flammable gases. J Hazard Mater 93:259–267
13. ANSI/ASHRAE (2007) Addenda a, b, c, d, e, f, g, and h to ANSI/ASHRAE Standard 34-Designation and safety classification of refrigerants, Atlanta, USA
14. Martin JJ, Hou YC (1955) Development of an equation of state for gases. AIChE J 1:142–151
15. Arora CP (2009) Refrigeration and Air conditioning, 3rd edn. Tata McGraw-Hill, New Delhi
16. Poling BE, Prausnitz JM, O'Connell JP (2001) The properties of gases and liquids. McGraw-Hill, New York
17. American Society of Heating (2017) ASHRAE handbook fundamentals. Refrigerating and Air-Conditioning Engineers, Atlanta, USA
18. Dalkilic AS, Wongwises S (2010) A performance comparison of vapour-compression refrigeration system using various alternative refrigerants. Int Commun Heat Mass 37:1340–1349
19. Bolaji BO, Huan Z (2014) Performance investigation of some hydro-fluorocarbon refrigerants with low global warming as substitutes to R134a in refrigeration systems. J Eng Thermophys Rus 23:148–157
20. Arora RC (2010) Refrigeration and air conditioning. Phi Ed., New Delhi
21. Baskaran A, Mathews PK (2012) A Performance comparison of vapour compression refrigeration system using Eco friendly refrigerants of low global warming potential. Int J Sci Res 2:1–8
22. ARI Standard 540 (2004) Performance rating of positive displacement refrigerant compressors and compressor units. Air Conditioning and Refrigeration Institute, Arlington, VA, USA
23. Wang F, Fan X, Lian Z, Wang F, Zhang X (2012) Performance assessment of heat pumps using HFC125/HCs mixtures. Int J Energy Res 36:1005–1014

Performance Assessment of a Solar Still Using Blackened Surface

Dillip Kumar Biswal

Abstract In this work, a blackened solar still has been fabricated and experimentally investigated for distillation of water. A tilted flat plate blackened-type solar still was designed for this purpose and fabricated. Solar still productivity is investigated by changing the input water quantity in the basin of the solar still. Indoor experimental testing was carried out. The effect of input water flow rate on the still productivity and purity of output water is investigated together. With the increase in quantity of input water, the solar still efficiency increases. The tests were conducted using the irradiance from a lamp array. It is found that 150 ml of distilled water is collected for a working period of 2 hours at 800 W input. An average overall efficiency of the designed solar still is found to be 7.95% for the 30 L of input water. With increase in water input to the solar still, its efficiency increases marginally. For the water input of 20 and 30 L, the solar still efficiencies are 7.41 and 7.95%, respectively. The quality of the distilled water is also determined experimentally and found to be satisfactory. Preliminary tests on the quality of distillate proved the possible use of the fabricated solar still for the production of distilled water for different uses in the college premises. By using the fabricated solar still, conductivity and TDS value of the sample water reduced significantly.

Keywords Solar still · Still efficiency · Blackened surface · Water output

1 Introduction

Along with food and air, the two important items which are absolutely necessary for human civilization are energy and water. Renewable energy, i.e., sun energy, can play a very vital role to meet a major part of our future energy needs. Survey shows that more than 90% of the water resources contain salts and harmful bacteria and cannot be used for drinking purposes. Many processes have been developed to purify

D. K. Biswal (✉)
Department of Mechanical Engineering, C. V. Raman Global University, Bhubaneswar 752054, Odisha, India
e-mail: dillipkumarbiswal@gmail.com

M. Ramgopal et al. (eds.), *Advances in Air Conditioning and Refrigeration*, Lecture Notes in Mechanical Engineering, https://doi.org/10.1007/978-981-15-6360-7_30

the excess amount of dissolved salts and harmful bacteria to make the water purified which can be used for drinking purpose.

A good method to obtain purified water is the distillation process. For large amount of purified drinking water, it is not feasible for the conventional distillation processes to meet the high demand. A source of energy, i.e., solar energy, which is available naturally can be used as energy input to the distillation of water processes. Therefore, solar distillation process used to purify the water seems to be a promising method and an alternative way for supplying small communities with purified water. There are several methods of water distillation process available using solar as the energy source. Simple equipment named 'solar still' can be used effectively and efficiently to provide purified water to a small community. Solar still system contains a deep basin containing water which is to be purified covered with a transparent material [1]. Although, the solar still system requires large area for installation and initial investments cost high, its operation and maintenance cost is low. In previous research, several types of solar still designs have been proposed by different authors, and many designed have found useful application all over the world. However, it is the best solution for inaccessible areas and small communities with lack of drinking water. In this work, a tilted flat plate blackened-type solar still was fabricated and used for distillation of water, and effect of various factors on the still productivity is investigated experimentally.

2 Literature Review

A double-condensing chamber solar still (DCS) had been designed by Tiwari et al. [2] in which the water is being transformed in to vapor in the first chamber by solar radiation and the water vapor is passed into a second chamber through a vent. Experiments were carried out to measure the DSC and compared with a conventional solar still. A novel low cost solar still design which reduced the overall cost of the solar still by 20% has been proposed by Kudish et al. [3]. Yadav et al. [4] designed, fabricated and tested an asymmetric line-axis compound parabolic concentrating single-basin 'solar still.' Meukam et al. [5] proposed a single-compartment model and a two compartment-type solar still, and experiment has been conducted to find out its effectiveness. Different types of single and double-slope solar stills are constructed by Khalifa et al. [6], and tests were conducted with the help of preheated feed water to show the effect of solar still performance. The efficiency and output of the solar still improves by using preheated feed water. The thermal analysis of solar distillation has been carried out by Tiwari et al. [7] to optimize the inclination angle of the glass cover for maximum output. A single-basin solar still has been fabricated by Farid and Hamad [8] made up of galvanized steel sheet with an inclined glass cover. By increasing the ambient temperature of water and decreasing the wind velocity, it is observed that the productivity of the solar still increases. Different plates in the solar stills were used by Panchal et al. [9, 10] to improve the solar still efficiency, and they found that by using M. S. Plate and G. I. Sheet the solar still efficiency increases by 15 and 20%, respectively. The effect of copper and aluminum plates on purified water output is investigated experimentally by Hitesh et al. [11] at different water depths under the same climate conditions and found that the output varies with the variation of plates and increases with the increase in water depth. Sahoo and Subushi [12] used jute cloth to improve the productivity of a solar still. Kaviti et al. [13] shows that the productivity of a solar still is greatly affected by the inclination angle of the glass cover. In comparison to 23° and 30° glass inclination angles, it is found that 30° gives more productivity compared to the 23° inclined glass cover Dinesh et al. [14]. Rashidi et al. [15, 16] found that solar still productivity can be enhanced by using porous media and nanofluid. Jani and Modi [17] carried out experimental investigation on double-slope single-basin solar still to improve the productivity by incorporating circular and square cross-sectional hollow fins.

From the literature review, it is observed that to remove harmful stuffing from water, solar still can be used as one of the key equipment. With appropriate modifications in design of glass cover (angles to the horizontal) and basin, solar still effectiveness can be improved. With this aim, the present work is designed to exploit solar energy for elimination of harmful content from drinking water by using a solar still. The simple modifications to the solar still are done by the use of blackened absorbed surface. Due to some limitations, in this experiment, electrical energy is used instead of solar energy.

3 System Description and Experimental Investigation

Figures 1a, b shows a schematic diagram of the designed solar still. The experimental setup of the 'solar still' is shown in Fig. 2a, b. The 'solar still,' with a base area of 39 × 42 inch, is fabricated from 1.5 inch thickness of glass fiber mixed with resin and $CaCO_3$ (chalk powder). The inner part of the basin is painted with blackened base liner to form a matt black surface for better absorptivity or very less reflectivity

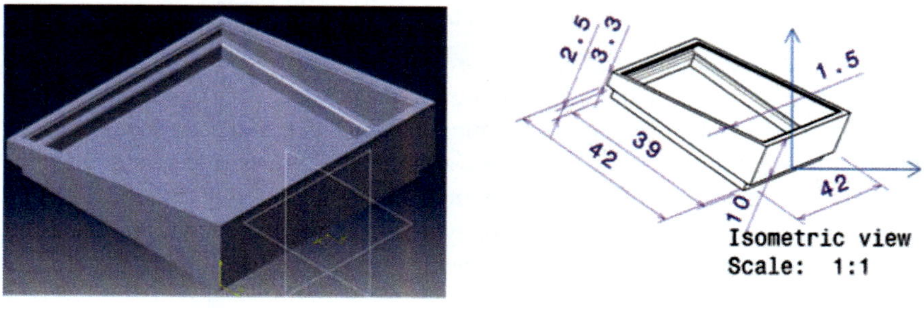

(a) (b)

Fig. 1 Schematic diagram of the designed solar still

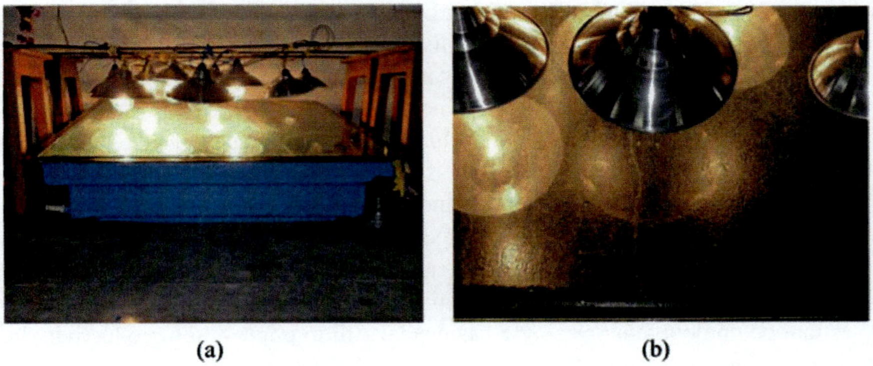

Fig. 2 Experimental setup of the solar still

and very less transmissivity. The 5 mm thick transparent glass cover of the solar still is placed at an angle of 9°74′ with the horizontal edge of the container to ensure maximum transmission of solar radiation into the still as well as enabling condensed vapor to filter down to the channel which is built in the still basin. The edges of the glass plate are sealed using rubber gasket to prevent heat losses and to make the basin air tight. The design incorporates a supply fill port at the upper half of the apparatus through which water is added into the still. The water in the basin was heated by the help of electrical energy, and the evaporated water was condensed near the cover. The output collector is kept along the lower edge of the glass cover and is used to collect the distilled water from the condensed water, and the output water is analyzed at Eastern Academy of Science and Technology, Bhubaneswar.

3.1 Result and Discussion

The area of the glass cover (1.138 m^2) is exposed to radiation, and transmissivity of the glass cover is taken from literature as 0.85 [8]. By using a thermometer, the initial temperature of the water is measured before putting into the solar still basin. Before keeping the glass cover in the solar still, the volume of water is also measured. Then, the measured amount of water is put into the solar still basin, and glass cover is placed above it to cover the basin. The volumetric output rate of the purified water from the solar still is obtained by collecting water in a jar and measured it with the help of a beaker. This experiment is conducted for estimating still efficiency for 20 and 30 L of water. Five quantities of electric bulb having 200 W were used in this experiment for 2 hours, to illuminate and heat the basin water instead of solar radiation, and the distilled water is collected from the basin. Due to external radiation, the calculated value of the input heat is taken as 800 W.

The efficiency of the solar still can be calculated by the formula [18]

$$\eta_{\text{still}} = \frac{m_v \times \text{Cp}_\text{w} \times (t_{\text{sat}} - \text{t}_{\text{wi}}) + m_\text{v} \times h_{\text{fg}}}{Q_{\text{in}}} \times 100 \quad (1)$$

where

Mass of the output water $m_\text{v} = 150$ ml $= 0.15$ kg

Specific heat capacity of water $\text{Cp}_\text{w} = 4.21$ kJ/kg K(Constant for water)

Latent heat of vaporization of water $h_{\text{fg}} = 2267$ kJ/kg

Temperature of saturated water $t_{\text{sat}} = 100\ °\text{C} = 373$ K

Temperature of input water $t_{\text{wi}} = 35\ °\text{C} = 308$ K

Transmissivity of glass $\tau = 0.85$

Area of the glass $A_\text{g} = 1.138\ \text{m}^2$

Area of water stored in still base $A_\text{s} = 1.1\ \text{m}^2$.

The solar still efficiencies are calculated for 20 and 30 L of water of operation and are calculated to be 7.41 and 7.95%, respectively. It is found that with increase in water input to the solar still, its efficiency increases marginally because the heat capacity increases with increasing water contents in the solar still basin whereas heat rejected from the basin remains constant.

Figure 3 shows the comparison between tap water and distilled water for pH value, conductivity and TDS value, respectively. From Fig. 3a for sample 1, it can be seen that, the pH value of distilled water was found to be 4.81 whereas it should be in between 6.5 and 7.5. The conductivity and TDS value for sample 1 also varied whereas these should be within 11 μs and less than 1 ppm, respectively. The conductivity was measured using an electronic conductivity meter, which generates a voltage difference between two electrodes submerged in the water. The drop in voltage due to the water resistance is used to calculate the conductivity per centimeter. This resulted due to the exposure of resulted water to the atmosphere (when the distilled water comes in contact with carbon dioxide, CO_2, it gets contaminated and turns acidic in nature). The variation was also caused due to the presence of dust particles and chemicals in the paint. The base of the solar still was not washed properly at first testing, and the water kept in the base remained unchanged for 15 days, which also had an effect on the pH, conductivity, TDS and temperature of water.

Therefore, to avoid such variations, the stored water was drained out and the base was cleaned thoroughly and wiped properly to avoid any contamination. Then, fresh water was poured into the base though the inlet channel. Again, the bulbs were illuminated, and the process was repeated. A pipe was connected to the outlet of the base through a connector to collect the distilled water in a bottle and was sealed properly to avoid contamination. The collected distilled water was taken in a beaker, and tap water was taken in another beaker to be tested in the environmental engineering laboratory of Eastern Academy of Science and Technology, Bhubaneswar.

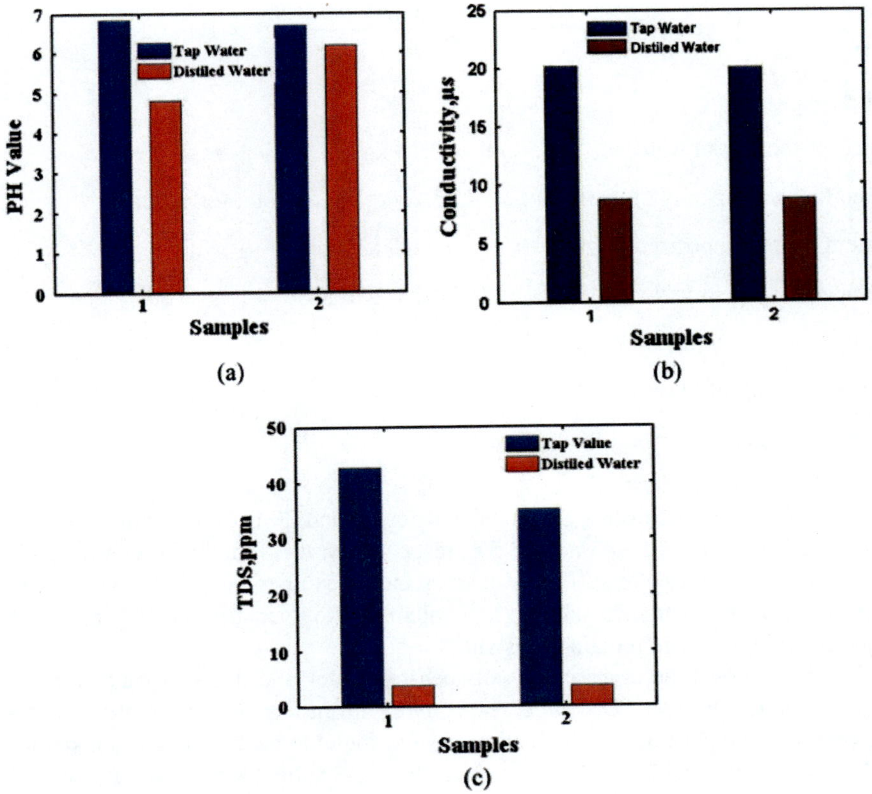

Fig. 3 Comparision between tap water and distilled water. **a** pH value. **b** Conductivity **c** TDS value

3.2 Measurement Error

The measurement errors allied with the solar still performance have been calculated by using sequential perturbation techniques [20]. As it a function of volume of purified water output, temperature of input water, area of glass cover and heat input to the solar still, it can be represented as

$$M_{\mathrm{error}} = f(m_{\mathrm{v}}, t_{\mathrm{wi}}, A_{\mathrm{g}}, Q_{\mathrm{in}}) \tag{2}$$

By considering the measurement errors in volume of purified water output (1%), temperature of input water (0.5%), area of the glass cover (0.1%) and heat input to the solar still (0.1%), the solar still efficiency was supposed to be accurate within ±2%.

3.3 Comparison of productivity of different solar still

As in this paper, due to some limitations, in this experiment, we used electrical energy as the input source instead of solar energy. We have calculated the input electrical energy as 800 W and considered to be the equivalent solar intensity and calculated the solar still efficiency. Compared to the result of Sahoo et al. [19], the authors reported the still efficiency was 7.28, 7.78 and 8.1%, respectively, for 10, 15 and 20 L of water, respectively, in mode I. Here in this work, we got 7.41 and 7.95% efficiency for 20 and 30 L of water, respectively. This is mainly because heat input and time duration is less compared to the previous work. In future work, the efficiency of the still can be increased by modifying different parameters which affect the still efficiency and using proper instruments for the measurement purpose.

4 Conclusion

In the present work, an attempt has been made to fabricate a tilted flat plate blackened-type solar still and the same has been used for the purpose of producing purified drinking water for a small community. The observations of the investigation may be summarized as follows.

- The pH value is found to be 6.2, which shows a good agreement with the previous results.
- By using the fabricated solar still, TDS value has been reduced around 90% compared to the sample taken from tap water (untreated).
- With increase in water input to the solar still, its efficiency increases marginally. For the water input of 20 and 30 L, the solar still efficiencies are 7.41 and 7.95%, respectively.

In future, the same setup can be experimented by using solar power as the heat input and compared with the present result.

References

1. Abdel-Rehim ZS, Lasheen A (2005) Improving the performance of solar desalination systems. Renew Energy 30:1955–1971
2. Tiwari GN, Kupfemann A, Aggarwal S (1997) A new design for a double-condensing chamber solar still. Desalination 114:153–164
3. Kudish AJ, Gale J, Zarmi Y (1982) A low cost design solar desalination unit. Energy Convers Manage 22:269–274
4. Yadav YP, Yadav AK, Anwar N, Eames PC, Norton B (1996) An asymmetric line-axis compound parabolic concentrating single basin solar still. In: Proceedings of world renewable energy congress, Denver, USA, pp. 737–740

5. Meukam P, Njomo D, Gbane A, Toure S (2004) Experimental optimization of a solar still: application to alcohol distillation. Chem Eng Process 43:1569–1577
6. Khalifa AJN, Al-Jubouri AS, Abed, MK (1999) An experimental study on modified simple solar stills. Energy Convers Manage 40:1835–1847
7. Tiwari GN, Thomas JM, Khan E (1994) Optimisation of glass cover inclination for maximum yield in a solar still. Heat Recov Syst CHP 14:447–455
8. Farid M, Hamad F (1993) Performance of a single-basin solar still. Renew Energy 3:75–83
9. Panchal HN, Shah PK (2012) Performance Improvement of Solar stills via experimental Investigation. Int J Adv Des Manuf Technol 5(5):19–23
10. Panchal HN, Shah PK (2012) Investigation on Solar stills having floating plates. Int J Energy Environ Eng 3(3):3–8
11. Hitesh N, Panchal PK (2014) Shah improvement of solar still productivity by energy absorbing plates. J Renew Energy Environ 1(1):1–7
12. Sahoo BB, Subudhi C (2019) Performance enhancement of solar still by using reflector-jute cloth-improved glass angle. J Eng Res 16(1):1–10
13. Dinesh K, Himanshu P, Zameer A (2013) Performance analysis of single slope solar still. Int J Emerg Technol Adv Eng 3(3):66–72
14. Kaviti AK, Yadav A, Shukla A (2016) Inclined solar still designs: a review. Renew Sustain Energy Rev 54:429–451
15. Rashidi S, Rahbar N, Valipour MS, Esfahani JA (2018) Enhancement of solar still by reticular porous media: experimental investigation with exergy and economic analysis. Appl Therm Eng 130:1341–1348
16. Rashidi S, Bovand M, Rahbar N, Esfahani JA (2018) Steps optimization and productivity enhancement in a nanofluid cascade solar still. Renew Energy 118:536–545
17. Jani HK, Modi KV (2019) Experimental performance evaluation of single basin dual slope solar still with circular and square cross-sectional hollow fins. Solar Energy 179:186–194
18. Tiwari GN, Dimri V, Singh U, Chel A, Sarkar B (2007) Compressive thermal evaluation of an active solar distillation system. Int J Energy Res 31(15):1465–1482
19. Sahoo BB, Sahoo N, Mahanta P, Borbora L, Kalita P, Saha UK (2008) Performance assessment of a solar still blackend surface and thermocol insulation. Renew Energy 33:1703–1708
20. Kline SJ, Mc Slintock FA (1953) Describing uncertainties in single-simple experiments. Mech Eng 3–8

Experimental Investigation of Thermal Performance of Solar Air Heater Having Hemispherical Fins on Absorber Plates

Sachindra Kumar Rout, Taraprasad Mohapatra, Chinmaya P. Mohanty, and Prasheet Mishra

Abstract The solar air heater with a hemispherical shape obstacle has been introduced, and an experimental study has been conducted to explore their performance over an extensive scope of running environments. The collector efficiency can be enhanced by increasing the heat transfer surface area by incorporating fins or obstacles to either over or under the absorbing surfaces. Different operating conditions have been chosen to run the experimental setup. Various measurable parameters like temperature of absorbing plate, temperature of inlet and outlet air, temperature of ambient air and solar radiation have been measured. The mass flow rates in the flow channel duct have been changed and measured to conduct the experiments for study the variation in performance for both finned and smooth absorbing plates. The study was concluded with calculation of the collector efficiency and reported a significant improvement in efficiency with hemispherical fins on absorber plates.

Keywords Solar air heater · Hemispherical · Collector

1 Introduction

Due to increase of population, the utilization of energy is very high, and hence, the researchers are focusing on the use of solar energy nowadays. The equipments used for solar energy have been improved with innovation of new technology and new methods. Solar air heater is one of the devices which have been significantly improved by various heat transferring approaches. Attaching obstacles or extended surfaces to the absorber plate is the most successful technique which enhances the heat transfer area by producing turbulence in the flow field. But one major problem associated with this approach is the pressure drop.

S. K. Rout · T. Mohapatra · P. Mishra
C. V. Raman Global University, Bhubaneswar 752054, Odisha, India

C. P. Mohanty (✉)
Vellore Institute of Technology, Vellore 632014, Tamil Nadu, India
e-mail: chinmaymohantymech@gmail.com

M. Ramgopal et al. (eds.), *Advances in Air Conditioning and Refrigeration*, Lecture Notes in Mechanical Engineering, https://doi.org/10.1007/978-981-15-6360-7_31

There are several research manuscripts available in the open literature on solar air heater supporting the research in this field. The absorber plate in solar air heater utilized various geometries like W-shaped ribs, W-shaped ribs with gap, V-shaped ribs, V-shaped ribs with gap, multi-V-shaped ribs and multi-V-shaped ribs with gaps, transverse ribs, and wire mesh screen. In European countries, solar air heaters play a major role in air-conditioning system. In recent years, the performances are tremendously increased by adopting various performance enhancement techniques. Though in India the ambient temperature is quite high as compared to other regions of the World of same global position, the use of solar air heater is not so high. But, nowadays this device uses solar energy in different sectors like agricultural and other process heating applications either by active or passive heating methods. Nowadays, these equipment are highly adopted by Asian countries for drying proposes and hot water generation proposes.

In our research work, we have used active heating method which can produce hot air for various purposes by using double pass solar air heater with obstacles in top of plate as compared to other conventional techniques.

There is a wide range of the application of solar air heaters in the field of agricultural and Industrial sectors. Therefore, a lot of research is being carried out by different groups of people throughout the globe to enhance the efficiency of the solar air heaters by adopting various augmentation methods like escalating the rate of heat transfer, appropriate physical shape and improved heat transferring medium of high heat absorptivity, and use of various thermal energy storage materials.

A solar air heater with a special type of coating on its absorber plate was experimentally analyzed by Pramanik et al. [1]. They used longitudinal fins at under part of the absorber plate to provide more surface area of contact for getting hot air by use of the solar energy. Prashant Verma et al. [2] used a novel wire screen matrix which was placed inside the solar air heater and solved numerically to investigate the parametric study of various operating conditions on the thermal and hydraulic performance. Lakshmi et al. [3] conducted experimental study on solar air heater to study the heat transfer and fluid flow effect in the mixed mode on its performance to dry stevia leaves. In addition, Vijayan et al. [4] conducted an experimental study on a solar air heater for drying bitter gourd slices in the city of Coimbatore. They incorporated a porous bed sensible heat storage medium to the solar dryer, and it was operated at forced convection environment. Lamrani et al. [5] analyzed the energy and environment of an hybrid solar using TRNSYS. This hybrid solar dryer can reduce the energy consumption and can decrease by about 34% of CO_2 emissions annually. A review report by Mustayen et al. [6] presented a detail study on the performance of many solar dryers and the factors that affect their efficiency.

In the present work, a comparative experimental work has been performed between smooth and finned solar air heater. A new type of fin with hemispherical shape is introduced and found a significant increase in the performance of solar air heater with negligible pressure loss. This paper demonstrates both design and fabrication of a couple of different types of solar air heater, known as smooth plate solar air heater and finned solar air heater.

Fig. 1 Fabrication of hemispherical fins and absorber plate

2 Experimental Study

The details about experimental setup are described in this section. It consists of wooden box fitted with absorber plate, glazing, air handling unit (fans), insulations and data accusation systems for recording the data during conducting the experiment. There are two types configuration of experimental setup: one is without fin and another set up is with hemispherical fins. The absorber plate is made of aluminum sheet as shown in Fig. 1. The absorber plate is fitted with a wood frame. Both inlet and outlet are situated on side of the frame. The ambient air enters to the system in upper section, and hot air goes out at the lower part. The fins are placed in a zigzag manner to provide a better air contact with the fins. The solar intensity during daytime was measured using a pyranometer. The fans were fitted with electrical controllers which can be used to change the speed of fan; as a result, we can control the mass flow rate of working fluid, that is, air. Thermocouples were attached at various locations like inlet and outlet of the set up to measure air temperature, along the absorber plate to measure plate temperature and outside the setup to measure surrounding temperature (Fig. 2).

3 Results and Discussions

The experimental data are recorded concurrently for both the finned and smooth absorbing plates. The limits of mass flow rate and Reynolds number investigated have been given in Table 1.

The collector energy efficiency can be evaluated with heat gain and solar radiation equations.

$$\text{Amount of useful energy absorbed by air (working fluid)} = Q_u = \dot{m}C_p(T_o - T_i) \quad (1)$$

$$\text{Incident energy on the absorber plate due to solar radiation} = Q_c = I * A_c \quad (2)$$

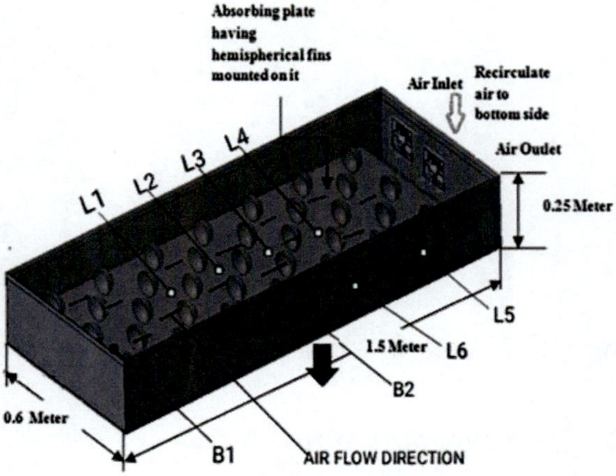

Fig. 2 Experimental setup

Table 1 Operating limits of flow parameters

Serial number	Parameters	Limits
1.	Mass flow rate ($\dot{m}$)	9×10^{-3} to 4×10^{-2} (kg/s)
2.	Reynolds number	2000–4000

$$\text{Thermal efficiency of the collector} = \eta_{\text{th}} = \frac{Q_{\text{u}}}{A_{\text{c}} I} = \frac{\dot{m} C_{\text{p}} (T_{\text{o}} - T_{\text{i}})}{A_{\text{c}} I} \quad (3)$$

4 Ambient Temperature

Figure 3 illustrates the change in the ambient air temperature and variation of solar radiation intensity from 10 am to 5 pm, where the temperature at 10 am was 32 °C and reached about 39 °C at 2 pm and then dropped to 33 °C at 4 pm on March 28, 2018. It was recorded 995 W/m^2 the intensity of radiation 991 W/m^2 at 2 pm on the same day.

Figure 4 presents hourly variation of air temperature of solar air heater (SAH) with a hemispherical shape fin and without fin at exit. The maximum temperature of at exit of solar air heater without fin was 45.2 °C, and with fin, it was recorded 57.46 °C at 2 pm. The leaving air temperature is higher with fin compared to without fin because

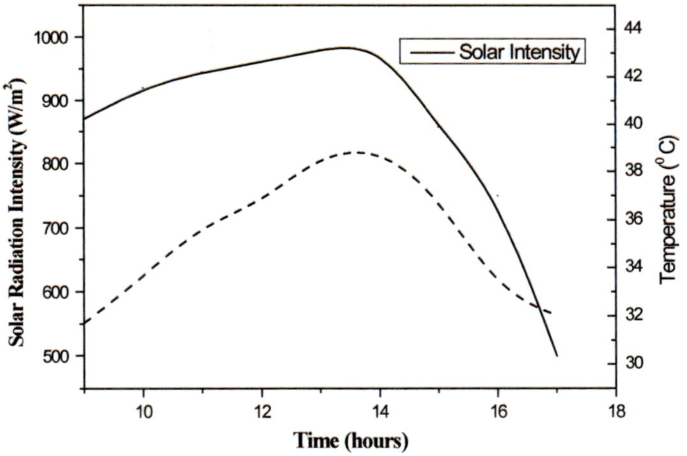

Fig. 3 Change of ambient air temperature and solar radiation intensity (Hourly)

Fig. 4 Air temperature variation of fluid (Hourly)

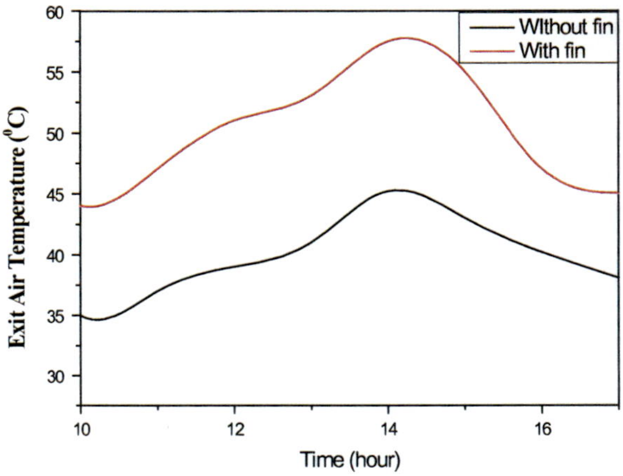

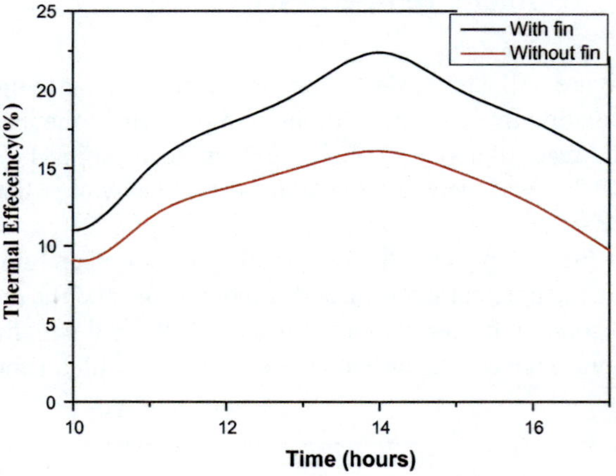

Fig. 5 Change of thermal efficiency (Hourly)

the surface area exposed to flowing air is increased due to adding of hemispherical fins and also thermal efficiency increases which is shown in Fig. 5. The absorber plate is the hottest part in the air heater, and it reached 73 °C at 2 pm. Thermal efficiency for without fin solar air heater varies with respect to solar intensity of the day which is 16.1% without fin whereas 22.4% for with fin attached. The average thermal efficiency for both the solar air heater is 12.8 and 17.45%, respectively.

5 Conclusion

In this experimental work, a novel shape hemispherical fin is introduced to the solar air heater. The performance of a solar air heater in terms of thermal efficiency, with a hemispherical shape fin absorber is experimentally presented and compared with smooth absorber plate. The outcomes from the experiments are listed below:

- Ambient air temperature is an important factor to get good air temperature distribution;
- The solar collector provided hot air exceeding 74 °C at its outlet;
- The air temperature at the outlet is slightly lower compared to the rest of the chamber;
- The leaving air temperature of the solar air heater with the hemispherical shape fins is higher than solar air heater having smooth absorber plate;
- The maximum thermal efficiency 16.1% without fin whereas 22.4% for finned absorber plate, respectively.

References

1. Pramanik RN et al (2017) Performance analysis of double pass solar air heater with bottom extended surface. Energy Procedia 109:331–337
2. Verma P, Varshney L (2015) Parametric investigation on thermo-hydraulic performance of wire screen matrix packed solar air heater. Sustain Energy Technol Assess 10:40–52
3. Lakshmi DVN et al (2019) Performance analyses of mixed mode forced convection solar dryer for drying of stevia leaves. Sol Energy 188:507–518
4. Vijayan S, Arjunan TV, Kumar A (2020) Exergo-environmental analysis of an indirect forced convection solar dryer for drying bitter gourd slices. Renew Energy 146:2210–2223
5. Lamrani B, Khouya A, Draoui A (2019) Energy and environmental analysis of an indirect hybrid solar dryer of wood using TRNSYS software. Sol Energy 183:132–145
6. Mustayen AGMB, Mekhilef S, Saidur R (2014) Performance study of different solar dryers: a review. Renew Sustain Energy Rev 34:463–470

A Study on Dual Cycle Based on VCR-VAR System

P. Ankit Subudhi and Santosh Kumar Panda

Abstract Refrigeration system plays an important role in cooling/heating process in day-to-day life. Efficient and better design of a refrigeration system is useful for saving cost, energy and needful for the future application. The individual vapour compression refrigeration (VCR) and vapour absorption refrigeration (VAR) system have some drawbacks, so the research is carried out for dual cycle based on VCR-VAR system. NH_3–$LiNO_3$ VCR and VAR system in the single circuit allows more enhancements of efficiencies than individual VCR and VAR system-based cycles. The absorption cycles works with NH_3–H_2O system, and the vapour compression cycle works with NH_3. The absorption cycles produces pure ammonia refrigerant, and the existing ammonia uses as a refrigerant for mechanical vapour compression plants.

Keywords Compression–absorption · Absorber · Generator · COP · Mass transfer · Heat transfer

1 Introduction

Recent trends of the cooling and air-conditioning methods are used with vapour-compression cycles, because it is simplicity in the construct and design with faster cooling device employing low cost compared to (Vapour Absorption Refrigeration) VAR system. Basically, VCR system works as a reverse heat engine, so heat energy is taken from a low temperature reservoir and delivers into a high-temperature reservoir. Based on the second law of thermodynamics, heat energy was unable to transfer from a low-temperature to a high-temperature reservoir. Refrigerant is defined as working fluids which transfer heat form one component to another which undergoes phase changes. The large amounts of heat energy are release due to phase change of refrigerant with the latent heat of vaporization of the refrigerant which changes in pressure and control by absorbing the refrigerant. The refrigeration cycle has a

P. Ankit Subudhi · S. K. Panda (✉)
National Institute of Science and Technology, Berhampur, Odisha, India
e-mail: santosh.panda@nist.edu

M. Ramgopal et al. (eds.), *Advances in Air Conditioning and Refrigeration*, Lecture Notes in Mechanical Engineering, https://doi.org/10.1007/978-981-15-6360-7_32

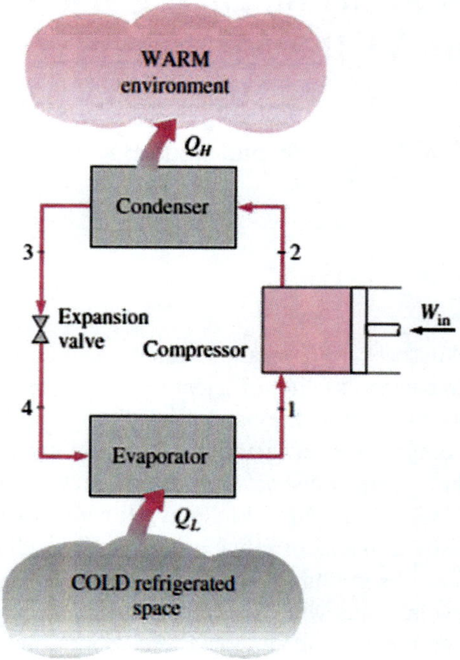

Fig. 1 Schematic diagram of vapour compression refrigeration cycle [3]

high temperature of reservoir at around room temperature acting as a source and a low-temperature reservoir at around less than atmosphere act as sink and maintain the refrigerant temperature to low in the cycle. The most common include air, ammonia, Freon-11 (R11), R12, R22, and R-134a. The refrigerant enters as low pressure into the compressor as a vapour having a low temperature and compressed in adiabatic compression process. The refrigerant leaves the compressor at high pressure and high temperature. The high pressure, high-temperature vapour condenses in the condenser by releasing the heat to the coolant. The condenser is in contact with the source of the refrigeration system. The liquid refrigerant expands with throttling valve from high-pressure to low-pressure vapour. The low-pressure, low-temperature refrigerant evaporates in the evaporator which acts as a sink and generates refrigerating effect. The low-pressure and low-temperature refrigerant leaves the evaporator as a refrigerant and is taken back into the compressor for recirculate, to beginning of the cycle (Fig. 1).

1.1 Vapour Absorption Refrigeration System

The VAR system may be used in both the domestic and large industrial refrigerating plants with the use of waste and renewable energy. The VAR system uses heat energy, instead of mechanical energy as in VCR systems, In the VAR system, an absorber,

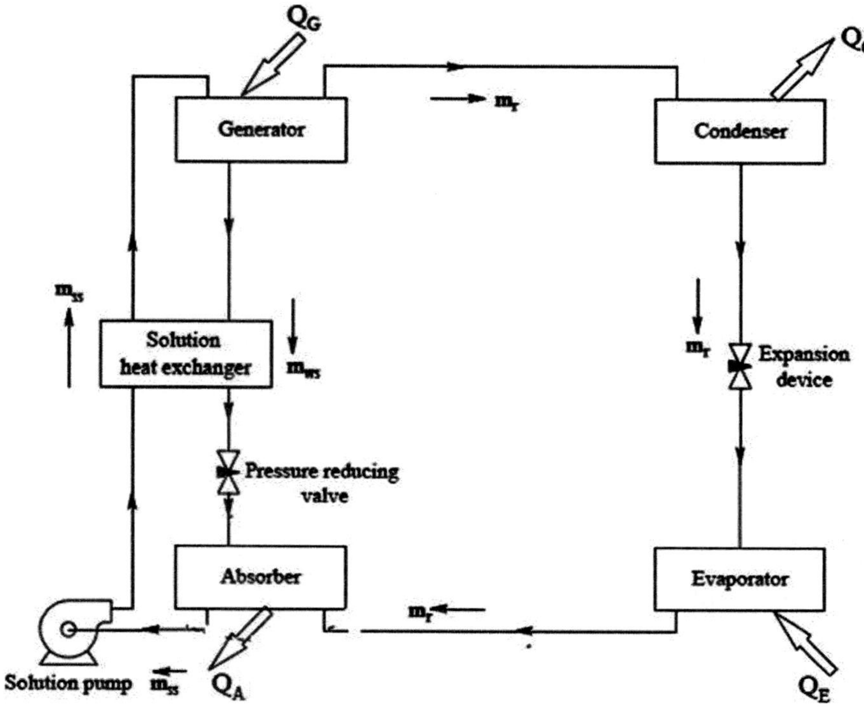

Fig. 2 Vapour–absorption refrigeration schematic diagram

pump, solution heat exchanger, generator, and pressure-reducing valve (PRV) replace the mechanical compressor. These components in VAR system perform the same function as that of a compressor in VCR system. Figure 2 shows a schematic diagram of VAR system. The refrigerant enters the evaporator as a low temperature and low pressure and creates the refrigeration effect, converts to vapour and enters to the absorber. Refrigerant vapour enters the absorber where it mixes with low concentration solution in refrigeration and converts to high concentration solution in refrigeration and releases the absorption mixing energy in the core channel. The absorption mixing energy removes from the absorber with circulating cooling water in the outside passage of the absorber. The strong solution is pumped with the help of solution pump to the generator through the SHX. The refrigerant vapour generated in the generator at high temperature with higher side pressure on the application of the heat energy supplied by hot circulating water is supplied; as a result, refrigerant vapour is generated at high pressure. After the refrigeration vapour releases, the solution converts to low concentration in refrigeration and is sent back to the absorber for further absorption process by following the path of low temperature side SHX and PRV. In SHX, the weak solution which comes out form the generator gives the heat to strong solution comes out from the absorber. This high-pressure refrigerant vapour is condensed in the condenser by removing the heat through of condensation by the coolant. The high-pressure condensed refrigerant liquid expands in the throttled expansion device to a low-pressure refrigerant liquid and fed to the evaporator to complete the VAR refrigerant cycle.

2 Compression-Absorption Refrigeration Systems

The dual VCR-VAR systems are considered as alternatives to VCR and VAR systems due to some of their speciality in operation features. This VCR-VAR enables to create a environment issue as the refrigerant–absorbent working fluid combination used such as ammonia–water and lithium bromide–water, which are well-known ozone friendly and have very low global warming potential (GWP). These VCR-VAR system have better COP, better capacity control with lower operating pressure in the system due to use of working fluid mixtures, compared to VCR systems. The VCR-VAR system has been shown in Fig. 3. The compressor used as fluid mixer has been added to the absorber to bring the high temperature vapour and the weak solution in refrigeration into thermal equilibrium. This phenomena is helpful to increasing the temperature of the absorber by increasing the inlet temperature of weak solution from the heat of superheated. After compressor process, liquid and vapours enter through different passages into the absorber, where liquid is distributed on the inner surface of the tube as a falling film and vapours are made to flow in the main channel. A phase separator has been used after generator to separate of unwanted absorbent and allows pure refrigerant vapours for the condenser after generation process.

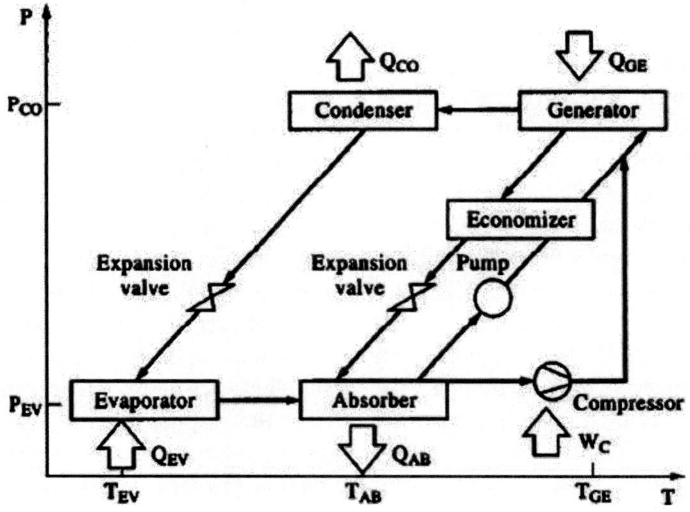

Fig. 3 Compression and absorption refrigeration system [1]

A VARS system operates with ammonia/lithium nitrate refrigerant–absorbent working fluid combination combined with mechanical vapor compression refrigeration system (VCR) working with ammonia as working fluid used as a combined cycle to increase the efficiencies than individual compression or absorption cycles by carrying out a thermodynamically simulation study [1]. The schematic of the cycle used by the authors is presented in Fig. 3. The cycle design is analysis over a range of refrigerant and absorption as fluid combination for the initial to final stage of dual cycle. The power generation and distribution effects were considered in deriving the heat energy ratios for each individual components of the dual combined cycle and overall COP of the individual and dual of VCR-VAR cycle, which will be helpful to meet the cooling needs of various countries. In VCR-VAR dual systems, the refrigerant works in a superheated condition as a working fluid at the compressor discharge and by the addition of the heat the refrigerant enters as a strong solution before it enters the generator. The heat transfer rate and the composition of the strong solution depend on the working mixture used, one and the compressor efficiency. Reciprocating compressors are taken to consideration for ammonia for more discharge and operating pressure. These systems have cylinder head with cooling system to improve the smooth and efficient operation to avoid damage due to excessive superheating and causing oil quality degradation. However, with synthetic oils, it might be possible to make better use of this heat. The simulation study considered generator temperature from 100 to 130 °C, condenser temperature from 20 to 40 °C, evaporator temperature from −20 to 0 °C, absorber temperature from 20 to 40 °C, and compression proportion 0 to 1. The coefficient of performance (COP) of the dual system was increased with the compression proportion and COP of Carnot for cooling (CCL) giving more COP compared to the other system such as enthalpy-based cooling (ECL) and individual system. The reported result found that it is possible to achieve up to a 10% increase in overall efficiency using dual absorption–compression refrigeration systems compare to the individual cycle.

An experimental study were [2] carried out for dual system, combination of enthalpy-based cooling system (ECL) and VARS with the ammonia/lithium nitrate fluid combination. The dual cycle used with the input energy ratio for combined hybrid system with more operate more efficiently in developing countries compare to individual developed system ones. The experiment is carried with putting a pre-cooler after the expansion valve for creating more cooling effect in the evaporator. A 7 kWh prototype model of a combined hybrid system was design and fabricated and to conduct the experimental results are reported in the paper. The performance of the experimental study represented with actual flow ratio (FRA) or thermodynamic flow ratio, second flow ratio and compression proportion and determine the COP, heat effectiveness ratio between COP of absorption system and ECL system and components loads like evaporator load, generator load. Maximum COP of the dual system was achieved using 90% compression and 10% absorption. The heat energy is not required for the absorption section, because of higher compressor ratio, and the heat is supplied by the superheated ammonia at the compressor outlet.

A 400 kW dual cycle of VCR-VAR system was considered for evaluating heat and mass transfer analysis(HMT) for the application of water chilling application [4].

Heat transfer coefficients and pressure drops in the absorber and desorber have been evaluated on the effects of the mass flow rate of weak solution, desorber pressure, and vapor area distribution and on the COP, cooling capacity of each individual components and absorber heat load. Three different configurations of the dual cycle used system, having increase in SHX area 17, 23, and 30%, have been studied. The result been reported that the 10–30% COP of the system increased with the larger relative SHX area and maintaining low rate of mass flow of weak solution.

An experimental study has been analyzed [5] for a single-stage dual VCR-VAR system for simultaneous heating and cooling analysis of the cycle components with R22 and E181 (DMETEG) as the working fluid combination. Firstly, the paper describe about the components and working principles of the dual cycle which consists the components like desorber, an absorber, a compressor, a solution pump, expansion valve and heat exchanger. A solution mixture of refrigerant and absorbent circulates through the desorber, absorber, solution pump, heat exchanger, and the expansion valve, while the refrigerant vapor flows through the condenser, evaporator, expansion valve, and compressor. The refrigeration effect is created at the desorber as the heat flows into the desorberas a part of the vaporization of refrigerant from the solution mixture. The refrigerant vapor is maintained by compressed to the absorber pressure by the compressor. The rest of solution in the desorber is pumped by the solution pump to increase the absorber pressure. The high-pressure refrigerant from the compressor and the weak solution from the solution pump are mixed to create the absorption process in the absorber. Exothermic process heat between the solution and refrigeration creates high temperature in the absorber, and the heat is removed with the circulating cooling water: The solution from the absorber flows through the expansion valve to decrease the pressure drops of the desorber. The SHX works between the weak and strong solutions to exchange the heat to improve the system performance. The experimental system components and schematic design based on the absorber results got from the theoretical result. The comparison study shows a good agreement with experimental and theoretical analysis of the dual system which gives better agreements.

In generator–absorber–exchange absorption compression (GAXAC) used as absorption cycle of 3.514 kW and generator–absorber–exchange (GAX) as a VCR cycle a combined cycle are shown in Fig. 4. Consider for numerical simulation using ammonia-water as working fluid combination considered [6]. The operating parameter of the simulation study consider as evaporator temperatures varying from −10 to 5 °C, condenser outlet temperatures from 20 to 40 °C. The absorber outlet temperatures were varied from 20 to 40 °C. The generator temperatures were varied from 110 to 150 °C. The numerical study found that lower-pressure ratio of the cycle has been optimized for better COP and COP effects on the operating temperature of components such as generator, condenser, absorber and evaporator, on the COP of the cycle determined. It is reported that for a given value of desorber and absorber temperatures difference, the maximum COP estimated for the effect to the optimum operating pressure ratio is independent of the temperatures of condenser, absorber and evaporator. The maximum parameter for the COP is reported as operating parameter for desorber temperatures 110–150 °C with approach temperature 14 °C at for

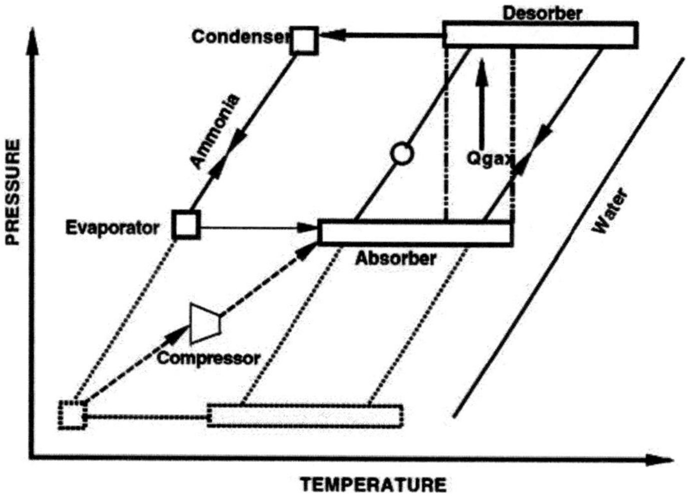

Fig. 4 Schematic diagram of of GAX and GAXAC cycle [6]

every pressure ratios are found to be 0.94–1, respectively. A comparison report of GAXAC and standard GAX cycle was carried out and found that GAXAC cycle has 26% better COP than the standard GAX cycle.

A comparison study of the COP of single-stage VAR, standard GAX, and GAX-based VAR-VCR cycles on the variation of condenser temperature were shown in Fig. 5. The graph show that increase in condenser temperature increases in COP, and enhances in the COP of single-stage VAR is minimum but it increases for the standard GAX cycle and more increases for the GAXAC cycle. The increases an amount of 26% in the cooling COP of GAX absorption–compression cycle compare to the cooling COP of standard GAX cycle. The paper also reported that the COP increases with increase in the low operating pressure ratio and generator temperature and degrades with increase in the absorber and condenser temperature. The optimized study of the cycle was also reported based on the optimum pressure ratio and the condenser, absorber and evaporator temperatures.

3 Conclusion

A dual cycle review study presented in this is paper based on the database available. Various research reports a combined dual cycle of VAR and VCR with a comparison study also based on individual cycle based on COP for various working fluid. The review gives a comparison study of individual system and dual system on compression proportion (function of the lower side pressure to higher side pressure) effect on COP of the system. The report also gives that COP and cooling effects of the cooling

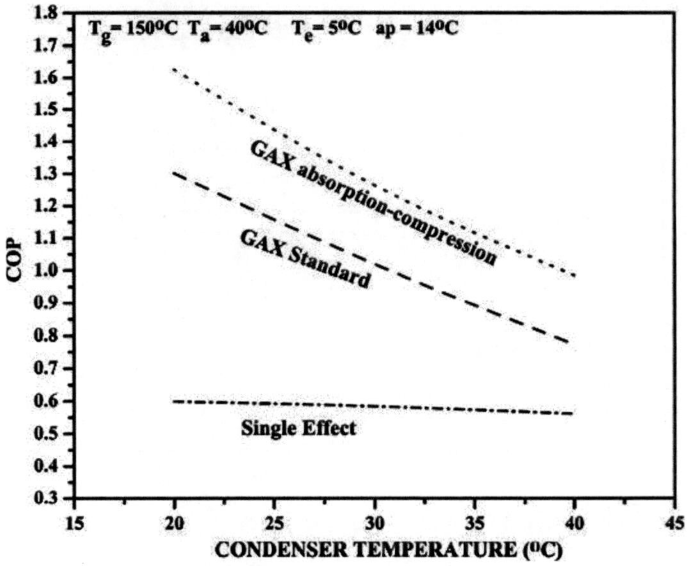

Fig. 5 Variation of the COP with condenser temperatures for dual cycle [6]

system is better in terms of increasing 10–30% in dual combination cycle compare to individual system.

References

1. Ayala R, Heard CL, Holland FA (1997) Ammonia/lithium nitrate absorption/compression refrigeration cycl. Part I. simulation. Appl Therm Eng 17(3):223–233
2. Ayala R, Heard CL, Holland FA (1998) Ammonia/lithium nitrate absorption/compression refrigeration cycl. Part II. Experimental. Appl Therm Eng 18(8):661–670
3. Cengel YA, Boles MA (2011) Thermodyancmics: an engineering appoach, 7th edn. McGraw-Hills, New York
4. Pratihar AK, Kaushik SC, Agarwal RS (2010) Simulation of an ammonia-water compression-absorptionrefrigeration system for water chilling application. Int J Refrig 33:1386–1394
5. Satapathy PK, RamGopal M (2008) Experimental studies on a compression–absorption system for heating and cooling applications. Int J Energy Res 2008(32):595–611
6. Rameshkumar A, Udayakumar M (2008) Studies of compressor pressure ratio effect on GAXAC (generator–absorber–exchange absorption compression) cooler. Appl Energy 85:1163–1172

Heat and Mass Transfer-Enhancement Technique Used for Vapour Absorption Refrigeration System

Raj Barun Raul and Santosh Kumar Panda

Abstract The vapour absorption refrigeration system (VARS) as a future cooling–heating system to meet the energy crises and to replace the mechanical vapour compression systems (VCR) also utilized the waste and renewable energy. The different renewable energy sources uses as the driving source for the VARS are waste energy from the industrial exhaust heat or solar thermal energy to efficiently run, meet the cooling load capacity, and have lesser impact on the climate and the environment. Nowadays, the cost of electric power raising high and the change in climate require the system for more efficient design to make the system more compatible size which guides to save energy. Absorber is a vital component for a VARS, heat and mass transfer (HMT) analysis concern. Better design of absorber impacts on the absorption process, coefficient of performance (COP) of the VARS which reduces the shape and size absorber and generator sizes. The objective of this paper is to identify various enhancement techniques used for heat and mass transfer like absorber and generator and summarize the system performance, on experimental and numerical studies. The enhancement techniques are classified into passive and active methods; twisted tape, swirl generator, axial guide vane, internal grooved, internal and external fins, rotation of tube are the geometrical device used as a passive methods, and use of additives, nano-fluids, induced pulsation by cams, reciprocating plungers, vibrators, use of magnetic fields, etc., are the passive technique. Finally, the present literature database contains suggestions for future work for the absorber and generator for enhancement in the HMT in absorbers and performance of the VARS.

Keywords VARS · Absorber · Heat transfer · Mass transfer · Absorption process · Enhancement technique

---

R. B. Raul · S. K. Panda (✉)
National Institute of Science and Technology, Berhampur, Odisha, India
e-mail: santosh.panda@nist.edu

M. Ramgopal et al. (eds.), *Advances in Air Conditioning and Refrigeration*, Lecture Notes in Mechanical Engineering, https://doi.org/10.1007/978-981-15-6360-7_33

1 Introduction

Depletion of natural resources caused due to increase in energy consumption requirements day to day with depletion of freshness air availability, ozone layer depletion (ODP), and global warming (GWP) causes a matter of concern for the living being. The increase in energy consumption is in part of the improvement of our standard of living and the society development of sectors as building, industry, transport and other facility. The building construction sector consumes maximum amount of energy compared to the other sources based on global energy consumption scenario. Air-conditioning point of view mechanical VCR system consumes high-grade electrical energy and widely used technology for the use of refrigerant. The uses of refrigerant are that it produces high impacts on the environment issues like of the ODP and GWP. The need to adopt ODP and climate-friendly technologies is the policies for the replacement of old technologies with alternative available energy-efficient servicing.

VARS have been recognized as the alternative and future refrigeration system for the mechanical VCR. It also provides high potential for energy saving compared to conventional electrically driven VCR technologies, which operated with environmentally friendly working fluids. The major drawbacks for the VARS are high initial investment, less COP, difficulty in compatible size, complicated in design, etc. Last few years, researchers keen to the interest in the development of new VARS by carrying various numerical and experimental studies to improve their performance and make comparison with the mechanical compression systems. The absorber is considered the important component of VARS due to the complexity in physics of HMT process with the absorption process and the absorption rates further enhanced by the application of passive and active enhancement techniques which allow for the absorber size to be compatible, energy efficient and cost saving. The enhancement techniques are classified into passive and active methods; twisted tape, swirl generator, axial guide vane, internal grooved, internal and external fins, rotation of tube are the geometrical device used as a passive methods, and the use of additives, nano-fluids, induced pulsation by cams, reciprocating plungers, vibrators, use of magnetic fields, rough and extended surfaces, etc., are the passive techniques. This paper provides a literature review for the enhancement of the HMT of with various refrigerant and absorbent working fluids combination.

1.1 Absorption Refrigeration System Technology

Figure 1 shows VAR system which is similar to the mechanical VCR comparing with the working principle. The compression between the VAR and mechanical VCR lies in the way of compressing the refrigerant vapour from a lower pressure zone to a high pressure zone, Fig. 1. The VAR system, the compressor is replaced by a thermal operated compressor. This subsystem consists mainly of a generator, a

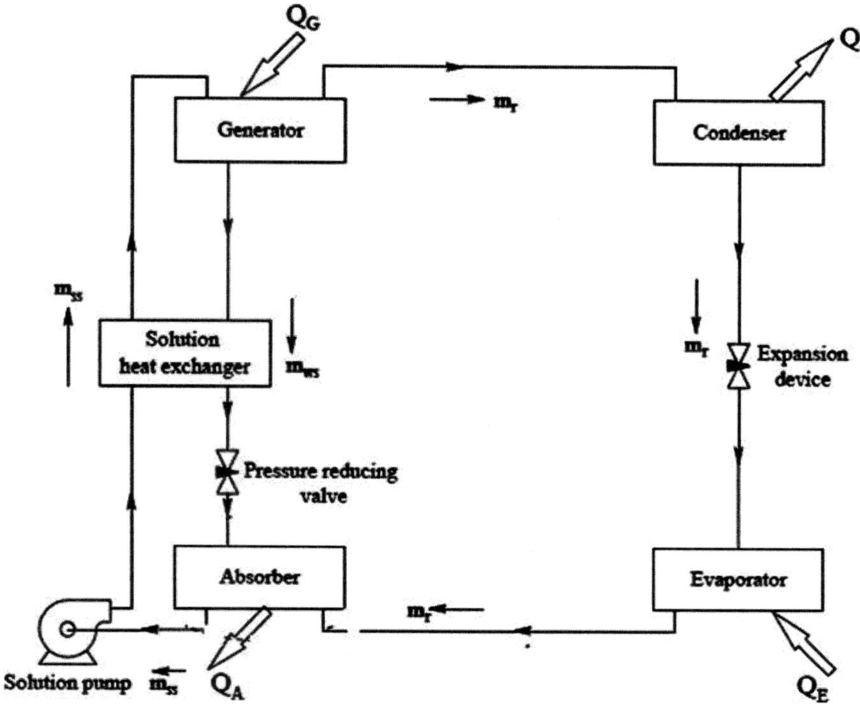

Fig. 1 Schematic diagram for VAR system

solution heatexchanger (SHX), solution pump, expansion or pressure reduction valve, and absorber. In the absorber, the refrigerant in vapour is injecting from the evaporator which will mix with the weak composition in solution coming from generator through the SHX and pressure reduction valve. The weak solution is converted to strong composition in the solution in term of concentration of the refrigerant at the flow progresses through the absorber. The solution pump is used to pump the strong solution to the generator through the SHX. The SHX is used to exchange the heat between low-temperature strong solution and high-temperature weak solution which will help in generation of the refrigerant in generator. The generation refrigeration vapours supply to the condenser for cooling process to repeat the absorption cycle. The COP of VAR system is defined as the relation between the ratio of the cooling or heating effect and the amount of heat power input or consumption by pump. The major advantage of the VAR system is less electricity power consumption by the solution pump, fewer moving parts leading to less noise and less maintenance, use natural refrigerant like water and ammonia which creates less impact on environmental; the energy sources uses are waste energy or renewable energy like solar, geothermal energy.

1.2 Working Fluids

Many refrigerant-absorption fluid combinations are used for VAR system. Ammonia-water (NH_3–H_2O) and water-lithium bromide (H_2O–LiBr) are the traditional working fluid combination for VAR system. The working pair of mixtures NH_3–H_2O and H_2O–LiBr have proved to have better characteristics and largely used fluid working pairs on the VAR cooling or heating systems. The NH_3–H_2O VAR system is mainly used for refrigeration applications due to the low NH_3 freezing point. NH_3 provides better thermodynamic and thermo-physical properties as refrigerant. NH_3 is environmentally friendly and emissions control of GWP. NH_3–H_2O VAR systems require high generator operation temperatures (more than110 °C) to keep the refrigerant cooling capacity at low evaporation temperatures. The disadvantages of the NH_3–H_2O system include toxicity and incompatibility with copper-based alloy and metals; this mixture is environmentally friendly with a lower cost. A few range of fluid combinations have been studied by many investigators for bubble dynamics, and as well as VAR system study like, viz. air-water, glycerol-air, methanol-air, ethanol-air, etc., were used for bubble dynamics studies and ammonia-water, water-lithium bromide, R134a-DMF, etc., for absorber studies.

1.3 Absorber

Absorbers are classified into three conventional types based on modes of design and operation: Falling film absorber (FFA), spray absorber (SA), bubble absorber (BA), and membrane-based absorbers (MBA). The FFA is most commonly used in the VAR system reported in the literature database. BA is used in VAR-based chillers that provide cooling effect the SA type absorber is new to the researchers. In the case of MBA, the few studies have just been reported in the literature recently. Detail description of the absorber size, shape, design configurations are presented in the upcoming sections.

1.4 Falling Film Absorber

FFA also is classified into two types based on the position of operation: horizontal tube and vertical tube. In horizontal tube configuration, the solution enters into the absorber through a mass distributor located in the upper parts of the absorber as shown in Fig. 2a. The solution flows as a film into a FFA parallel to the cooling water tube surfaces. The solution leaving the mass distributor has contact with the saturated vapour at the vapour–liquid interface phase of refrigerant which occupies the volume of the absorber. The weak solution follows the path along the falling film, which travels from one tube to another, follows the bundle of tubes, hence allows more

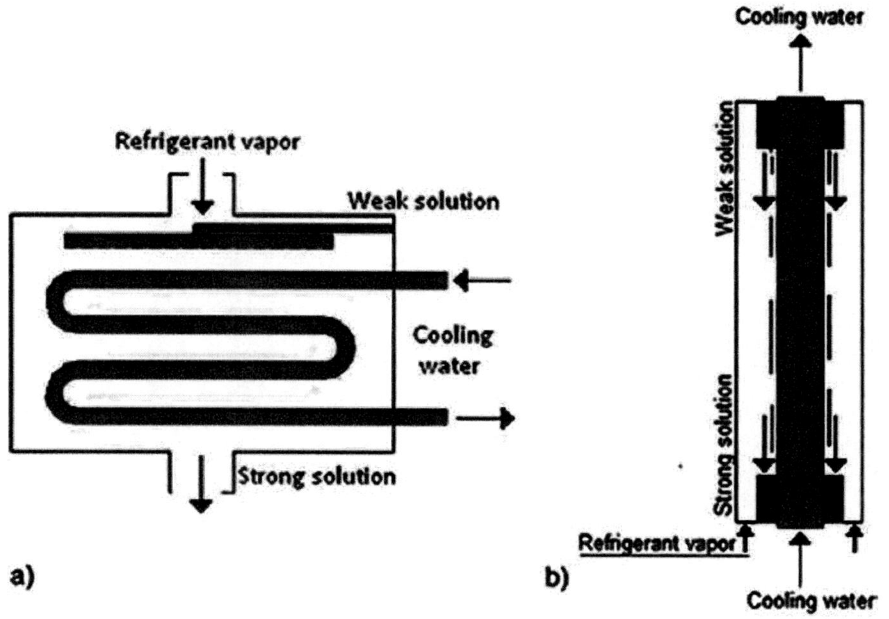

Fig. 2 Falling-type absorber. **a** horizontal-type. **b** Vertical-type [1]

refrigerant absorption in the solution. The strong solution in refrigerant leaves the absorber through the lower parts. Vertical-type FFA configuration is shown in Fig. 2b, which is counter flow-type configuration. The cooling water flows enter from the bottom and move to upward along the tubes, and the weak solution flows from the top and moves downward through the outer tube, meanwhile the refrigerant vapour flows in the counter current path for the improvement of HMT analysis. This configuration study is further extended for the design of air-cooled absorbers, in which the solution and refrigerant flow inside the tube. H_2O–LiBr working fluid combination are used in horizontal-type FFA configurations. Due to the insignificant pressure drop involve; the design is more complicated. Also due to the low operating pressures, solution mass distribution limitations and insufficient wetting of the cooling water tube surface area are the affects parameter for HMT analysis.

1.5 Spray Absorber

The SA contains an adiabatic chamber and a solution sprayer, Fig. 3a. The weak solution in refrigerant from the generator is introduced into the SA where it gets contact with the refrigerant in vapour phase come out from the evaporator. After the absorption process completed, the solution mixture, which becomes strong in refrigerant, leaves the absorber through the bottom part of the adiabatic chamber. SA

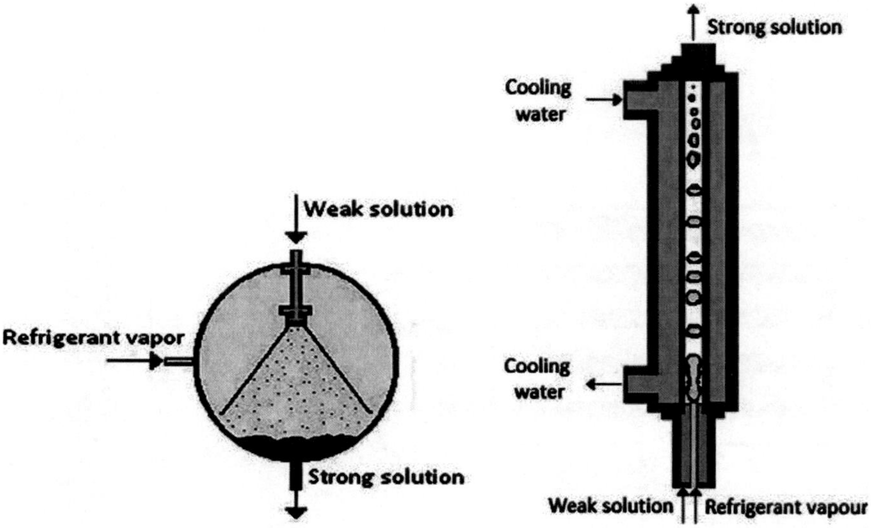

Spray Absorber[1] Bubble Absorber[1]

Fig. 3 **a** Spray-type absorber. **b** Bubble-type absorber

used to disperse the solution into the chamber with fine drops to enhance the HMT through mixing with the reduction on the solution resistance. Figure 3a represents SA occurs in an adiabatic chamber showing the HMT process. The process allows to make use of additional necessary components to enhance sub-cool absorption process in the weak solution. The recirculation of the strong solution throughout the adiabatic SA were creates an important role in order to increase absorption process and to make the solution equilibrium and stable conditions.

1.6 Bubble Absorber

A vertical position BA is shown in Fig. 3b. The basic principle of BA configuration is that the pure refrigerant enters from the bottom and mixes with the weak in refrigerant solution which also enters from the bottom, moves upward direction and come out as a strong solution in refrigerant at the upper part of the absorber. The refrigerant in vapour state comes in the absorber in bubble mode and is absorbed by the solution mixture as it rises upwards against the flow of the weak solution and converted strong solution after mixing with refrigerant solution. The mixing heat energy is released during the absorption of the refrigerant vapour, which is extracted by the circulating a cooling fluid which flowing outside channel of the BA. According to the available literature database, BA provides higher HMT coefficients during the absorption process in comparison to FFA.

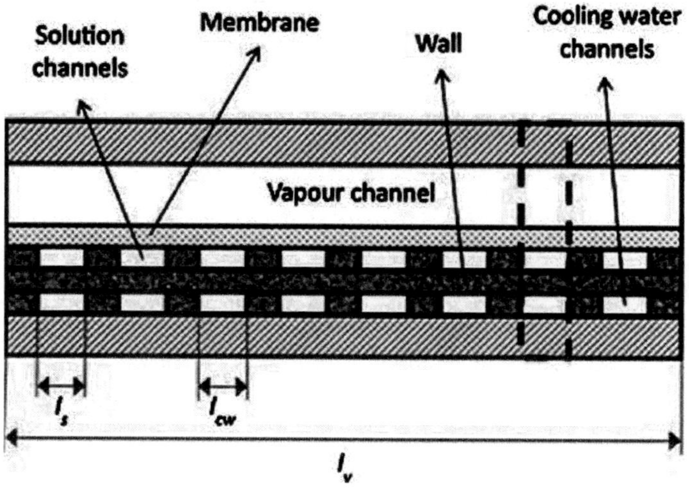

Fig. 4 Membrane-type absorber

1.7 Membrane Absorber

MBA plays an important role on the VAR system study. MBA consist of a hydrophobic micro-porous-based membrane located between solution and refrigerant vapour flow channel as shown in Fig. 4. The membrane provides the refrigerant vapour to mixes with the weak solution which travels in opposite direction to each other and the cooling water flow next channel to the solution flow channel which is separated by conductive wall.

The passive intensification of the HMT phenomenon implementing in the absorbers by applying three different techniques such as mechanical treatments of surface or surface designs, mixing of additives in working fluids, and nanotechnology. The resistance to the HMT is found on the solution side which is also a mixing techniques are applied for enhance the mixing of liquid and vapour phases, which enhances the HMT processes. The research on the technique is the interested topic for the absorbers and the VAR systems on HMT analysis.

1.8 Enhancement Technique

The HMT phenomena takes place in the outer surface or the vapour–liquid interface of the falling film absorbers; hence, improvement in the surface properties or mechanical treatments of the surface increases the HMT performance. Improvement in the surface includes scratched or increases in roughness of a surface, curvature tubes, fluted tubes, micro-finned tubes, corrugated plates, finned plates, etc. HMT in bubble absorber takes place in the inner surface of the tubes or channels where the absorbent solution and refrigerant mixing process takes place. Therefore, surface treatments in the inner surface of the bubble absorbers involve by the use of internal extended surfaces or corrugated surfaces such as internal grooves, fins, baffle plates. Some enhancement technique also increases the HMT phenomena by providing tapes, swirl generator (axial and radial), multiple injection, etc. Microchannel tubes or compactness of the inner surface and constrained designs can be used in case of membrane absorbers for the enhancement. A review of the theoretical, experimental, and numerical studies available in the open literature based on design of the absorbers with enhanced surface designs for absorption systems is presented in the following section.

The HMT performance is evaluated [13] for three different types of bubble absorbers with combination of desorber–absorber heat exchanger absorption cycles for a compact designs for generator–absorber heat exchanger applications (GAX). The compact designs tested were spiral flutes, which involved roughness, and internal spacers. An experimental study for compact heat exchanger with enhanced surface was used as falling film absorber [9] on NH_3–H_2O absorption processes. Off strip fin is used between two plates for absorber design on the solution flow area, and rectangular plain fins are used on coolant side. The performance study analyses on the effects are like vapour and solution flow and temperature, and inlet concentration on HMT rates. The study found that solution Nusselt number was affected by solution and vapour flow rate of the falling film, and Sherwood number was more affected by the vapour flow effects more in a comparison to the solution flow. Finally, the authors also developed correlations for HMT coefficients for this design. A vertical tube falling film absorber was used for experimental study [7], with different enhance areas for heat transfer in a bubble mode a narrow plate with different corrugation angles (30° and 60°). The plates examined on the corrugation angle affected to the heat transfer performance of the falling absorber and compared with plain surface plate which is three times more.

A H_2O–LiBr falling film absorber considers evaluating the performance analysis [17] with implemented of tubes with different shapes of extended surfaces such as bare tube, floral-shaped tube, twisted floral tube, and bumping bare tube. The experiments studies were conducted with the effect of the parameters such as solution and cooling water flow rates. The authors also reported that tubes wettability of the surface improves the HMT between the solution mass flow and the refrigerant. The effect of solution and refrigerant vapour flow rates on a plate-type heatexchanger consider with NH_3–H_2O bubble absorber with three different types of plates evaluated [12]. The different plates are considered such as hair-lined plate treated by laser, smooth plate, and plate treated by sand paper, and proposed a correlations for HMT analysis in terms of the Nusselt and Sherwood number.

Experimental study was conducted [10, 15] with the effects of micro-channel based H_2O–LiBr falling film absorber for analysis of the performance and reported that the wettability increased the performance, and the hatched tubes shows two times better than the smooth tube. Experiments were carried [6] based on falling film tube absorbers which consists of 4 columns, 60 horizontal tubes on each column, and tested with and without a screen mesh and calculated the thermal load of each components. The study also compared the performance on with and without screen mesh and reported the enhanced result found with the mesh. A horizontal film absorber was considered for performance analysis NH_3–H_2O [9] with different plate surfaces. Porous medium was used for guiding the refrigerant and weak solution side where as the coolant flowed in counter-current flow to enhance the HMT rate. A HMT study NH_3–H_2O working pair for on a corrugated plate bubble absorber was considered for experimental work [2]. L-type plate heat exchanger was used as the absorber having three channels where refrigerant vapour entered in bubble mode from the bottom of the middle channel in parallel with the solution flow. Experiments were performed varying parameters such as solution inlet temperature, solution flow rate, inlet solution concentration, absorber pressure, and cooling water inlet temperature. Bubble absorber with an internal smooth tube is used NH_3 absorption process in solution mixture of NH_3–$LiNO_3$ as working pair [1]. A comparison study based on experimental result was reported between the smooth surface tube and corrugated plate as per the maximum absorption rate which was up to 1.7 times higher than that of the smooth tube. NH_3–H_2O VAR system was examined [3] on a double pipe absorber to study the absorption performance with a helical static mixer in both the central channel for solution and annular channel for the coolant. The study reported that an improvement of 20% in vapour absorption performance and 31.6% in heat load in a comparison of smooth pipes. Many report like [16] evaluated the HMT rate in a H_2O–LiBr falling film absorber with a stainless steel mesh packing structure along the mixing absorption process. As many as forty stainless steel mesh screens are consisted of folded as longitudinal troughs inserted between the horizontal tubes for the advanced surface design. Based on the experimental results, the solution mass transfer by 17.2% mesh increased compared to that of the bundle of tubes without the mesh. A falling film absorber with M–W mesh guider inserts was used [4] for the HMT coefficients which were increased around 33.4 and 55.4% compared to the bare tubes. A [15] H_2O–LiBr falling film plate absorber designed with offset-strip fins manufactured form sand blast used for absorption process, conducting experiment with the effect of the solution temperature, solution flow rate, and the cooling water temperature.

1.9 Nano-Particles in Working Fluids

Nanotechnology is considered for HMT enhancing technique for the various components of VAR systems. The working fluid of VAR system combined with nano-particles is called binary nano-fluid or binary mixture. The implementation of nano-particles increases the thermal conductivity (k) of the working fluids used on conventional VAR system and enhances the mixing process on absorption. Nano-fluids [6] used for increases in the k of the solution with the increase in the temperature for enhance of the heat transfer rate and vapour absorption rate to increase in mass transfer. The nano-particles [5] used for the heat transfer enhancement with the enhanced of thermal conductivity but did not significant result on the enhancement in the convective heat transfer. Study of mass transfer enhancement has been considered for the optimized of the volume fraction with improvement k using nano-particles [12].

2 Conclusions

The present literature study reviewed the various enhancement techniques used for HMT analysis on VAR system based on published literature database. At first, the different types of absorbers such as falling film spray and bubble absorber with their construction, design, and working were classified on experimental studies on the absorption process, for various operating conditions to found the absorption characteristics for different working fluid combinations. Then, the various techniques used for HMT analysis were described for vapour absorption process using advanced surface designs, additives, and nano-fluids, or nano-particles were reviewed. The studies also focused the effects of the operating conditions such as solution flow; temperature, concentration, pressure, and coolant temperature, while others were based on geometrical configuration such as tube spacing, the number of tubes, and diameters others. The performance of VARS has also improves with the complex absorber designs and enhanced heat and mass transfer in working fluids by adding surfactants and nano-particles.

References

1. Amaris C, Bourouis M, Vallès M (2014) Effect of advanced surfaces on the ammonia absorption process with NH3/LiNO3 in a tubular bubble absorber. Int J Heat Mass Transfer 72:544–552
2. Cerezo J, Bourouis M, Vallès M, Coronas A, Best R (2009) Experimental study of an ammonia-water bubble absorber using a plate heat exchanger for absorption refrigeration machines. Appl Therm Eng 29:1005–1011
3. Cerezo J, Best R, Chan L, Romero R, Hernandez J, Lara F (2018) A theoretical-experimentalcomparison of an improved ammonia-water bubble absorber by means of ahelical static mixer. Energies 11:56
4. Chen Y, Cao R, Wu J, Yi Z, Ji G (2016) Alternate heat and mass transfer absorption performances on staggered tube bundle with M-W corrugated mesh guider inserts. Int J Refrig 66:10–20
5. Ding Y, Chen H, He Y, Lapkin A, Yeganeh M, Siller L et al (2007) Forced convective heat transfer of nanofluids. Adv Powder Technol 18(6):813–824
6. Das S, Putra N, Thiesen P, Roetzel W (2003) Temperature dependence of thermal conductivity enhancement for nano fluids. J Heat Transf 125:567–574
7. Goel N, Goswami Y. Experimental verification of a new heat and mass transfer enhancement concept in a microchannel falling film absorber. J Heat Transf 129(2):154–61
8. Helbing U, Würfel R, Fratzscher W (2000) Comparative investigations of non-adiabatic absorption in film flow and bubble flow. Chem Eng Technology 23:1081–1085
9. Jenks J, Narayanan V (2008) Effect of channel geometry variations on the performance of a constrained microscale-film ammonia-water bubble absorber. J Heat Transf 130(11):1–9
10. Kang YT, Akisawa A, Kashiwagi T (1999) Experimental correlation of combined heat and mass transfer for NH_3-H_2O falling film absorption. Int J Refrig 22:250–262

11. Kim JK, Park CW, Kang YT (2003) The effect of micro-scale surface treatment on heat and mass transfer performance for a falling film H_2O/LiBr absorber. Int J Refrig 36:575–585
12. Krishnamurthy S, Bhattacharya P, Phelan PE (2006) Enhanced mass transport in nanofluids. Nano Lett 6(3):419–423
13. Lee KB, Chun BH, Lee JC, Lee CH, Kim SH (2002) Experimental analysis of bubble modein a plate-type absorber. Chem Eng Sci 57:1923–1929
14. Merrill TL, Setoguchi T (1995) Perez-Blanco H. Passive heat transfer enhancement techniques applied to compact bubble absorber design. J Enhanc Heat Transf **2**(3):199–208
15. Mortazavi M, Isfahani RN, Bigham S, Moghaddam S (2015) Absorption characteristics of falling film LiBr (lithium bromide) solution over a finned structure. Energy 87:270–278
16. Park CW, Kim SS, Cho HC, Kang YT (2003) Experimental correlation of falling film absorptionheat transfer on micro-scale hatched tubes. Int J Refrig 26:75–763
17. Wu J, Yi J, Chen Y, Cao R, Dong C (2015) Yuan S. Enhanced heat and mass transfer inalternating structure of tubes and longitudinal trough mesh packing in lithiumbromide solution absorber. Int J Refrig 53:34–41
18. Yoon JI, Kwon OK, Moon CG (1999) Experimental investigation of heat and mass transfer in absorber with enhanced tubes. J Mech Sci Technol 13(9):640–646

Heat Transfer Analysis of Clay Pot Refrigerator Adopting Curvature Effect

Abhijit Date, Kaushal Prasad, Akshay Shirsat, and Roshan Mayekar

Abstract Preservation of food is major concern in India. In rural areas where electricity is not reached, clay pot is best option for preservation. The steady-state performance of clay pot is already carried out by considering cylindrical shape in the literature. But, in actual practice shape of the clay pot is curved; hence, it is necessary to study the curvature effect on the performance of clay pot refrigerator. For particular cooling load, the refrigeration temperature and geometrical dimensions can be predicted under various ambient conditions. Introduction of curvature in model reduces size of pot for the same refrigeration temperature. For a parabolic geometry, approximately 7–18% higher efficiency can be achieved as compared to cylindrical shape at same surface area for given atmospheric relative humidity and temperature. Also, overall size of model can be reduced. Hence, more compact model can be used for same heat transfer.

Keywords Clay pot refrigerator · Curvature effect · Cylindrical shape · Efficiency

Nomenclature

A	Surface area (m^2)
B	Spalding number
B_i	Biot number
b	Clay wall thickness variable (m)
C_p	Specific heat (J/kg K)
g	Mass transfer coefficient (kg/m^2 s)
H	Pot height (m)
h	Enthalpy of mixture (J/kg)
k	Thermal conductivity (W/m K)
K_p	Permeability (m^2)

A. Date (✉) · K. Prasad · A. Shirsat · R. Mayekar
Mechanical Engineering Department, Finolex Academy of Management and Technology, Ratnagiri 415639, India
e-mail: abhijitdate1@gmail.com

M. Ramgopal et al. (eds.), *Advances in Air Conditioning and Refrigeration*, Lecture Notes in Mechanical Engineering, https://doi.org/10.1007/978-981-15-6360-7_34

K_{hy}	Hydraulic conductivity (m/s)
L_e	Lewis number (–)
	Evaporation rate (kg/s)
M	Mass (kg)
P	Pressure (N/m^2)
Q	Heat transfer (W)
T	Temperature (°C)
l	Bulge length (m)

Greek symbols

α	Heat transfer coefficient (W/m^2 K)
β	Volumetric coefficient (K^{-1})
ω	Mass fraction
η_{th}	Thermodynamic efficiency
ε	Emissivity
λ	Latent heat (J/kg)
Φ	Relative humidity

Suffix

a	Air or atmospheric
cl	Clay
cold	Inner pot environment
dp	Dew point
e	Energy conservation principle
eff	Sand + water
i	Inner pot
in	Radiation + convection
load	Cooling load
m	Mass conservation principle or mean
M	Transferred substance state
mean	Mixed mean
nc	Natural convection
o	Outer pot
rad	Radiation
ref	Reference value
v	Vapor
w	Water or w-w state
∞	Ambient condition

1 Introduction

Preservation of food is major concern in India. In rural areas where electricity is not reached, clay pot is best option for preservation. Clay pot gives refrigeration effect on the principle of evaporative cooling, and analysis of cylindrical shape clay pot is already carried out in past. In order to make this analysis more general or practical, curvature of clay pot has to be considered. For the simplicity, parabolic shape is considered in order to study the effect of curvature of pot on various output parameters like inside cold temperature (T_{cold}) and efficiency of pot. As stated and explained in [1], in order to overcome difficulty in defining COP of refrigerator efficiency, it is defined in following way.

$$\eta_{\text{th}} = \frac{T_{\infty} - T_{\text{cold}}}{T_{\infty} - T_{\text{dp}}} \tag{1}$$

The refrigeration temperature inside pot depends upon various parameters like inner and outer radius, height, effective conductivity of insulation, cooling load, etc. which are studied in detail further. Apart from these parameters, introduction of bulge length (l) also had a great influence on efficiency and more importantly on size of pot. Any kind of curved clay pot can be converted into equivalent cylindrical model and analyzed as mentioned in [1].

As mentioned earlier, heat and mass transfer analysis of clay pot is carried out by considering cylindrical geometry. The analysis is done using Reynolds flow model to find out the functional dependence between input parameters like geometrical parameters, ambient conditions and heat transfer parameters and output parameters that are efficiency and inside cold temperature [2]. Numerical calculation is done for a reference case like assuming values of geometrical and heat transfer parameters. For this reference case, efficiency and inside cold temperature are evaluated at various ambient conditions. Ambient condition that is studied in [1] is percentage of relative humidity, e.g., 10, 20, 30% and so on and environmental temperatures 35, 40 and 45 °C. It is concluded in [1] that maximum efficiency of pot is obtained at 40% relative humidity.

In this paper, the effect of the curvature on the performance of clay pot refrigerator as well as parametric variation is studied. Figure 1 shows the assumed geometry of clay pot which is analyzed in [1].

2 Modeling of Clay Pot Refrigerator

2.1 Parabolic Model

Figure 2 shows modified model of the refrigerator considered for analysis. The traditional clay pots have irregular or undefined curved shape. But for the simplicity,

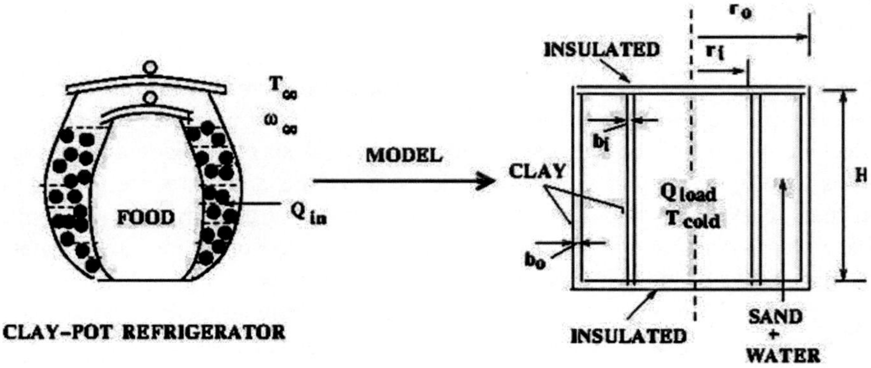

Fig. 1 Clay pot refrigerator and assumed model [1]

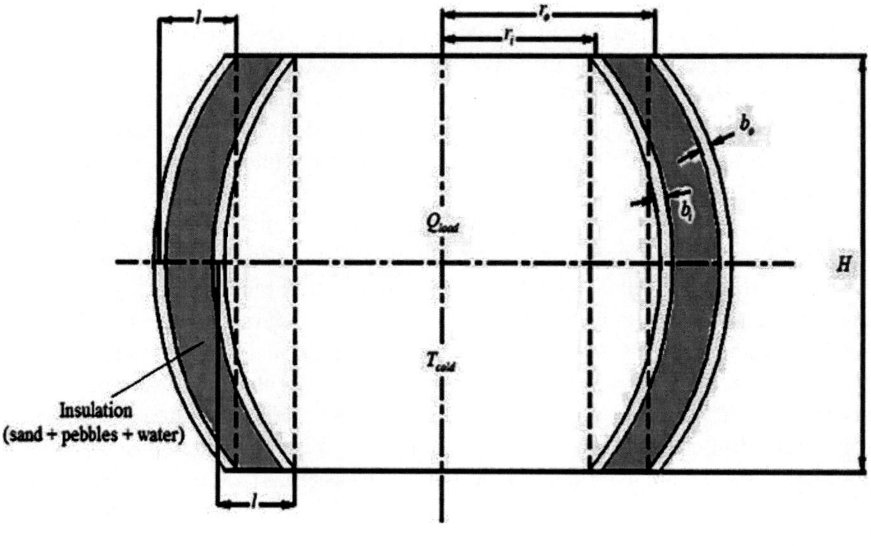

Fig. 2 Developed parabolic model

parabolic curvature is considered. Hence, assumed model is combination of cylinder and parabolic curve [3–6]. The cylinder has inner radius r_i and outer radius r_o with height H. For the parabolic curve, bulge length is assumed as l. The parabola passes through outer end points of cylinder. Inner and outer clay pot thicknesses are b_i and b_o. Thermal conductivity of clay used to made clay pots is k_{cl} and effective thermal conductivity of the insulation (sand + pebbles + water) is k_{eff}. Then, the relationship between output parameters and input parameters will be given by,

$$(\eta, T_{cold}) = f\left\{(T_\infty, \varphi_\infty), (r_i, r_o, H, b_i, b_o, l), (k_{cl}, k_{eff}, \alpha_i, \alpha_o, \varepsilon), k_{hy}\right\} \quad (2)$$

Table 1 Assumed input parameters

Geometric parameters (m)	Specific heat (J/kg K)	Thermal conductivity	Hydraulic conductivity	Others
$r_i = 0.075$ $r_o = 0.15$ $b_i = b_o = 0.005$ $H = 0.3$ $l = r/4$	$C_{p_a} = 1005$ $C_{p_v} = 1880$ $C_{p_w} = 4186$	$k_a = 0.027$ $k_{cl} = 1.5$ $k_{eff} = 2$	$k_{hy} = 8 \times 10^{-9}$ m/s	$Q_{load} = 1$ W $\upsilon_a = 16.5 \times 10^{-6}\,\text{m}^2/\text{s}$ $T_{ref} = 0\,°\text{C}$ $\lambda_{ref} = 2503$ kJ/kg

2.2 *Assumed Input Parameters*

See Table 1.

2.3 *Conversion of Curved Pot into Equivalent Cylindrical Pot*

For any type of curvature surface construction of its equivalent cylindrical model is carried out by taking integration of that curve over that height and dividing this by height. Mathematical procedure is as follows

As shown in Fig. 3, the equation of parabola is,

$$y^2 = kx \tag{3}$$

Since the parabola passes through pt. $\left(r, \frac{h}{2}\right)$ value of k is $k = -\frac{h^2}{4l}$.

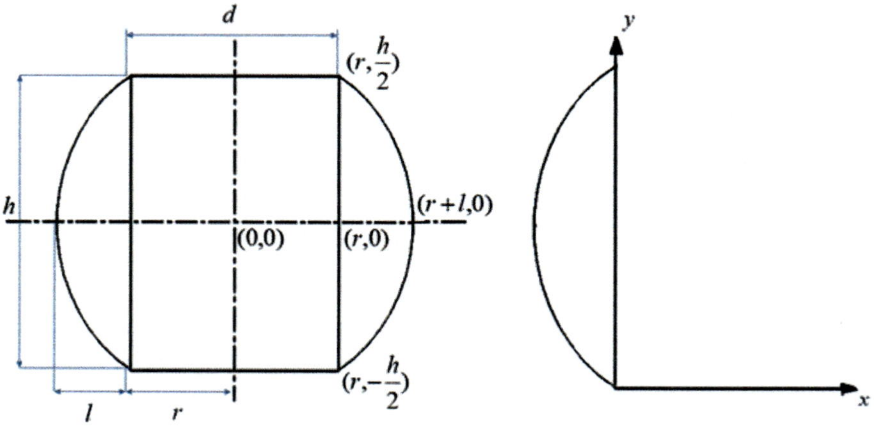

Fig. 3 Geometrical model

$$\int_{-h/2}^{h/2} x.dy = 2\int_{0}^{h/2} x.dy \tag{4}$$

$$2\int_{0}^{h/2}\left(l - \frac{4ly^2}{h^2}\right).\mathrm{d}y = \frac{2lh}{3} \tag{5}$$

Now, approximate line that represents this parabola can be given as,

$$e = \frac{\text{Area under the curve}}{\text{Base length}} \tag{6}$$

$$\therefore e = \frac{\left(\frac{2lh}{3}\right)}{h} = \frac{2l}{3} \tag{7}$$

For parabolic curve equivalent radius is (Fig. 4),

$$\begin{aligned} r_{e_o} &= r_o + e \\ r_{e_i} &= r_i + e \end{aligned} \tag{8}$$

The rest procedure for calculation of efficiency and T_{cold} temperature is same as stated in [1].

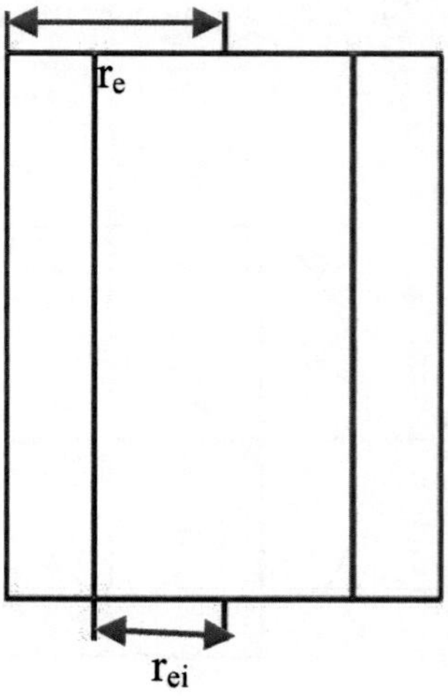

Fig. 4 Equivalent cylindrical model for parabolic profile

Table 2 Validation of parabolic model with cylindrical

Parabolic model	Cylindrical model
$T_w = 33.287\ °C$	$T_w = 33.287\ °C$
$T_o = 33.1362\ °C$	$T_o = 33.1362\ °C$
$T_i = 30.7843\ °C$	$T_i = 30.7843\ °C$
$T_{cold} = 32.9572\ °C$	$T_{cold} = 32.9572\ °C$
$Q_L = -13.7915$ W	$Q_L = -13.7915$ W
$Q_{in} = 19.0471$ W	$Q_{in} = 19.0471$ W
COP = 0.0525	COP = 0.0525
$\alpha_{rad} = 6.7439$	$\alpha_{rad} = 6.7439$
$\eta_{th} = 0.5663$	$\eta_{th} = 0.5663$
$g = 0.003196$	$g = 0.003196$
$\alpha_{nc} = 3.2911$	$\alpha_{nc} = 3.2911$
$B = 0.0075$	$B = 0.0075$

2.4 Validation of Model

Assumed parabolic model is combination of cylinder and parabola. Additional parameter in this model is bulge length (l). Hence, this model can be validated by making this bulge zero and comparing results with cylindrical model results in [1]. Following are the results obtained by keeping bulge zero in parabolic model for reference case $T_\infty = 40\,°C$, $\varphi = 50\%$, $Q_{load} = 1$ W.

From Table 2, the difference between values of parabolic and cylindrical model is negligible. Hence, the assumed parabolic model is correct for curvature heat transfer.

3 Result and Discussion

3.1 Comparison of Cylindrical and Parabolic Model with Same Surface Area

For comparing heat transfer performance of two models, surface area of outer pot should be same. Hence, for reference case of $r_i = 0.075$ m, $r_o = 0.15$ m and $H = 0.3$ m inner and outer surface areas are $A_o = 0.2827\,m^2$ and $A_i = 0.1414\,m^2$. For these same surface areas, construction of parabolic model is as follows.

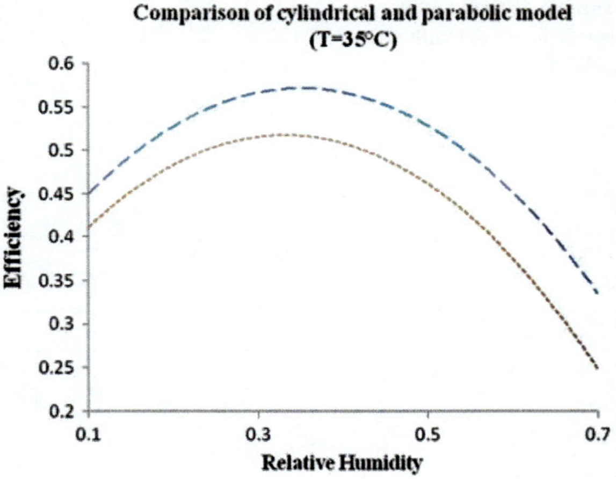

Fig. 5 Efficiency versus relative humidity (35 °C)

For construction, we have to keep two parameters fixed. So, height is shortened. Hence, height and bulge are

$H = 0.2$ m & $1 = 0.0375$ m	
$A_o = \pi \times D_{e_o} \times H$	$A_i = \pi \times D_{e_i} \times H$
$\therefore 0.2827 = \pi \times D_{e_o} \times 0.2$	$\therefore 0.1414 = \pi \times D_{e_i} \times 0.2$
$\therefore D_{e_o} = 0.45\,\text{m}$	$\therefore D_{e_i} = 0.225\,\text{m}$
$\therefore r_{e_o} = 0.225\,\text{m}$	$\therefore r_{e_i} = 0.1125\,\text{m}$
$\& \; r_{e_o} = r_o + \frac{2}{3}1$	$\& \; r_{e_i} = r_i + \frac{2}{3}1$
$\therefore 0.225 = r_o + \frac{2}{3} \times 0.0375$	$\therefore 0.1125 = r_i + \frac{2}{3} \times 0.0375$
$\therefore r_o = 0.2\,\text{m}$	$\therefore r_i = 0.0875\,\text{m}$

Taking these values of geometric parameter, results are evaluated and compared with original cylindrical model. Variation of efficiency versus relative humidity is plotted for both model. It is shown in Figs. 5, 6 and 7.

It is clearly seen from Table 3 that efficiency increases about 4.5% of cylindrical shaped clay pot (Figs. 6 and 7).

3.2 *Effective Heat Transfer Per Unit Surface Area*

From Table 4, it comes to know that heat transfer rate per unit outer surface area for parabolic shape is approximately 3% greater than that of cylindrical shape. Figures 8, 9 and 10 show that heat transfer rate per unit outer area is highest at 10% relative humidity and goes on decreasing with increase in humidity.

Table 3 Comparison of two models with same surface area

ϕ_∞	COP		T_{cold}		η_{th}	
$T_\infty = 35$ °C						
	Parabolic	Cylindrical	Parabolic	Cylindrical	Parabolic	Cylindrical
0.1	0.026561	0.027648	18.50182	19.84609	0.456045	0.418887
0.2	0.031462	0.03271	21.18979	22.3792	0.524341	0.479182
0.3	0.037973	0.039429	23.7417	24.78445	0.557502	0.505866
0.4	0.047009	0.048745	26.17216	27.07656	0.564232	0.506427
0.5	0.060313	0.062445	28.49594	29.2708	0.541858	0.477303
0.7	0.121002	0.124787	32.91169	33.46015	0.330879	0.243979
$T_\infty = 40$ °C						
ϕ_∞	COP		T_{cold}		η_{th}	
	Parabolic	Cylindrical	Parabolic	Cylindrical	Parabolic	Cylindrical
0.1	0.022035	0.022934	20.25779	21.83914	0.527523	0.485268
0.2	0.02624	0.027275	23.50042	24.88825	0.605165	0.554263
0.3	0.031804	0.033015	26.53491	27.74122	0.643859	0.586177
0.4	0.03949	0.040935	29.38718	30.42357	0.654801	0.590857
0.5	0.050727	0.052502	32.07894	32.95717	0.636873	0.566261
0.7	0.10117	0.104299	37.10516	37.70604	0.442487	0.350641
$T_\infty = 45$ °C						
ϕ_∞	COP		T_{cold}		η_{th}	
	Parabolic	Cylindrical	Parabolic	Cylindrical	Parabolic	Cylindrical
0.1	0.01859	0.019345	21.82995	23.67315	0.598859	0.55122
0.2	0.022254	0.023126	25.68313	27.2891	0.684865	0.627926
0.3	0.027073	0.028095	29.23344	30.61942	0.728462	0.664425
0.4	0.033682	0.034902	32.52439	33.70635	0.743527	0.673084
0.5	0.043271	0.044767	35.59464	36.58778	0.730289	0.653176
0.7	0.08549	0.088102	41.24017	41.90198	0.554789	0.457136

3.3 Parametric Variation

Variation in output parameters (efficiency and temperature inside pot) can be studied in by keeping either ambient temperature (T_∞) constant or ambient relative humidity (ϕ_∞) constant. The important parameters that give comparative variation in efficiency are inner and outer radius (i.e., r_i & r_o), height (H), effect on efficiency which is studied by varying these parameters ±20%.

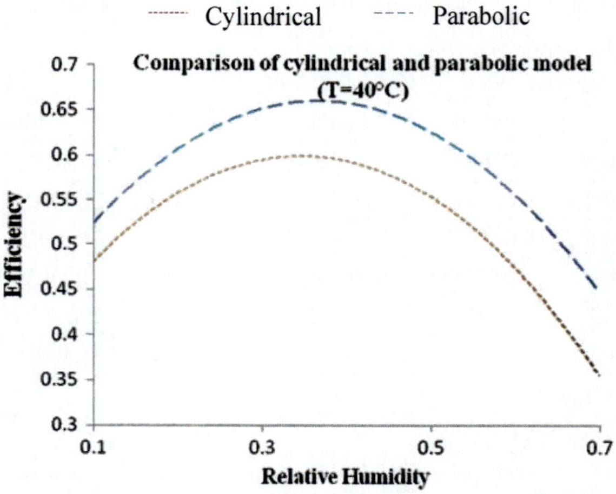

Fig. 6 Efficiency versus humidity (40 °C)

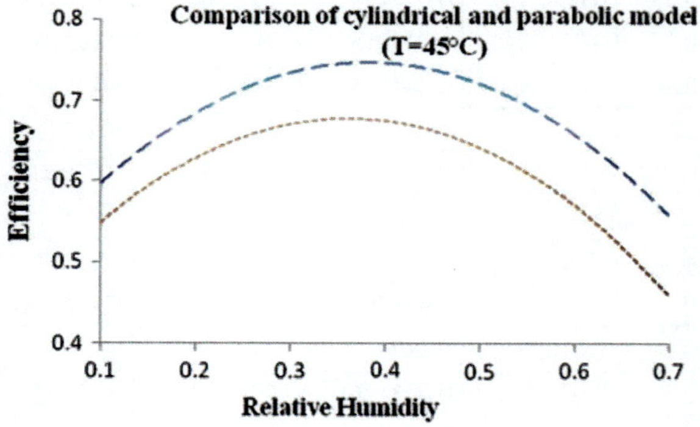

Fig. 7 Efficiency versus humidity (45 °C)

Table 4 Heat transfer per unit outer surface area

ϕ_∞	T35		T40		T45	
	Parabolic	Cylindrical	Parabolic	Cylindrical	Parabolic	Cylindrical
0.1	129.6179	124.3847	156.9692	150.681	186.7118	179.2942
0.2	108.8786	104.5894	131.2481	126.1319	155.3898	149.3991
0.3	89.60305	86.16287	107.6672	103.5883	127.0994	122.3495
0.4	71.69941	69.02013	86.02529	82.86282	101.4675	97.79734
0.5	55.10339	53.10169	66.18539	63.82843	78.19857	75.46651
0.7	25.69218	24.80566	31.42199	30.37314	37.83366	36.60716

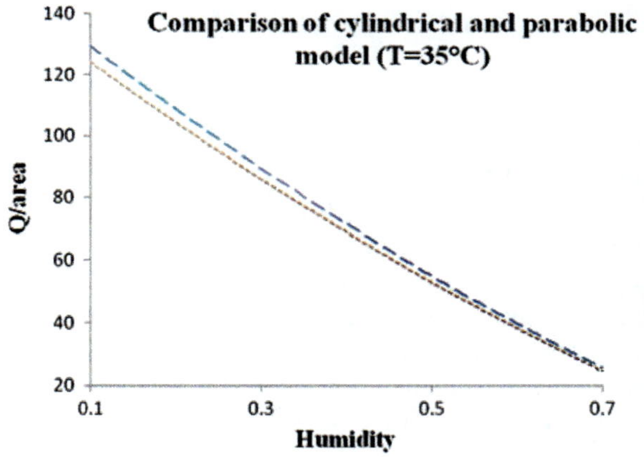

Fig. 8 Q/area versus humidity (35 °C)

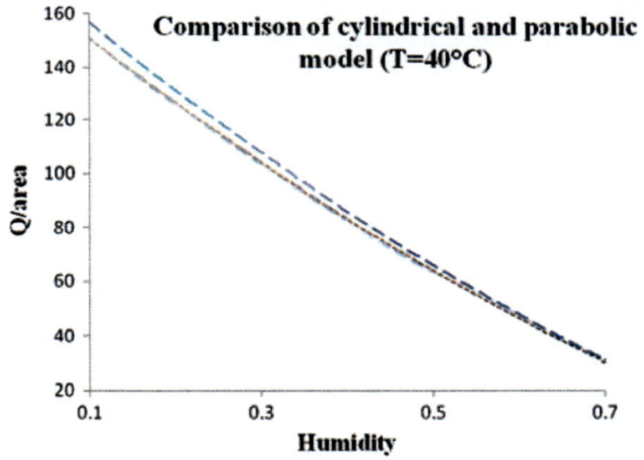

Fig. 9 Q/area versus humidity (40 °C)

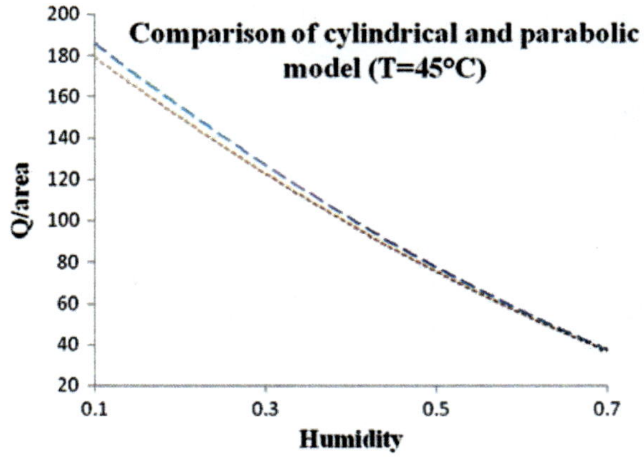

Fig. 10 Q/area versus humidity (40 °C)

3.4 *Ambient Temperature Constant* (T_∞)

Change in efficiency and inside cold temperature is plotted with variation of input parameters for three ambient temperatures, viz. 35, 40 and 45 °C for common relative humidity of 50%.

Figures 11 and 12 show that increase in inner radius efficiency decreases and inside cold temperature increases. This variation is due to decrease in insulation thickness. As inner radius increases keeping outer radius constant, difference between outer and inner radius decreases. That means overall insulation goes on reducing increasing conductive heat transfer from inside of pot to environment. This increases inside cold temperature reducing the efficiency of pot.

Figures 13 and 14 show that increase in outer radius efficiency increases and inside cold temperature decreases. This is because of increase in insulation. With increase in outer radius keeping inner radius constant, difference between outer and inner radius increases. This results in increasing overall insulation thickness,

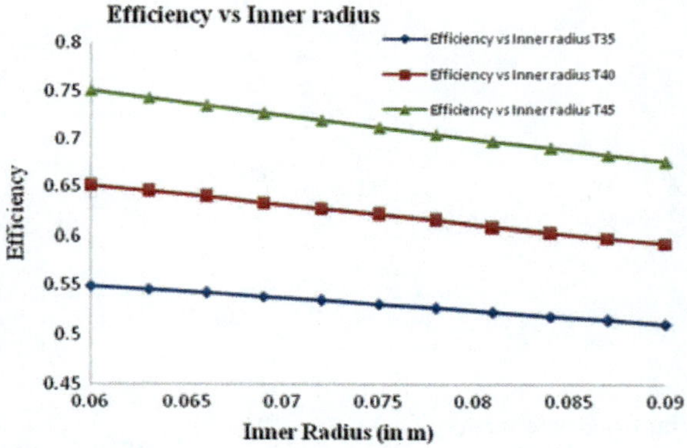

Fig. 11 Efficiency versus inner radius

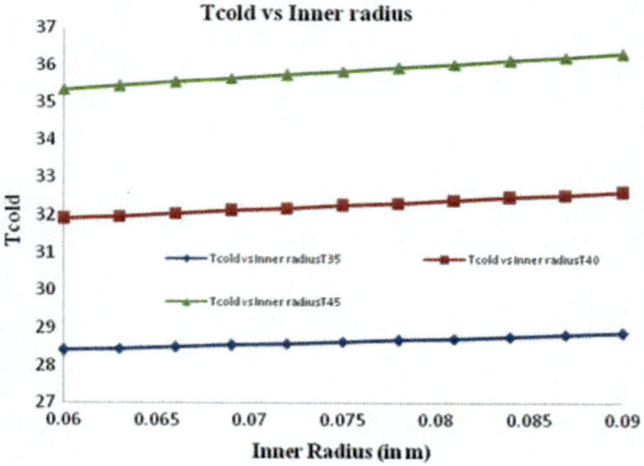

Fig. 12 T_{cold} versus inner radius

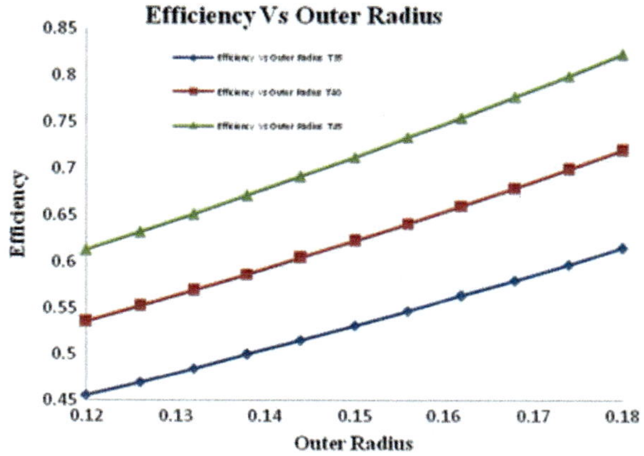

Fig. 13 Efficiency versus outer radius

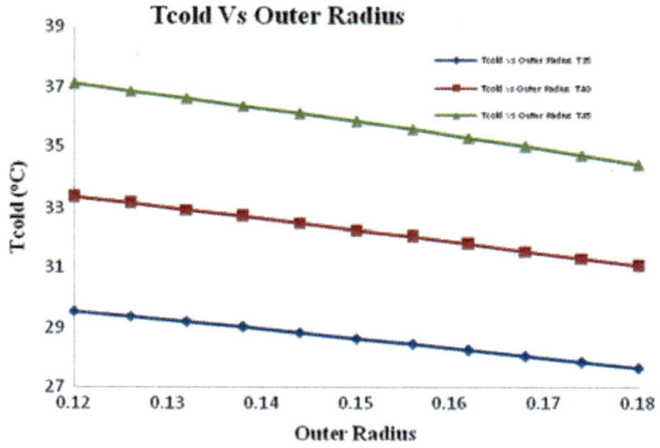

Fig. 14 T_{cold} versus outer radius

thereby reducing conductive heat transfer across walls. Because of this, inside cold temperature reduces increasing efficiency of pot.

Increase in height efficiency increases, and cold temperature inside pot decreases as shown in Figs. 15 and 16. This is because with increase in height outer surface area of pot increases increasing conductive heat transfer rate from outer surface of pot to environment. All this results in increase evaporative cooling effect.

3.5 Ambient Relative Humidity Constant (ϕ_∞)

Change in efficiency and inside cold temperature is plotted with variation of input parameters at various humidity values, viz. 10, 20, 30, 40, 50 and 70% keeping common ambient temperature 40 °C.

Efficiency is evaluated at various values of inner radius. This variation is shown in Fig. 17. Efficiency at all relative humidity values goes on decreasing for 0.06 to

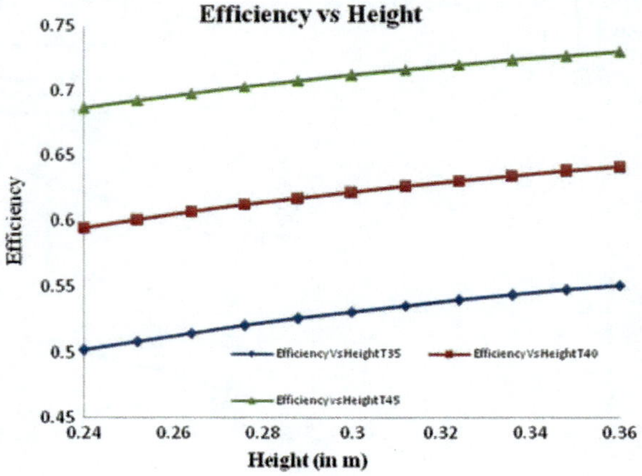

Fig. 15 Efficiency versus height

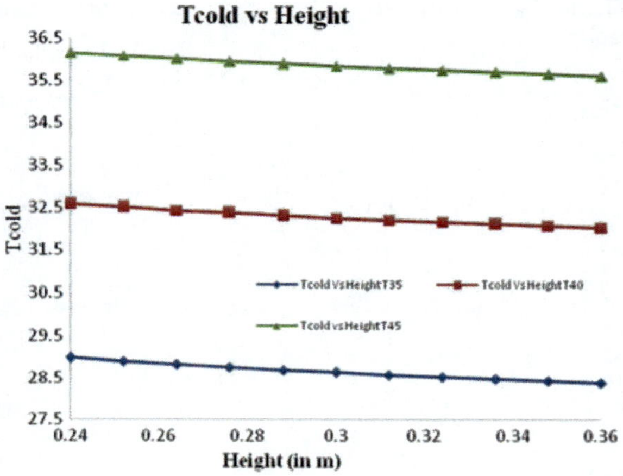

Fig. 16 T_{cold} versus height

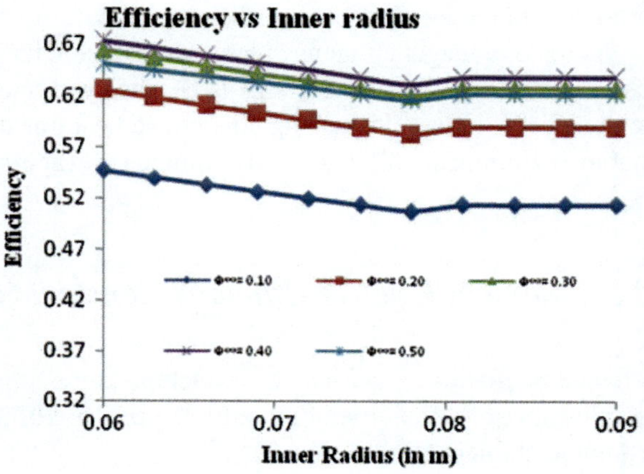

Fig. 17 Efficiency versus inner radius

approx 0.08 m, increases slightly and remains constant up to 0.09 m

Clearly seen from Fig. 18 that increase in outer radius efficiency goes on increasing. This is because of increase in insulation thickness of clay pot. Rate of increase in efficiency is maximum for ambient relative humidity 40% and minimum at 10 and 20%.

As shown in Fig. 19 increase in height efficiency increases and temperature decreases slightly. As height increases, outer surface area of pot increases resulting in increase in evaporative cooling effect. This achieves more cold temperature and increases efficiency. Rate of increase in efficiency is maximum at 70% relative humidity. There is very slight increase efficiency at 10, 20 and 30% relative humidity. Medium rate of increase in efficiency is at 40 and 50% relative humidity.

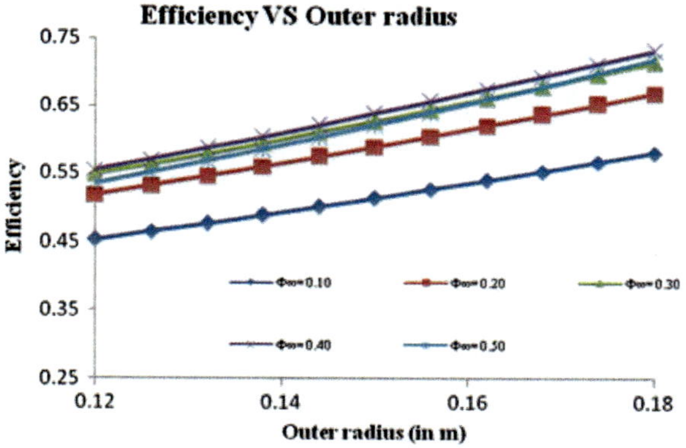

Fig. 18 Efficiency versus outer radius

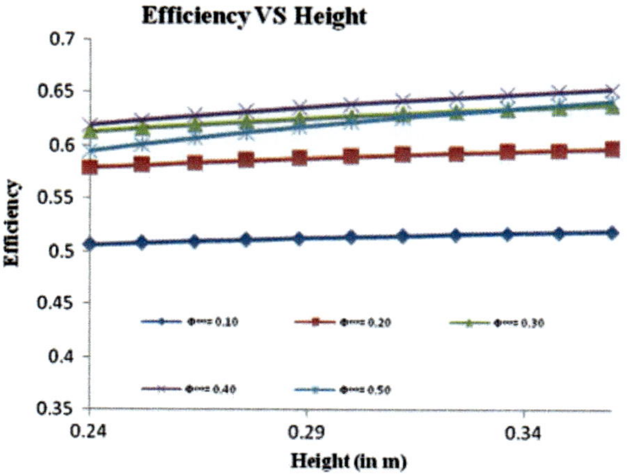

Fig. 19 Efficiency versus height

4 Conclusions

In this paper, the clay pot refrigerator is analyzed by considering the parabolic shape. The main conclusions are as follows:

1. For a parabolic geometry, 7–18% higher efficiency is achieved as compared to cylindrical shape at same surface area for given T_∞ & ϕ_∞. Also, overall size of model reduces. Hence, more compact model can be used for same heat load.
2. The efficiency decreases by approximately 1% with increase in inner radius by 6 mm and increases by approximately 5% with increase in outer radius by 2 mm.
3. The effect of increasing height results in increase in efficiency by 1.5% for every 4 mm increment in height.
4. Efficiency of pot decreases by approximately 1% with every 3 mm increment in inner bulge and increases approximately by 5% with every 3 mm increment in outer radius.

References

1. Date AW, Damle RM (2015) Heat and Mass Transfer Analysis of a Clay-Pot Refrigerator. J Energy Heat Mass Transf 37:11–25
2. Date AW (2012) Heat and mass transfer analysis of a clay-pot refrigerator. Int J Heat Mass Transf 55:39773983
3. Date AW, Convective heat and mass transfer, NPTEL Online Video Course, Lecture 34. http://www.nptelvideos.in/2012/12/convective-heat-and-mass-transfer.html
4. Incropera FP, DeWitt DP (2002) Fundamentals of heat and mass transfer, 5th edn. Wiley, New York
5. Nield DA, Bejan A (2005) Convection in porous media. Springer, New York
6. Cengel YA, Ghajar AJ (2007) Heat and mass transfer: fundamentals and applications. McGraw-Hill Education, 2 Penn Plaza, New York

CFD Analysis of Heat Transfer in Liquid-Cooled Heat Sink for Different Microchannel Flow Field Configuration

Balaji Kumar Choudhury and Manoj Kumar Gouda

Abstract The thermal control of electronic devices is become an essential due to heavy uses in the modern world. To overcome this difficulty, the utilization of microchannel heat sink is an important device to cool the electronic circuits. So it is required for better understanding of the fluid flow and heat transfer in microchannel heat sink. Last two decades, a lot of investigation was carried out to improve the performance of the microchannel heat sink. The current paper deals with the study of conjugate heat transfer of a silicon-based microchannel heat by making a three-dimensional model. The 3-D model of microchannel heat sink consists of a silicon substrate of 10 mm length having rectangular cross section with different geometries. The impact of geometry on the distribution of temperature in the microchannel heat sink is presented and discussed by taking constant heat source and constant pumping power. This model was validated by comparing the obtained results with previously published papers on geometric optimization of a microchannel heat sink with liquid flow. A suitable geometric was found out by considering heat transfer, fluid flow keeping in mind of manufacturing the microchannel.

Keywords Numerical simulation · Heat transfer · Microchannel · Microelectronic cooling

1 Introduction

There may be a requirement for improvement of technology to improve the performance of a machine but sometimes it's difficult to get the desired result. Sometimes with improved technology also there may be some degraded output. Now days in usual day to day life we come across many electronics devices. In automobile, aircraft and industry also many electronics components are used for control and system monitoring. The electronics components in a small area results overheating. Overheating can result in the failure of the system. So scientists are trying to find greater efficient

B. K. Choudhury (✉) · M. K. Gouda
Parala Maharaja Engineering College, Berhampur, Odisha, India
e-mail: balaji.me@pmec.ac.in

M. Ramgopal et al. (eds.), *Advances in Air Conditioning and Refrigeration*, Lecture Notes in Mechanical Engineering, https://doi.org/10.1007/978-981-15-6360-7_35

mechanism to lower the thermal impact of all those electronic devices. With the development of micro electro mechanical system (MEMS) in engineering, micro-channels are extensively used for heat transfer applications.

Now a day, the modern micro electro-mechanical devices, advanced very large scale integration technology and the related micro channel, has led to appreciable increases the packing densities and heat generation within these devices. It is a matter of serious concern for the safety as well as reliability of the electronic devices. For continuous operation of the electronic devices the heat must be removed else the electronic devices may be failed due to temperature rise in the circuits. Hence, it is essential to develop a new methodology by which a large amount of heat fluxes can be removed with in less time. One important solution for this is to utilize micro-channel heat sinks with optimized geometry.

To design an effective microchannel heat sink, essential information of the characteristics of the heat transfer and fluid flow in microchannel are essential. The concept of micro channel heat sink at first proposed by Tuckermann and Pease [1] in 1981, used water inside a silicon micro channel because the running liquid, to deplete electricity from an electronic chip. They stated that under laminar flow heat transfer may be improved by means of decreasing the channel height right down to microscale. Li et al. [2, 3] had studied 3-D governing equations for fluid flow and heat transfer the use of a partial dimensionless evaluation for minimum pumping power. The dimensionless velocity and dimensionless temperature had been calculated, and the general thermal resistance in the microchannel heat sink is analyzed. Knight et al. [4] had explained the governing equations of fluid dynamics and combined conduction/convection heat transfer in a micro channel heat sink in dimensionless form for each laminar and turbulent fluid flow. From their investigation he concluded that when the pressure drop via the channels is small, laminar region yield lower thermal resistance than the turbulent region. Toh et al. [5] had investigated that the 3-dimensional fluid flow and heat transfer phenomena of inner heated microchannel. The regular, laminar flow and governing equations are solved the usage of a finite-extent approach. They founded temperature of the water is increased, leading to a lower viscosity and therefore smaller frictional losses. Wu and Cheng [6] has been performed an experiment to measure the friction factor of laminar flow of water and the experimental information is found to be in right agreement with the analytical solution. Peng and Peterson [7] has been investigated that the single-section pressured convective heat transfer and fluid flow characteristics of water in rectangular microchannel was investigated experimentally. The outcomes indicated that the geometric configuration had a large impact at the single-section convective heat transfer and the fluid flow characteristics. Choi and Cho [8] had studied the impact of the element ratio of square channels at the cooling traits of paraffin slurry float with a linear array of discrete heat transfer. They found that the element ratio of the rectangular channel is one of the major factor out of four experimental parameters. Tunc and Bayazitoglu [9] had investigated that convective heat transfers in a rectangular microchannel. The results display a comparable behavior to preceding research on circular microchannel. The values of the Nusselt number varies widely for different element ratios. Knight et al. [10] presented a paper that for fully developed fluid flow in closed finned channels. They concluded that the dimensionless thermal resistance in terms of the number of channels. Wei and Josh [11] had investigated with smaller inlet flow velocity in a microchannel that requires much less pumping power to do with a positive rate of heat transfer. Li and Peterson [12] had numerically investigated by developing a complete 3-D model of silicon based microchannel considering the conjugate heat transfer. The 3-D heat transfer model has been carried out to optimize the geometric structure of micro heat sinks. Under a regular pumping power of 0.05 W for a water-cooled micro heat sink, the optimized geometric parameters of the structure as determined through the model have a pitch of 100 μm, channel width of 60 μm and channel height of 700 μm. Fisher and Torrance [13] had concluded that the average thermal resistance reduced by varying the channel boundary, channel width, and wall thickness with concerning pressure drop and pump work.

2 Methodology

2.1 3-D Conjugate Heat Transfer Modeling in Microchannel Heat Sink

A3-D model of rectangular micro channel heat sink was developed as shown in Fig. 1. A uniform heat flux was supplied at the bottom of the heat sink. This highest quality layout changed by means of minimizing the thermal resistance between the heat sink and the coldest factor in the coolant and is referred to as the global thermal resistance while subjected to a precise heat input and pumping power.

To compute global thermal resistance value, the pumping power is considered as a constant value. Under one of these constraints, a highest quality evaluation could be achieved by means of comparing the thermal resistance for exclusive configurations,

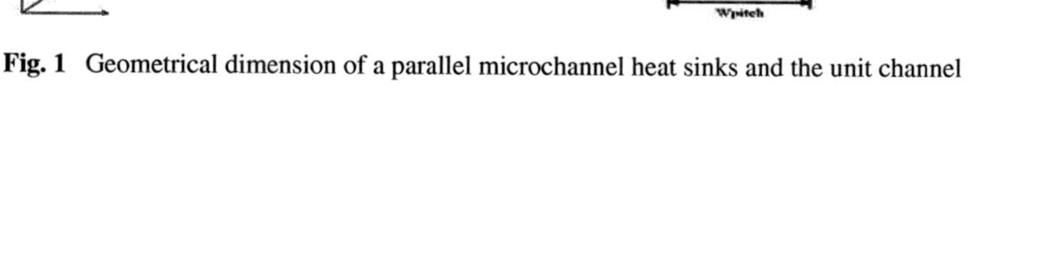

Fig. 1 Geometrical dimension of a parallel microchannel heat sinks and the unit channel

to decide how each of the dimensional parameters influences the general thermal overall performance. This pumping power (in Watt), is defined as

$$\bar{P} = V.\Delta P = N u_{\mathrm{m}} A_{\mathrm{c}} \Delta P \tag{1}$$

where, V = Total volume flow rate m^3/s, N = Number of channel, u_m = Mean velocity of the channel m/s, A_c = Area of the cross-section of the channel m^2, ΔP = Pressure drop across the channel Pa.

The given Fig. 1 shows that, in the current investigation the dimension of the heat sink is 10 mm × 900 μm × 10 mm ($L_x \times L_y \times L_z$). The thicknesses of the silicon wafers evaluated vary primarily based upon the intensity of the channel being investigated. In order to simplify the mathematical formula, the subsequent assumptions have been made.

- The fluid flow through the channel is laminar and the hydrodynamic fluidflow is fully developed. This assumption is based upon the small hydraulic diameters and low flowrates.
- The thermodynamic and hydrodynamic properties of materials are assumed as a constant with a reference temperature of 300 K.

Due to the symmetrical structure of the heat sink, the computational domain made like a rectangular channel. If the gravitational force and the heat dissipation due to viscosity are overlooked for steady, incompressible, laminar fluid flow, the governing equations for the liquid flows are

Continuity equation is

$$\frac{\partial u}{\partial x} + \frac{\partial v}{\partial y} + \frac{\partial w}{\partial z} = 0 \tag{2}$$

Momentum equation is

$$\rho\left(u\frac{\partial u}{\partial x} + v\frac{\partial u}{\partial y} + w\frac{\partial u}{\partial z}\right) = -\frac{\partial P}{\partial x} + \mu\left(\frac{\partial^2 u}{\partial x^2} + \frac{\partial^2 u}{\partial y^2} + \frac{\partial^2 u}{\partial z^2}\right) \tag{3}$$

$$\rho\left(u\frac{\partial v}{\partial x} + v\frac{\partial v}{\partial y} + w\frac{\partial v}{\partial z}\right) = -\frac{\partial P}{\partial y} + \mu\left(\frac{\partial^2 v}{\partial x^2} + \frac{\partial^2 v}{\partial y^2} + \frac{\partial^2 v}{\partial z^2}\right) \tag{4}$$

$$\rho\left(u\frac{\partial w}{\partial x} + v\frac{\partial w}{\partial y} + w\frac{\partial w}{\partial z}\right) = -\frac{\partial P}{\partial w} + \mu\left(\frac{\partial^2 w}{\partial x^2} + \frac{\partial^2 w}{\partial y^2} + \frac{\partial^2 w}{\partial z^2}\right) \tag{5}$$

where (x, y, z) are the orthogonal components of the body force field. And the energy Equation is

$$\left(u\frac{\partial T}{\partial x} + v\frac{\partial T}{\partial y} + w\frac{\partial T}{\partial z}\right) = \alpha\left(\frac{\partial^2 T}{\partial x^2} + \frac{\partial^2 T}{\partial y^2} + \frac{\partial^2 T}{\partial z^2}\right) \tag{6}$$

2.2 Boundary Condition

The velocity is zero at all walls of the channel except the channel inlet and outlet.The velocity inside the channel is assumed to be uniform.

$$\text{at the inlet } x = 0,\ P = P_{\text{in}},\ v = 0,\ w = 0 \tag{7}$$

$$\text{at the outlet } x = L_{\text{x}},\ P = P_{\text{out}},\ V = 0,\ w = 0 \tag{8}$$

and no slip condition ($u = 0$, $v = 0$, $w = 0$) is assumed at the inner surface.

The adiabatic boundary conditions are applied to all the boundaries of the solid area of the micro channel except the bottom wall of the, channel inlet and channel outlet in which a steady heat flux is implemented. The liquid temperature at the inlet of the channel is equal to a given consistent inlet temperature, i.e.

$$\text{At } x = 0,\ \text{if } (y, z) \in \text{channel } T_{\text{l}} = T_{\text{in}} \text{ Else, } -\lambda_{\text{w}}\frac{\partial T_{\text{W}}}{\partial x} = 0 \tag{9}$$

$$\text{At} x = L_X \text{ if}(y, z) \in \text{channel} - \lambda\frac{\partial T_{\text{l}}}{\partial x} = 0 \text{ Else, } -\lambda_{\text{w}}\frac{\partial T_{\text{W}}}{\partial x} = 0 \tag{10}$$

$$At\ y = 0, -\lambda_{\text{w}}\frac{\partial T_{\text{W}}}{\partial y} = q_{\text{w}} \tag{11}$$

$$At\ y = L, -\lambda_{\text{w}}\frac{\partial T_{\text{W}}}{\partial y} = 0 \tag{12}$$

$$\text{At } z = 0, -\lambda_{\text{w}}\frac{\partial T_{\text{W}}}{\partial z} = 0 \tag{13}$$

$$\text{At } z = W_{\text{pitch}}, -\lambda_{\text{w}}\frac{\partial T_{\text{W}}}{\partial z} = 0 \tag{14}$$

The mean velocity u_{m} can be determined analytically ($U_{(x,y)} = u_{(y,z)}/u_{\text{m}}$) and the Reynolds number can be expressed as

$$R_{\text{e}} = \frac{\rho u_{\text{m}} D_{\text{h}}}{\mu} \tag{15}$$

where, hydraulic diameter is

$$D_{\text{h}} = \frac{2HW}{H + W} \tag{16}$$

Fanning friction issue is a dimensionless variety used as a nearby parameter in continuum mechanics calculation.

$$f = \frac{H^2 + W^2}{(H + W)^2} \frac{24}{R_e} \tag{17}$$

For fully-developed, laminar flow, $R_e.f$ is a constant [3] for a particular channel and thus the velocity is only a function of the geometry of the channel.

$$R_e.f = \gamma \tag{18}$$

$$\text{Where } \gamma = 4.7 + 19.64\,\text{G} \tag{19}$$

$$G = \frac{H^2 + W^2}{(H + W)^2} = \frac{\alpha^2 + 1}{(\alpha + 1)^2} \tag{20}$$

where, α is the aspect ratio of the channel, α = H/W

$$\text{Pressure drop across the channel; } \Delta P = f \frac{4L_x}{D_h} \left(\frac{1}{2} \rho u_m^2 \right) \tag{21}$$

We can determine the velocity field by using the above equation.

A grid independence test has been carried out to test which mesh length gives correct result.

$$u_m = \sqrt{\frac{D_h^2 \bar{P}}{2\gamma \mu L_x \text{NWH}}} = \sqrt{\frac{2\alpha \bar{P}}{\gamma \mu (\alpha + 1)^2 \text{NL}_x}} \tag{22}$$

3 Results and Discussions

The numerical calculation were carried out for the dimension of 10 mm × 900 μm × 10 mm ($L_x \times L_y \times L_z$) for the heat sink channel with the hydrodynamic conditions $\bar{P} = 0.05$ W. The temperature distribution was developed in this section. The overall thermal resistance can be calculated as

$$R_T = \frac{\max(T_w(x, y, z)) - T_{l,in}}{q_w L_x L_z} \tag{23}$$

Here the reference temperature of the water hasbeen chosen as 300 K. Because the thermo-physical properties like the liquid viscosity are temperature based. The velocity and Reynolds numbers are distinct under the same pressure drop conditions.

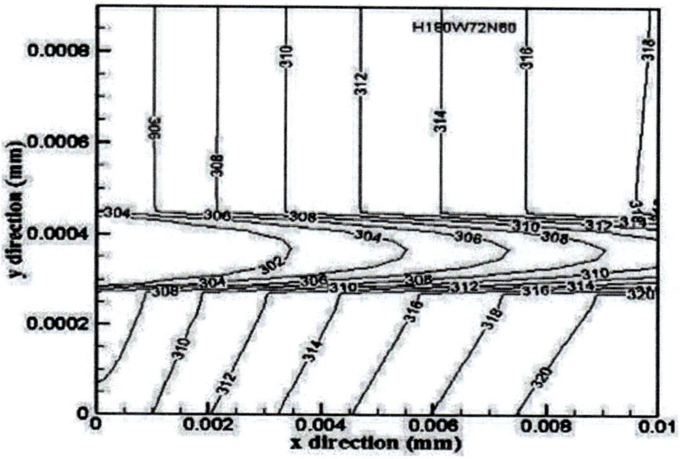

Fig. 2 Cross-sectional temperature distribution of the MCHS at $Z = W_{\mathrm{pitch}}/2$ for $\beta = W/W_{\mathrm{pitch}} = 0.72$ and $\alpha = H/W = 2.5$

3.1 *Temperature Contour*

Initially the liquid temperature at the inlet is distributed uniformly (at 300 K). Due to the assumption of hydrodynamic fully developed flow, the temperature profiles developed as shown in the figures. The maximum temperature is observed the channel outlet. This is due to the low velocity of the drift and maximum temperature of fluid at exit of channel. The figures show that, the silicon wafer extract more heat through the channel walls of the heat sink at $Z = W_{\mathrm{pitch}}/2$ and $x = L_x$ for $N = 60$, $\beta = W/W_{\mathrm{pitch}} = 0.6$ and 0.72 is presented as an illustration of the temperature distribution. The variation of temperature in the bulk liquid and substrate are represented by variations of the values of the different line contours. The temperature distribution is clearly shown in Figs. 2, 3, 4 and 5 and this exhibit the highest temperature occurs at the bottom of the substrate near the outlet of the channel heat sink.

3.2 *Model Validation*

In order to validate the numerical results, it is compared with the numerically obtained thermal resistance, as presented in the published literature [3]. The results of this comparison are as shown in Fig. 6 which is in a very good agreement with the present investigation.

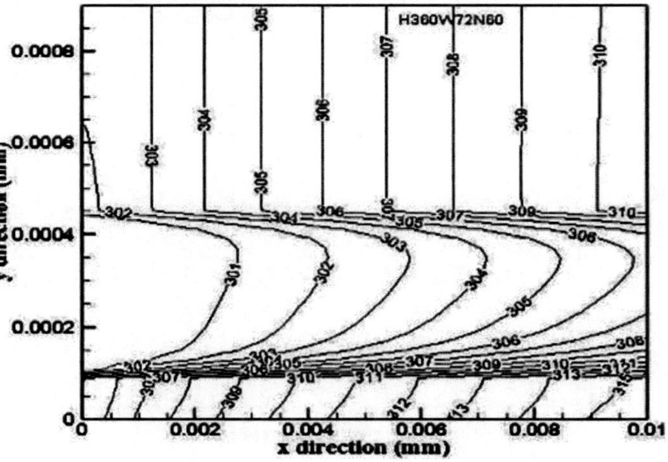

Fig. 3 Cross-sectional temperature distribution of the MCHS at $Z = W_{\text{pitch}}/2$ for $\beta = W/W_{\text{pitch}} = 0.72$ and $\alpha = H/W = 5$

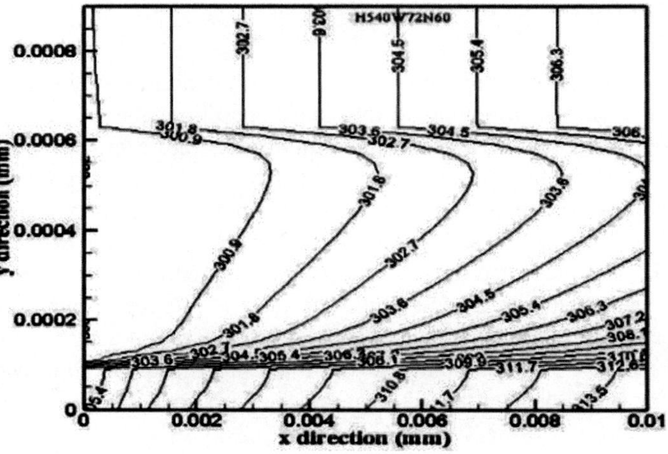

Fig. 4 Cross-sectional temperature distribution of the MCHS at $Z = W_{\text{pitch}}/2$ for $\beta = W/W_{\text{pitch}} = 0.72$ and $\alpha = H/W = 7.5$

3.3 *Overall Thermal Resistance*

Microchannel heat sink with liquid cooling performance is investigated by calculating thermal resistance which depends on inlet temperature of fluid and maximum wall temperature of heat sink. Figures 7, 8, 9 and 10 illustrate a comparison of the thermal resistance at the outlet among the different number of channels with the variation of $\beta = W/W_{\text{pitch}}$ and a given constant pumping power, $\bar{P} = 0.05$ W and constant height of the channel, $H = 180$, 360, 540 and 720 μm respectively.

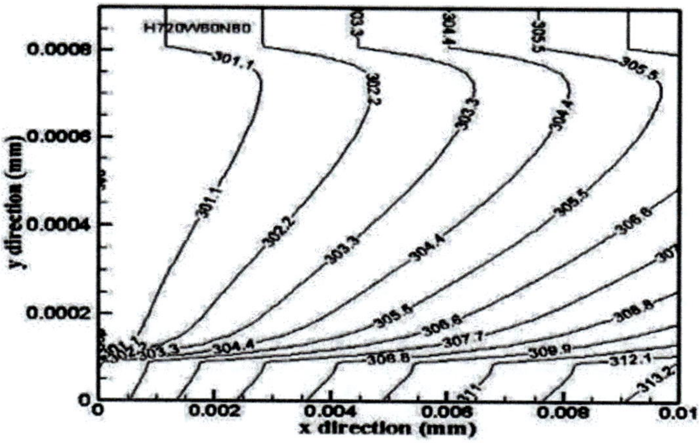

Fig. 5 Cross-sectional temperature distribution of the MCHS at $Z = W_{pitch}/2$ for $\beta = W/W_{pitch} = 0.6$ and $\alpha = H/W = 12$

Fig. 6 Comparison of outlet thermal resistance for numerical calculation in this work with numerical result of reference at $\beta = W/W_{pitch} = 0.6$ and $N = 100$

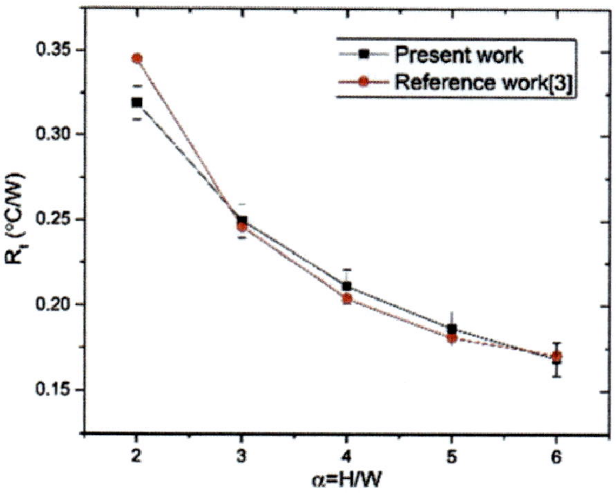

In Fig. 7 for $H = 180$ μm, it is clear that by increasing the values of $\beta = W/W_{pitch}$ the sidewall becomes thinner which results higher the heat transfer rate due to increase in passage cross-sectional area. The lowest thermal resistance occurs at $N = 60$, 80, 100, 120, 140 and $\beta = W/W_{pitch} = 0.72$ for $H = 180$ μm under $\bar{P} = 0.05$ W. Basically, the thermal conductivity inside the silicon is very excessive in comparison to the value of water. The convective thermal resistance is inversely proportional to area of cross-section (A) and convective heat transfer coefficient (h). When the convective heat transfer coefficient increases the overall thermal resistance decreases. A close inspection [3] exhibits that once the component ratio (α) decreases, the mean velocity will increase. Thus, in Fig. 7, when the channel height is held constant and $\beta = W/W_{pitch}$ increases for a particular no. of channels, the aspect ratio decreases

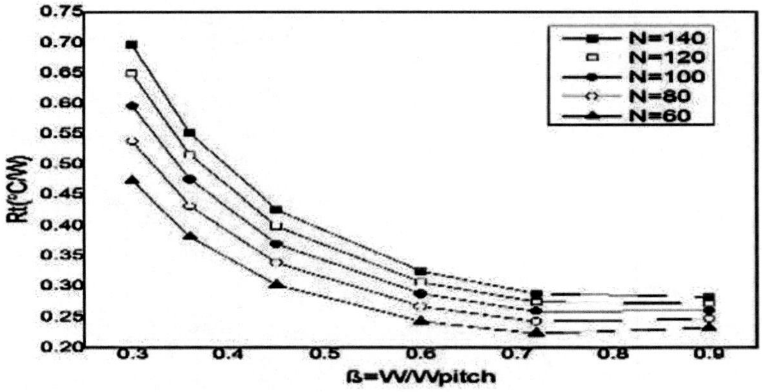

Fig. 7 Comparison of the outlet thermal resistance for different no. of channels with variation of $\beta = W/W_{\text{pitch}}$ and height of the channel $H = 180\ \mu\text{m}$

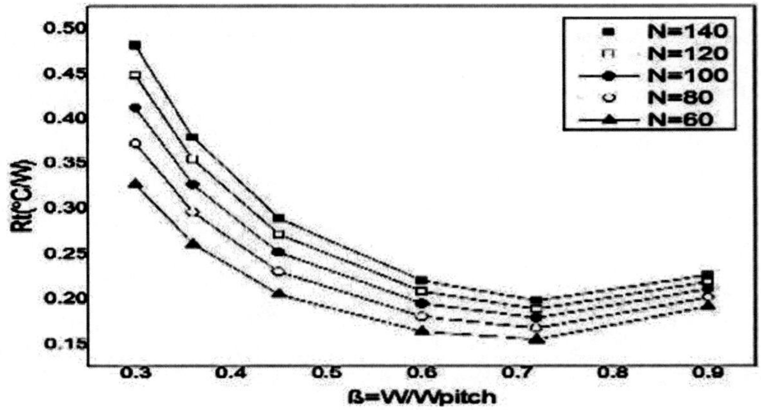

Fig. 8 Comparison of the outlet thermal resistance for different no.of channels with variation of $\beta = W/W_{\text{pitch}}$ and height of the channel $H = 360\ \mu\text{m}$

and the heat transfer area will increase. With those outcomes, the convective thermal resistance decreases with increase in $\beta = W/W_{\text{pitch}}$.

These results are looking different from the previously published papers [3]. The second reason for this difference is a result of balance among the flow resistance and the heat transfer surface area for distinct number of channels. The flow resistance has a negative impact on the thermal resistance that is low velocity but increasing the heat transfer surface area has a positive impact on the thermal resistance. Flow resistance will higher when the channel size is small alternatively more channel increase the heat transfer surface area. Third reason for the difference is a result of previously discussed the axial heat conduction will produce bad impact on the thermal resistance when the sidewall of the channel gets shrinking. The minimum thermal resistance

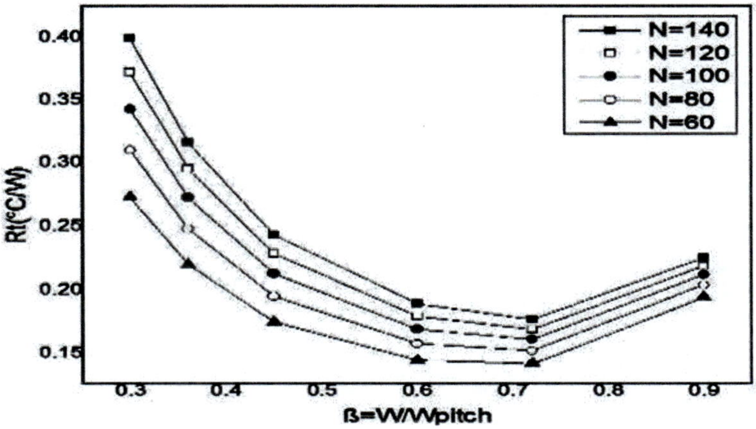

Fig. 9 Comparison of the outlet thermal resistance for different no. of channels with variation of $\beta = W/W_{\text{pitch}}$ and height of the channel $H = 540\ \mu\text{m}$

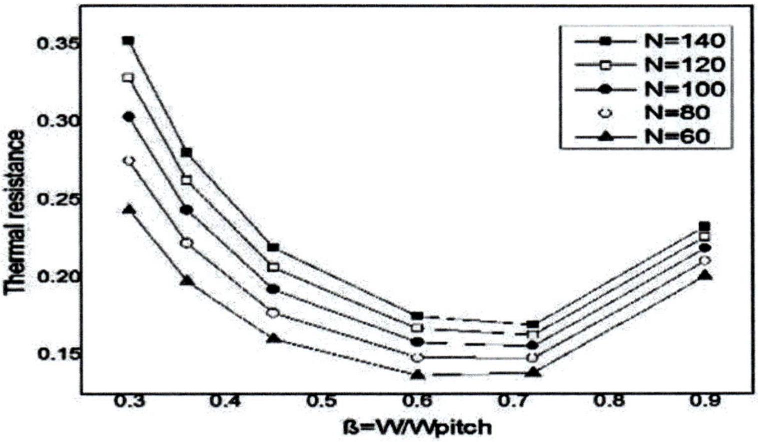

Fig. 10 Comparison of the outlet thermal resistance for different no. of channels with variation of $\beta = W/W_{\text{pitch}}$ and height of the channel $H = 720\ \mu\text{m}$

is mainly compromise between the heat transfer area of liquid flow, flow resistance and the heat conduction in the substrate of the wall.

In Fig. 7 the lowest thermal resistance occurs at $N = 60$ and $\beta = W/W_{\text{pitch}} = 0.72$ for $H = 180\ \mu\text{m}$ under $\bar{P} = 0.05\,\text{W}$. Higher the channel, the higher the flow resistance which causes lower the velocity and results in adecrease in heat transfer rate.

Similar tendency was observed in Figs. 8, 9 and 10 for $H = 360$, 540 and 720 μm but with a different slope for each curve due to more contribution from heat conduction and little contribution from flow convection. In Figs. 8 and 9, for $H = 360\ \mu\text{m}$,

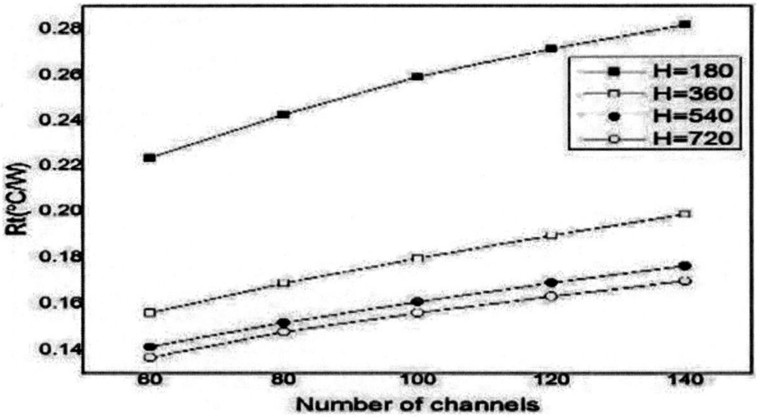

Fig. 11 Comparisonof thermal resistance and no. of channels when $H = 180, 360, 540$ and $720\ \mu m$ respectively

the lowest thermal resistance was found at $N = 60$ and $\beta = W/W_{pitch} = 0.72$ when $\bar{P} = 0.05\,W$. Whereas in Fig. 10 for $H = 720\ \mu m$, the lowest thermal resistance occurs at $N = 60$ and $\beta = W/W_{pitch} = 0.6$, when $\bar{P} = 0.05\,W$.

A comparison of minimum thermal resistance for different number of channels is as shown in Fig. 11. The figure shows that for a specified number of channels, when the channel height is maximum it gives lower thermal resistance which is compared for $H = 180, 360, 540$ and $720\ \mu m$.

4 Conclusion

The present 3-D computational domain was investigated for different flow field configurations of microchannel heat sink. Here the overall thermal resistance was calculated and analyzed in the microchannel heat sink. Minimization of the thermal resistance for a microchannel heat sink with silicon wafer under low pumping power and the flow lies within the fully developed laminar region only. From the computational points of view, the most suitable variety of channels become observed to be $N = 60$ with channel height $H = 720\ \mu m$ and a ratio of $\beta = W/W_{pitch} = 0.6$ gives minimum thermal resistance. But from the manufacturing points of view, the most suitable variety of channels become observed to be $N = 60$ with channel height $H = 360\ \mu m$ and a ratio of $\beta = W/W_{pitch} = 0.72$.

References

1. Tuckerman DB, Pease RFW (1981) High-performance heat sinking for VLSI. IEEE Electron Device Lett EDL 2:126–129
2. Li J, Peterson GP, Cheng P (2004) Three-dimensional analysis of heat transfer in a micro heat sink with single phase flow. Int J Heat Mass Transf 47:4215–4231
3. Li J, Peterson GP (2006) Geometric optimization of a micro heat sink with liquid flow. IEEE Trans Compon Packag Technol 29(1):145–154
4. Knight RW, Hall DJ, Goodling JS, Jaeger RC (1992) Heat sink optimization with application to micro channels. IEEE Trans Compon Hybrid Manufact Technol 15(5):832–842
5. Toh KC, Chen XY, Chai JC (2002) Numerical computation of fluid flow and heat transfer in micro channels. Int J Heat Mass Transf 45:5133–5141
6. Wu HY, Cheng P (2003) Friction factors in smooth trapezoidal silicon micro channels with different aspect ratios. Int J Heat Mass Transf 46(14):2519–2525
7. Peng XF, Peterson GP (1996) Convective heat transfer and flowfriction for water flow in microchannel structures. Int J Heat Mass Transf 39(12):2599–2608
8. Choi M, Cho K (2001) Effect of the aspect ratio of rectangular channels on the heat transfer and hydrodynamics of paraffin slurry flow. Int J Heat Mass Transf 44(1):55–61
9. Tunc G, Bayazitoglu Y (2002) Heat transfer in rectangular micro channels. Int J Heat Mass Transf 45(4):765–773
10. Knight RW, Goodling JS, Hall DJ (1991) Optimal thermal design of forced convection heat sinks-analytical. ASME J Electron Package 113(3):313–321
11. Wei XJ, Joshi Y (2003) Optimization study of stacked micro-channel heat sinks for micro-electronic cooling. IEEE Trans Compon Packag Technol 26(1):55–61
12. Li J, Peterson GP (2007) 3-Dimensional numerical optimization of silicon-based high performance parallel microchannel heat sink with liquid flow. Int J Heat Mass Transf 50:2895–2904
13. Fisher TS, Torrance KE (2001) Optimal shapes of fully embedded channels for conjugate cooling. IEEE Trans Adv Package 24(4):555–562
14. Madou MJ (2002) MEMS fabrication. In: Gad-el-Hak M (ed) The MEMS handbook, vol. 16, no. 1. CRC, Boca Raton
15. Kays WM, Crawford ME (1990) Convective Heat and Mass Transfer. McGraw-Hill, New York

Effect of Speed of Condenser Fan Motor on Vapor Compression Refrigeration System

Punyabrata Acharya, Balaji Kumar Choudhury, and Sachindra Kumar Rout

Abstract The main aim of this paper is to analyze the effect of speed of condenser fan motor on various parameters of vapor compression refrigeration system. A detailed study has been done on the performance of evaporator and the heat absorbing capacity of refrigerant when there is drop in temperature of the cooling space. The compressor being the work-consuming component, its performance has also been analyzed. The graph has been plotted between different temperatures and other parameters to understand the process. The experiment was conducted several times to confirm the results and performance. It was carried out by changing the condenser fan motor (CFM) speed with mentioning the reasons. Commercially used and freely available software, Cool Pack, is used to evaluate the refrigeration cycle performance. This also utilized to plot the p–h diagram and also to get the properties of the refrigerant R134a.

Keywords Vapor compression refrigeration · Performance of condenser and evaporator · Condenser fan motor

1 Introduction

The vapor compression refrigeration cycle is a commonly used refrigeration system. This cycle is based upon capability of some liquids to absorb high amount of heat in terms of latent heat at a much lower temperature than the surrounding. The refrigeration effect produced by these liquid refrigerants is more as compared to the melting of ice. It has few advantages like refrigerating effect can be produced or controlled whenever required. Also the vaporized liquid refrigerant can be condensed, and that can again have converted into liquid state. By this way, the same liquid can be recirculated to operate in cycle and thereby producing the refrigerating effect. Hence, the

P. Acharya · B. K. Choudhury (✉)
Parala Maharaja Engineering College, Berhampur, Odisha, India
e-mail: balajichoudhury@gmail.com

S. K. Rout
C. V. Raman College of Engineering, Bhubaneswar 752054, Odisha, India

M. Ramgopal et al. (eds.), *Advances in Air Conditioning and Refrigeration*, Lecture Notes in Mechanical Engineering, https://doi.org/10.1007/978-981-15-6360-7_36

vapor compression refrigeration system uses a liquid refrigerant which evaporates and condenses readily at low temperature and this operates in a closed cycle.

The performance of a refrigerating system is expressed in terms of coefficient of performance (COP). It is defined as the ratio of refrigerating effect to the total work done on the system. Therefore the coefficient of performance [1] can be increased by increasing the refrigerating effect or by decreasing the total work on the system.

The major components of a vapor compression system [2–4] are evaporator, compressor, condenser and expansion valve. The diagram of the arrangement is as shown in Fig. 1. The low-temperature liquid refrigerant at A enters the evaporator and cools the space that is required by absorbing the latent heat from the refrigerant. Then, it converted into low-pressure vapor. Now this vapor refrigerant is at low temperature and low pressure (B), and it is compressed by the compressor to high temperature and pressure vapor at state C. This vapor is condensed into liquid in the condenser by utilizing a cooling fan. Thereafter, it is passed through the expansion valve. Here, the vapor is undergone isenthalpic expansion, and temperature and pressure become low. This is again recirculating to the space to be cooled. From this, it is concluded that the liquid refrigerant plays a vital role in taking the heat from the space by the evaporator and rejecting the heat outside by the condenser [5]. Taking the heat from the cooled space is directly proportional to the rejecting the heat outside. Several studies have been done on condenser [6] to improve the heat rejection by changing the geometrical parameters [7] to increase the COP. To increase the heat rejection from the condenser, a fan [8] is utilized (in case of air-cooled condenser). It is required to control the speed of the fan [9] and focus on the effect of this on other parameters of the vapor compression refrigeration system.

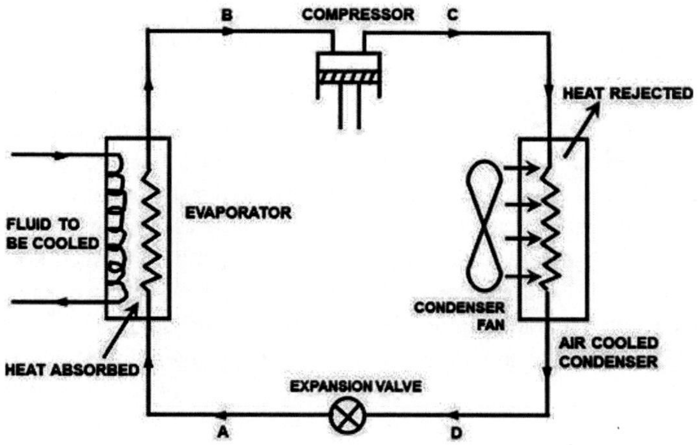

Fig. 1 Simple vapor compression refrigeration system

2 Experimental Details

The available refrigeration test rig is used for the experimental study of vapor compression system. The experimental setup [5] is designed in such a fashion that vapor compression cycle and its component can be easily understood which can record temperatures and pressures with time.

The experimental setup consists of a compressor, fan cooled condenser, drier, capillary tube, evaporator coil and water tank. All the components are mounted on a base stand. For measuring the temperature at the node points, resistance-type temperature sensors (Pt100 sensors) are attached and this is connected to a displayer. To measure the pressure, dial pressure gauges are used, while for the suction pressure compound pressure gauge is used. To measure the refrigerant flow rate, rotameter is used. An anemometer or wind meter is used to measure air speed through the condenser. To maintain uniform temperature throughout the water tank, a motor pump is used as a stirrer. Voltmeter, ammeter and energy meter are also mounted to account the electrical parameter like voltage, current and energy consumed. For the control, the VCR system a thermostat is used which senses the temperature of a system to maintain system's temperature near a desired set-point. It acts as a cutoff of the system; it switches off the compressor as soon as the temperature of the evaporator drops below set-point (Fig. 2).

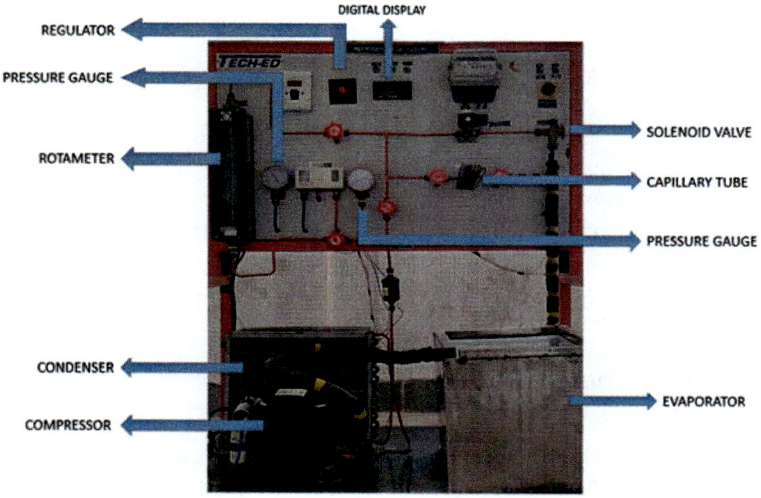

Fig. 2 Experimental setup

3 Procedure of Testing the Performance

The evaporator tank is filled completely. Thereafter, all the valves are checked to ensure the flow of refrigerant in right direction. Then, power is supplied to the stirrer. To start the experiment, the tank water temperature is raised to 40 °C by using an electric heater. The energy meter initial reading is taken. The flow of refrigerant is recorded by the rotameter. The pressure gauge reading is taken for suction side and discharge side. It is observed that temperature of water tank decreases due to the refrigerating effect. The readings are recorded at every 5 min of interval till the temperature of the load tank is became 5 °C. The above procedure is followed, and the experiment was performed at two conditions [8] as described below.

- Allowing condenser fan speed at low or minimum rpm.
- Allowing condenser fan speed at high or maximum rpm.

During the time, performance of all the components was studied and results were recorded. The enthalpy values were taken by the property table of Cool Pack software. Cool Pack [10] is commercially used and dedicated software for designing and optimizing the vapor compression refrigeration system. This software was used to evaluate the performance of the system. Also the properties (pressure, temperature, specific volume, enthalpy, etc.) of the refrigerant R134a were taken from the software property table. This software was also helpful to draw the cycle on p–h diagram. This is discussed in Results and Discussion.

4 Results and Discussion

Observations were done by concentrating on temperature at various locations of the system. And it is compared with varying condenser fan motor (CFM) speed from minimum to maximum.

4.1 Effect on Compressor Inlet and Exit Temperature

After starting the system, the evaporator temperature decreases with time. As the exit temperature of evaporator and compressor inlet temperature are almost same, so compressor inlet temperature also decreases with time as shown in Fig. 3. In case of minimum CFM speed, compressor exit temperature is higher than the maximum CFM speed (Fig. 4).

Compressor exit temperature mainly depends on the work done by the compressor. The compressor work can be given by $Wc = -vdp$. As the specific volume in case of maximum CFM speed is less than the case of specific volume with minimum CFM speed. So compressor has to do less work in case of maximum CFM speed. So it

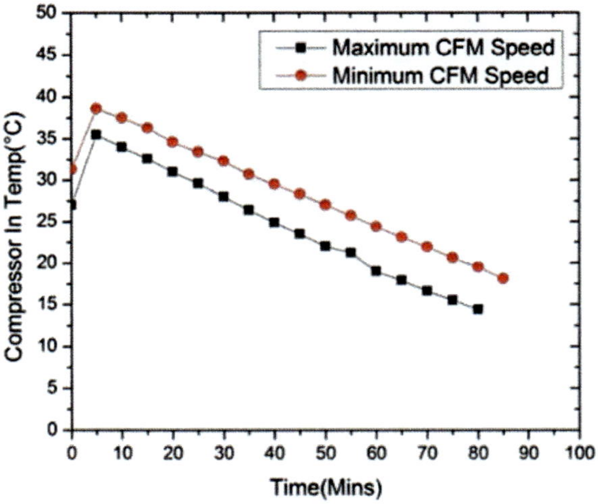

Fig. 3 Compressor inlet temp versus time

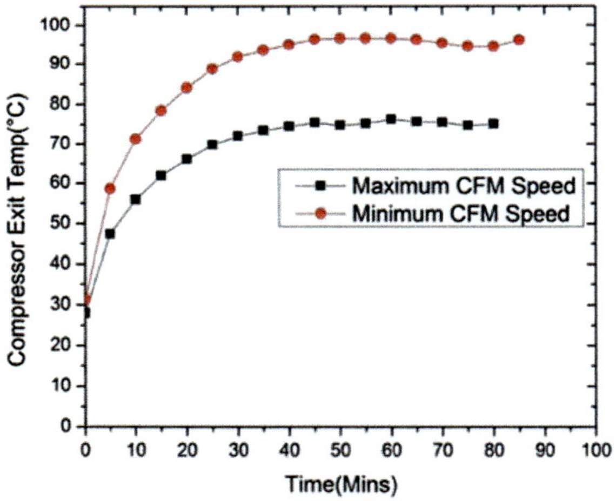

Fig. 4 Compressor exit temp versus time

results in decreasing the compressor work which results in reducing the compressor exit temperature. Also with increase in time, the specific volume in the individual case goes on increasing that is why the compressor exit temperature increases in both the cases.

4.2 Effect on Condenser Exit Temperature

Condenser exit temperature mainly depends upon amount sub-cooling the refrigerant undergoes which depends upon the velocity of the condenser fan. As with maximum CFM speed, we increase the forced heat transfer coefficient on air side. This results

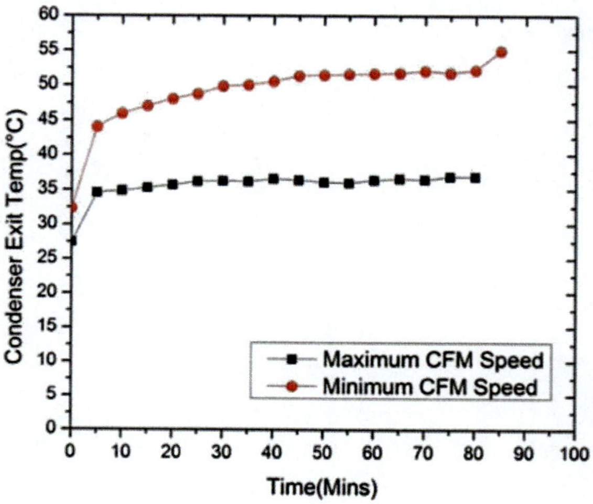

Fig. 5 Condenser exit temp versus time

in more heat drop in the condenser so the refrigerant temperature is less in the case with minimum CFM speed. But with increase in time the condenser outlet temperature increases because the compressor outlet pressure increases with respect to time (Fig. 5).

4.3 Temperature After Isenthalpic Expansion

The maximum CFM speed results in sub-cooling of the refrigerant before entering into the expansion valve or capillary tube. So after the isenthalpic expansion in capillary tube, the temperature reduces (Fig. 6).

4.4 Effect on Compressor Work

The compressor work remains almost constant for maximum CFM speed case, but for the case of minimum CFM speed, it almost increases with time. Also the compressor work with maximum CFM speed is always lower than compressor work with minimum CFM speed (Fig. 7). The compressor work mainly depends on inlet condition of the refrigerant, i.e., its specific volume. The compressor work can be given by $Wc = -vdp$. The specific volume in case of maximum CFM speed is less than the case of specific volume minimum CFM speed. So compressor has to do less work in case of with maximum CFM speed. So it results in decreasing the compressor work.

Fig. 6 Evaporator temp versus time

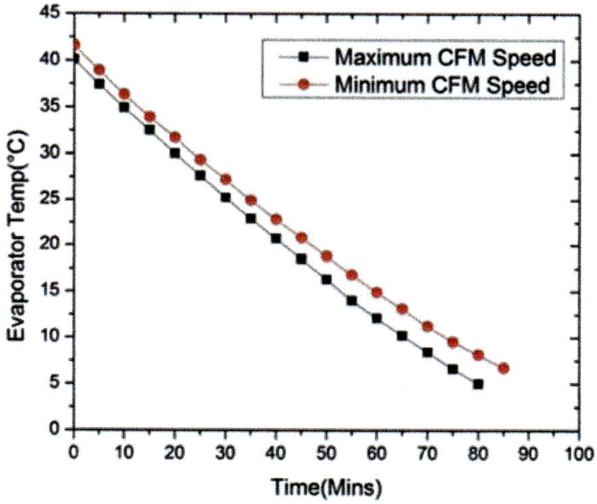

Fig. 7 Compressor work versus time

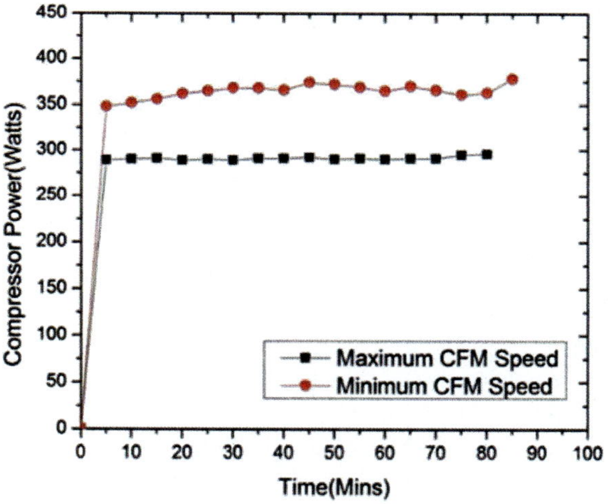

4.5 Effect on COP

Figure 8 shows the COP decrease with time, and COP for the case of with maximum CFM speed is higher than the COP of refrigerant minimum speed of CFM. COP of the system mainly depend on the refrigerating effect and compressor work done. As refrigerating effect is always higher for the case of with maximum CFM speed also the compressor work is lower (as explained in Fig. 7) so COP will be always high for the case of with maximum CFM speed condition.

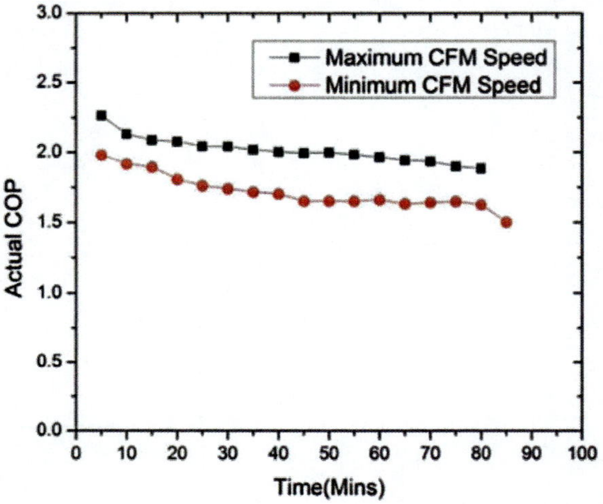

Fig. 8 COP versus time

4.6 *Comparison on p–h Chart*

The p–h chart is drawn using the software Cool Pack. The cycle with minimum CFM speed and maximum CFM speed is shown in Fig. 9.

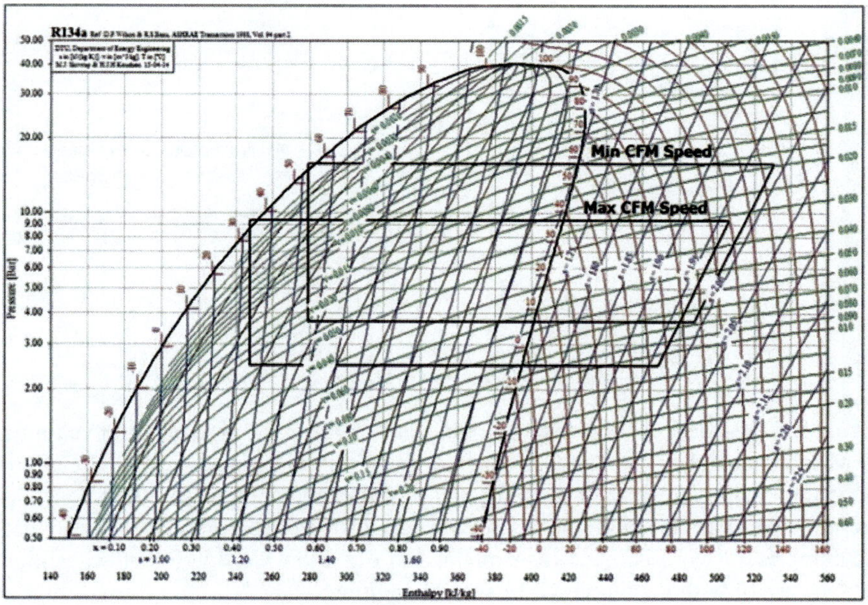

Fig. 9 p–h chart comparing two cycles

5 Conclusion

The experimental observations have been done on vapor compression refrigeration system. Performance of the system was found out as well as individual components like compressor, condenser and evaporator have been evaluated. It is observed that maximum CFM speed results better in terms of COP.

We found that with increasing CFM speed, the pressure head reduces which is very advantageous. The higher speed also ensures low heat loss which enhances the system performance. Lower speed of condenser fan makes the compressor exit temperature higher which requires a larger amount of cooling medium (air) to pass through the condenser. The most important observation was the rate of cooling of the space, and it cools at a faster rate with higher CFM.

Cool Pack software is used to check the performance of the system. This also utilized to plot the p–h diagram and also to get the properties of the refrigerant R134a.

The knowledge gained by experimental investigation can be utilized to design another VCR system, i.e., constant temperature bath/chillier. The constant temperature bath can be utilized as a vital component in the field of thermal engineering.

References

1. Pottker G, Hrnjak P (2015) Effect of the condenser subcooling on the performance of vapor compression systems. Int J Refrig 50:156–164
2. Arora CP (2002) Refrigeration and Air-conditioning. Tata McGraw-Hill, New Delhi
3. Miller R, Miller MR. Air conditioning and refrigeration. McGraw-Hill, New York
4. Incropera FP, De Witt DP. Fundamentals of heat and mass transfer, 4th edn. Wiley, New York, p 445
5. Qureshi BA, Zubair SM (2014) The impact of fouling on the condenser of a vapor compression refrigeration system: an experimental observation. Int J Refrig 38:260–266
6. Saechan P, Wongwises S (2008) Optimal configuration of cross flow plate finned tube condenser based on the second law of thermodynamics. Int J Therm Sci 47(11):1473–1481
7. Wright MF (2000) Plate-fin-and-tube condenser performance and design for refrigerant R-410A air-conditioner, M. Tech. thesis, Georgia Institute of Technology
8. Yu FW, Chan KT (2006) Improved condenser design and condenser-fan operation for air-cooled chillers. Appl Energy 83(6):628–648
9. Yu FW, Chan KT (2008) Optimizing condenser fan control for air-cooled centrifugal chillers. Int J Therm Sci 47(7):942–953
10. Jakobsen A, Rasmussen BD, Skovrup MJ, Andersen SE (2001) Cool pack tutorial—Version 1.46

Scope of Using Photovoltaic Cell to Power Electrical Units of Air-Conditioned Linke Hofmann Busch (LHB) Coaches Used in Indian Railways

Dilip Kumar Bagal, Abhishek Barua, Siddharth Jeet, Antarjyami Giri, Ajit Kumar Pattanaik, and Surya Narayan Panda

Abstract The fourth-largest railway network in the world, i.e. Indian Railways, operates more than 25,000 passenger trains everyday which consists of either ICF coaches or LHB coaches. Though the old ICF coaches generate the electricity to power the electrical things fitted inside it on its own, their productions have been stopped due to their ageing technology. They are now slowly replaced by modern Linke Hofmann Busch (LHB) coaches which are more reliable than the ICF coaches in almost every aspect except in the case of electricity generation. LHB coaches do not have self-electric generation systems; hence, they have to depend upon the End on Generation (EOG) or Head on Generation (HOG) system present in the train for supply of electricity inside the coaches. Nearly 90% of trains with LHB coaches are dependent on EOG system which consists of diesel generators which generates electricity for the train. This study investigates the use of solar photovoltaic (PV) modules which can be installed on top of train coaches and how it can be beneficial for an AC LHB coach. Many researchers have carried out their research for finding the capability of this type of electricity generation technique earlier. A comparative study has been carried out to find the economical way to supply electricity in the coach. This will not only help in the conservation of fuel, but also helps in significant reduction in noise and air pollution.

Keywords ICF · LHB · Photovoltaic cell · EOG · HOG · Indian railways

D. K. Bagal (✉) · A. Giri · A. K. Pattanaik
Government College of Engineering, Kalahandi, Bhawanipatna, Odisha, India
e-mail: dilipbagal90@gmail.com

A. Barua · S. Jeet
Centre for Advanced Post Graduate Studies, BPUT, Rourkela, Odisha, India

S. N. Panda
Birsa Institute of Technology Sindri, Dhanbad, Jharkhand, India

M. Ramgopal et al. (eds.), *Advances in Air Conditioning and Refrigeration*, Lecture Notes in Mechanical Engineering, https://doi.org/10.1007/978-981-15-6360-7_37

1 Introduction

Indian Railway invests crores of rupees for the operation of more than 20,000 passenger trains daily which consists of either ICF coaches or LHB coaches. Though the old ICF coaches generate the electricity to power, the electrical things fitted inside it on its own. They have been fitted with alternators which are driven by the rail wheel while the train is moving, thus producing electricity which is stored in batteries and used in the coaches for lighting and other purposes. This restricts the speed of train to 110 kmph only. Though this technology is green and does not require and other renewable sources to run, production of ICF coaches has been stopped due to their ageing technology and demand for a faster and more reliable coach. They are now slowly replaced by modern Linke Hofmann Busch (LHB) coaches which are more reliable than the ICF coaches in almost every aspect except in the case of electricity generation. Due to its design to achieve a high speed to 160 kmph, LHB coaches do not have self-electric generation systems as no alternator system has been present in them. Hence, LHB coaches have to depend on either End on Generation (EOG) or Head on Generation (HOG) system present in the train for supply of electricity inside the coaches. All LHBfied trains have EOG system, and nearly 90% of them are dependent on EOG system which consists of diesel generators which generate electricity for the train. Rest 10% LHB trains run on HOG system where electricity is directly taken from OHE wires by the electric locomotive, rectified, converted into direct current and supplied to the whole train for lighting and other use [1–3].

EOG system consists of diesel-powered generators which not only produces noise but also emits harmful exhaust gases mixed with black smoke throughout the journey of a LHB train. In India, more diesel consumption is done by the EOG diesel generators in full AC LHB trains rather than normal LHB trains with mixed type coaches. Vasisht et al. [1, 2] first proposed and conducted trail by fitting solar photovoltaic cell over LHB coach and presented its benefit and cost-saving techniques. Based on that research, this study emphasis on the use of fitting solar photovoltaic cell over AC LHB coach and to find out whether this technique is better than HOG technique or not. Usage of this technique by Indian Railways in the present time has also been discussed here.

2 Electrical Supply in AC LHB Coaches

As stated earlier, two types of electrical power supply system are used in Indian Railways, i.e. End on Generation (EOG) and Head on Generation (HOG) systems.

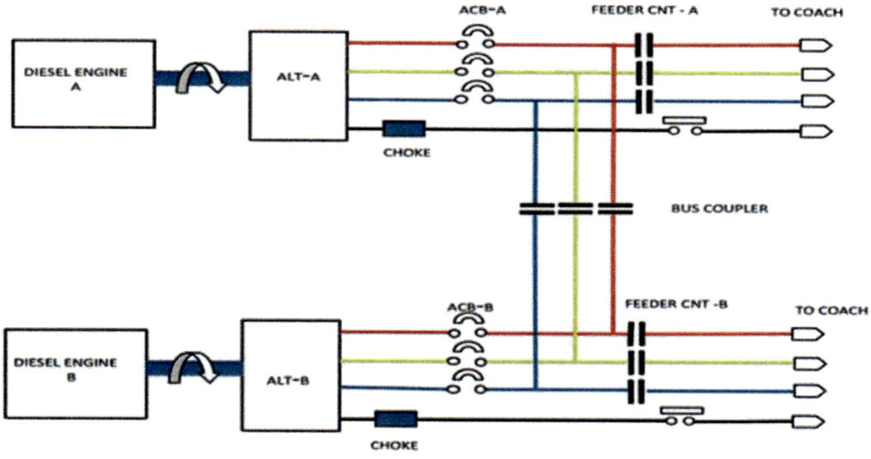

Fig. 1 Power car block diagram with EOG supply [3]

2.1 *End on Generation (EOG) System*

In EOG system, the electrical load, i.e. loads of fans, lights and air-conditioning units, are mentioned as "hotel load" in LHB coaches of express trains is powered by the power cars which are placed at both ends of a train rake. Inside the power car consists of two diesel alternator sets which generate three-phase (4 wires) power supply of 750 V 50 Hz, and the same is transferred to complete rake through two parallel cables feeders termed as Feeder-A and Feeder-B, running through the whole length of the train. A 60 kVA transformers present inside LHB coaches tap the electric power supply which is again transformed to 415 V for running electrical equipment working at this voltage and again converted into 110 V AC single phase for running electrical equipment working at that voltage. Figure 1 shows the power car block diagram with EOG supply [3].

Since the power cars are positioned at both end of train rake, the system is termed as End on Generation (EOG).

2.2 *Head on Generation System*

In HOG scheme, power is fed from the electric locomotive to the train to run hotel load of the whole train rake. First, the electric locomotive draws the power from the overhead electrical supply using its pantograph to traction transformer present inside it. This traction transformer consists of a hotel load winding of 945 kVA, at minor voltage of 750 V single-phase, which fluctuates with overhead electrical supply voltage deviations. This 750 V single-phase supply is transferred to hotel load converter which gives 750 V three-phase 50 Hz output power supply for running hotel load of entire rake [3]. Figure 2 shows HOG process layout. Figure 3 shows the power car block diagram with HOG supply.

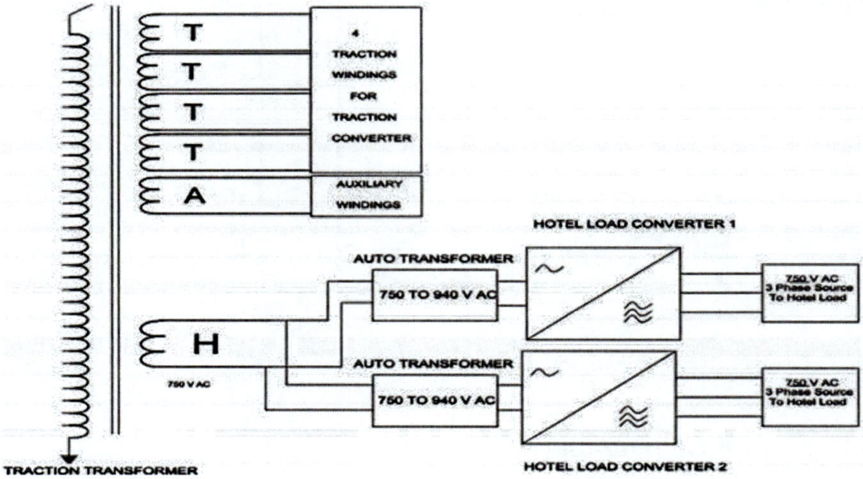

Fig. 2 HOG process layout [3]

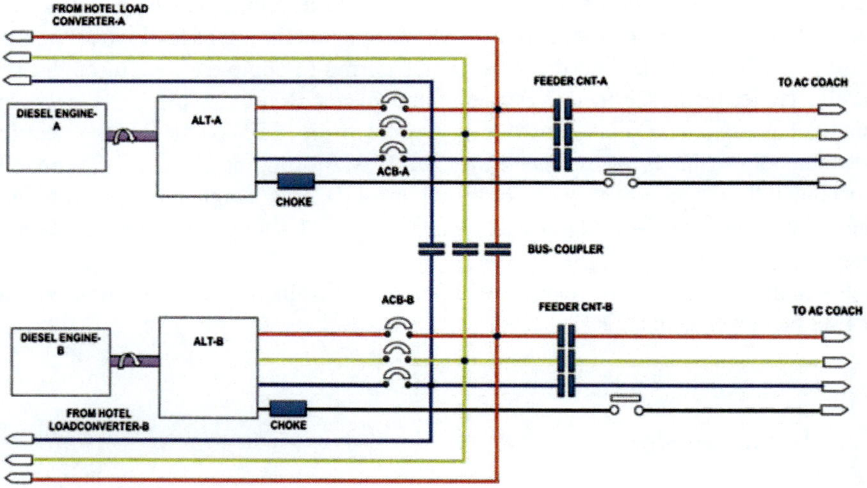

Fig. 3 Power car block diagram with HOG supply [3]

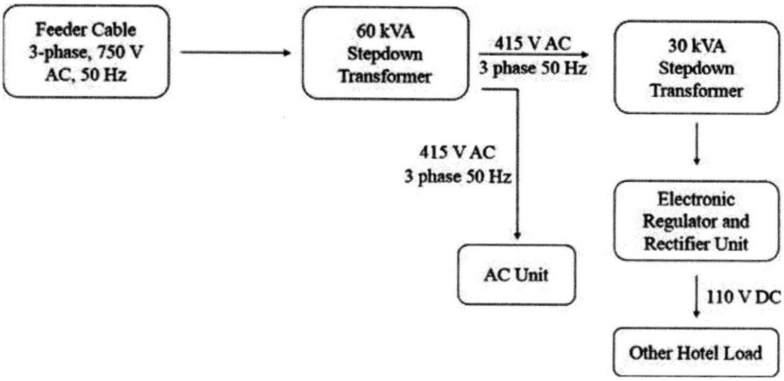

Fig. 4 Power supply layout of AC LHB coach [1]

2.3 *Electric Supply Inside AC LHB Coach*

Alternator sets generating three-phase (4 wires) power supply of 750 V 50 Hz is transferred to whole rake using two parallel feeder cables present inside the whole rake. Tapping of electric power supply is done at each coach using 60 kVA transformers present inside LHB coaches and is converted into 415 V for feeding the equipment working at this voltage and again transmuted into 110 V AC single phase for powering electrical equipment operational at that voltage. Compressor motors, heaters, condenser fan motors present in the evaporator of AC LHB Coach directly uses 415 V, three-phase supply. 110 V, three-phase AC power supply is used to run evaporator blowermotor. For other lighting loads, like lamp, fan, mobile charging socket, present inside coach, a 30 kVA step-down transformer is used for stepping down the voltage to 110 V DC, through an electronic regulator and rectifier unit by linking them among line and neutral on secondary side of this transformer. Figure 4 shows power supply layout of AC LHB coach [1].

3 Scope of Using Photovoltaic Cell to Power Electrical Units of AC LHB Coaches

In the present arrangement of operation of power cars, two of the power cars have a diesel tank with 3000 L capacity each which are entirely filled during the commencement of the journey. Out of two, only one power car is put to facilitate and the other is kept as a standby since the electricity generated by single alternator is enough to power the electrical load of entire train rake though the second power can be used according to the load requirement [4, 5]. Table 1 shows different types of electrical loads in AC LHB coach, and few loads among them are different in quantities for different types of AC LHB coaches as Indian Railways uses five different variants

Table 1 Different types of electrical loads in AC LHB coach [1]

Hotel load equipments	Rating (W)	Hotel load equipments	Rating (W)
Ac units	11,500	Lavatory lamp	20
Tubelight (variant 1)	25	Vacuum flush	50
Tubelight (variant 2)	40	Other loads	100
Emergency lamp	20	Power sockets for mobile phones	100
Night lamp	15		

of AC LHB coaches viz. AC 1st Class (24 seater), AC 2-tier (54 seater), AC 3-tier (72 seater), AC Chair car (78 seater), AC Executive class (56 seater).

For mounting solar photovoltaic cell on rooftops, it is necessary to contemplate the dimensions and built of AC LHB coach. Figure 5 shows a bezel-less AC three-tier LHB coach. It is not possible to use complete area of the roof surface for installing solar photovoltaic cell since clearances have to be left for easy access for various purposes and roof-mounted AC package unit (RMPU) on the rooftop. Hence, the actual area available for fixing solar photovoltaic cell is less. The total roof area shall be allocated into four equal parts parted by alleys, which is presented in Fig. 6. The power generated using solar cell would depend on time and direction of travel and will fluctuate. Since the roof of the coach is curved, installation of standard size solar cell panel available in market could not be practicable since AC LHB coach roofs are very curved. Graphical demonstration of the installation of solar photovoltaic cell panels on top of the roof of AC 3-tier LHB coaches would be comparable to the illustration in Fig. 7 (Fig. 8).

Vasisht et al. [1, 2] calculated and reported that around Rs. 15 lacs (approx.) have to be invested for a train with 19 coaches on which solar photovoltaic cell system would be mounted. Though this can be done with any kind of LHB coach, return of investment can be gained within 2 yr 6 m after proper operation and maintenance.

Fig. 5 A bezel-less AC Three-Tier LHB coach [4]

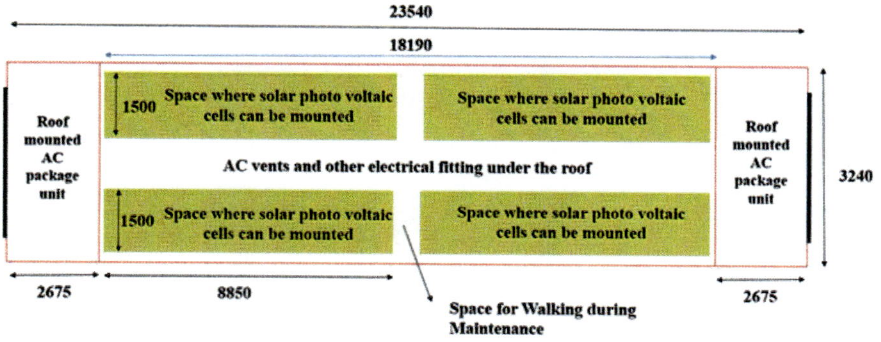

Fig. 6 Top view of AC LHB coach with the proposed design of fitting solar photovoltaic cell panels [1, 2]

Fig. 7 Graphical demonstration of an AC 3-Tier LHB coach mounted with solar photovoltaic cell panels [1, 2]

Annually 90,000 L approx. units of diesel will also be saved which in turn saves around Rs. 64 lacs (taking diesel rate as Rs. 71/L) on a train after the use of solar photovoltaic cell.

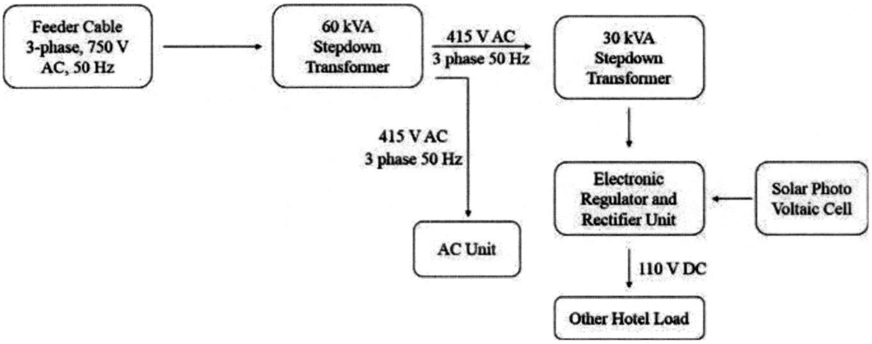

Fig. 8 Power supply diagram of an AC LHB coach with solar photovoltaic cell [1]

4 Comparison for Better Alternative of EOG: Solar Cell or HOG System

According to the report of Research Design and Standards Organisation (RDSO), through HOG system, annual energy cost for running hotel load will be around Rs. 80 lacs for a train with 19 AC LHB coaches. For running same train with EOG system, around Rs. 180 lacs are invested for diesel fuel which more than double of the investment of HOG system. This investment continues every year. Hence, the installation of solar photovoltaic cell on the AC LHB coaches can help in saving a lot of money since it is a one-time investment.

On July 14, 2018, Indian Railways launched its first solar-powered diesel electrical multiple unit (DEMU) train from Delhi Sarai Rohilla railway station in New Delhi to Farrukhnagar railway station in Haryana. Total 16 solar panels were fitted in six coaches producing 300 W_p from each panel. The solar panels were manufactured under "Make in India" initiative costing around Rs. 54 lacs. This train also has a power back-up for all electrical units which can run on batteries for around 72 h. About 17 units of power are generated by the solar panels in a day which powers the lighting system present inside the coaches. Figure 9 shows India's first solar-powered diesel electrical multiple unit train.

In December 2018, the Southern Railways fitted six non-AC coaches of the Coimbatore–Mayiladuthurai Jan Shatabdi Express train with photovoltaic panels at a cost of Rs. 15.2 lakhs, with the panels providing 4.8 kW per coach. In addition to the one coach that was fitted with the solar panels as a trial attempt in 2017, seven out of the 15 coaches are now powered by solar energy tapped through the photovoltaic panels. Figure 10 shows Coimbatore–Mayiladuthurai Jan Shatabdi Express train with photovoltaic panels.

Fig. 9 India's first solar-powered diesel electrical multiple unit (DEMU) train

Fig. 10 Coimbatore–Mayiladuthurai Jan Shatabdi Express train with photovoltaic panels

5 Conclusion

From this study, various conclusions can be drawn. Demonstration of the feasibility and scope of using solar photovoltaic cell on AC LHB coaches of trains was presented. The use of photovoltaic cell will benefit the Indian Railways by saving a lot of money spent on diesel fuel. The venture put into this assignment shall be mended in 3 yrs. There are unquestionably encounters involved in this system, but it is essential to highlight the profits on a larger scale whose influence would be incredible. According to the reports of RDSO, through HOG system, annual energy cost for running hotel load will be around Rs. 80 lacs for a train with 19 AC LHB coaches. For running the same train with EOG system, around Rs. 180 lacs are invested for diesel fuel which more than double of the investment of HOG system which continues every year. Hence, the installation of solar photovoltaic cell on the AC LHB coaches can help in saving a lot of money since it is a one-time investment. Some of the premium express trains which are fully air-conditioned run by Indian Railways, i.e. Garib Rath Express, Shatabdi Express, Rajdhani Express, Humsafar Express, AC Duronto Express, Tejas Express, Vande Bharat Express; also other prestigious Superfast Mail and Express trains can also be benefitted with this arrangement if the system can be developed and used.

References

1. Vasisht MS, Vishal C, Srinivasan J, Ramasesha SK (2014) Solar photovoltaic assistance for LHB rail coaches. Curr Sci 107(2):255–259
2. Vasisht MS, Vishal C, Srinivasan J, Ramasesha SK (2017) Rail coaches with rooftop solar photovoltaic systems: a feasibility study. Energy 118(1):684–691
3. Kesari OP, Deo SK. Head on Generation (HOG)—a step towards energy efficiency. Research Designs and Standards Organisation (RDSO). https://rdso.indianrailways.gov.in/works/uploads/File/Paper%20on%20HOG.pdf
4. Maintenance Manual for AC LHB Coaches (2013) Research Designs and Standards Organisation (RDSO). https://rdso.indianrailways.gov.in/works/uploads/File/Maintenance%20Manual%20for%20AC%20LHB%20Coaches.pdf
5. Maintenance Manual for LHB Coaches (2013) https://rdso.indianrailways.gov.in/works/uploads/File/Maintenance%20Manual%20for%20LHB%20Coaches(8).pdf

Thermoelectric Systems for Sustainable Refrigeration

Prasanta Kumar Satapathy

Abstract Today's concern over drastic environmental degradation and depletion of reserve of fossil fuel have made the scientists to search for some renewable energy sources and advanced thermodynamic systems so that irreversibility can be minimized. Thermoelectric system has a great potential to generate electricity in the range of some microwatt to 500 W on the principle of Seebeck effect. It can also generate refrigerating effect ranging from 5 mW to 500 W on the principle of Peltier effect. It utilizes the waste heat from industry, motor vehicle or the solar energy. The efficiency of this system is determined from non-dimensional parameter Figure of Merit, *ZT*. Review of the updated research papers reveals the development of thermoelectric materials which give higher values of *ZT* so that system efficiency is increased. Enhancement of *ZT* is done by doping the bulk material with some other suitable one in nanostructural form. This decreases the lattice thermal conductivity and increases the power factor. As a result, *ZT* value of a thermoelectric material increases. System performance can be enhanced by properly utilizing the available heat. This is made possible by integrating photovoltaic cells with thermoelectric generator. Minimizing the resistances to heat flow and current flow through the thermoelectric modules and thermal resistance matching for thermoelectric cooling system also enhances the performance of a thermoelectric system. Development of thermoelectric cooling system run by thermoelectric generator, known as thermoelectric self-cooling system, draws more attention for the research as this technology gives a sustainable cooling system.

Keywords Thermoelectric generators · Thermoelectric cooler · Figure of merit · Portable cooler · Thermal resistance

P. K. Satapathy (✉)
Mechanical Engineering, College of Engineering and Technology, Bhubaneswar, Bhubaneswar, India
e-mail: premdamayanti@yahoo.com

M. Ramgopal et al. (eds.), *Advances in Air Conditioning and Refrigeration*, Lecture Notes in Mechanical Engineering, https://doi.org/10.1007/978-981-15-6360-7_38

1 Introduction

Nowadays, most of the research activities are focused on the sustainable development. Sustainability is defined as the preservation of productive capacity for a foreseeable future. Productive capacity needs resources of material and energy, technology and conducive environment to lead a healthy life. When a society enjoys the developmental benefits preserving the productive capacity for the foreseeable future, it is said that it undergoes a sustainable development. A worldwide estimate shows that around 40% of energy is consumed by refrigeration and air-conditioning sectors. Preservation of goods and human comfort are the two most energy-consuming applications of refrigeration cycle. In view of the sharp declining of the fossil fuel and sudden rise in the environmental pollution, scientists and researchers are striving to develop new technologies so that energy consumption can be reduced or renewable energy can be used, and as a result environment pollution can be reduced. In this context, it is very much essential to discuss on the existing technology for refrigeration systems and their use. In this regard, a scientific approach towards adopting advanced and efficient technology for refrigeration and air-conditioning and conservative approach for the use of these systems will definitely lead us towards a sustainable development. After successful implementation of Montreal Protocol to phase out ozone-depleting substances like CFCs since 1987, Kigali amendment in 2016 came into force for phase out of HFCs, particularly high GWP HFCs. This amendment was specifically meant for climate system protection by reducing the use of high GWP HFCs and promoting the use of energy efficient technology. In short, it can be said that this amendment provides an opportunity for search of 'Sustainable Technologies' for refrigeration systems. It is normally perceived that sustainable system means the use of renewable energy and most preferably solar energy. Much of the research has been carried out, and feasibility study has been reported on vapour absorption refrigeration systems using solar energy. This paper presents a detailed review of the working of a thermoelectric refrigeration or cooling system and selection of an efficient thermoelectric materials pair which gives better performance than the existing materials pair in use. For thermoelectric cooler (TEC) to be operational in sustainable mode, electricity should be supplied through renewable energy source. In this discussion, thermoelectric generators (TEG) have been opted for the supply of electricity to TEC.

2 Working Principle

When electric current is passed through two dissimilar metals joined together, the following effects are observed [1].

(i) When emf is applied across the two ends, one end becomes hot and other end becomes cold depending upon the direction of current as shown in Fig. 1a and known as Peltier effect and expressed as $Q = \pi_{PN} I$, where π_{PN} is the Peltier

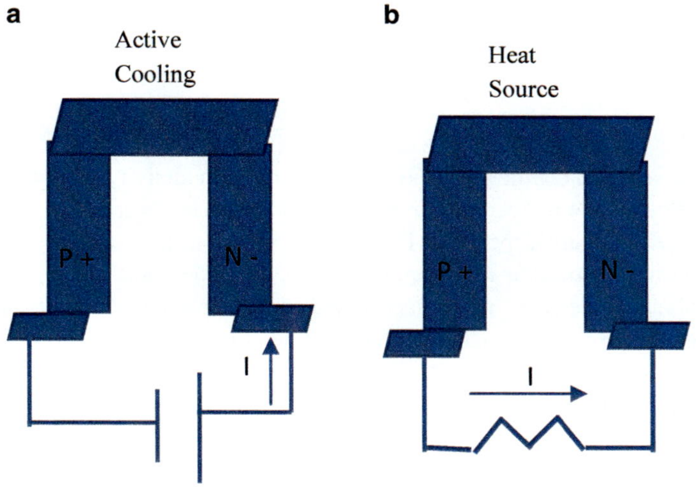

Fig. 1 Schematic diagram of **a** Thermoelectric cooler, **b** Thermoelectric generator

coefficient and I is the current passed through it. This is the working principle for thermoelectric cooling (TEC).

(ii) When two ends are maintained at different temperature, emf is generated as shown in Fig. 1b and known as Seebeck effect and expressed as $\Delta V = \alpha \Delta T$, where α is Seebeck coefficient. This is the working principle for thermoelectric generation (TEG).

(iii) There is reversible absorption or liberation of heat in a homogeneous conductor exposed to simultaneous temperature and electrical gradients, and it is expressed as $\tau = (Q/\Delta T)/I$.

(iv) Due to flow of current, there will be generation of heat known as Joulean effect and calculated from $Q = I^2 R$.

(v) Heat is conducted from hot end to cold end as per the Fourier law of heat conduction.

It is evident from the above phenomena that it is the two dissimilar materials, usually the semiconductors, properties determine the performance of a thermoelectric system. Also the system configuration and the availability of heat source and sink with minimal irreversibility during the heat transfer process increase the efficiency and the reliability of the system.

3 Thermoelectric Materials

Thermoelectric material works on the principle of Seebeck effect (for thermoelectric generation) and/or Peltier effect (for thermoelectric cooling). Its effectiveness is described on the basis of a parameter (Z) known as 'Figure of Merit' and more conveniently, is expressed in a non-dimensional form for performance evaluation as

$$ZT = \frac{\alpha^2 \sigma}{\kappa} T; \quad (1)$$

where α is Seebeck coefficient, σ is the electrical conductivity, κ is the thermal conductivity, T is the temperature in Kelvin. It is because Figure of Merit (Z) of a material is temperature dependent. From this expression, it is evident that a good thermoelectric material must have high Seebeck coefficient and electrical conductivity, combinedly called as 'power factor' and a low thermal conductivity. But these properties of any thermoelectric material contradict each other as shown in Fig. 2 [2]. However, thermoelectric material performance (ZT) and power factor can be enhanced by (i) decreasing the thermal conductivity specifically decreasing the lattice and bipolar conductivity, because thermal conductivity of a material is contributed by electron movement, lattice vibration and bipolar effect. If electron movement is reduced, the electrical conductivity will be reduced which is not desirable; (ii) increasing the Seebeck coefficient; (iii) increasing the electrical conductivity. This is only possible by impregnating the bulk material with suitable doping material or by layering the bulk material using nano-engineering techniques. Increase in carrier concentration decreases Seebeck coefficient, but increases the electrical conductivity, thermal conductivity as well. Maximum value of ZT is obtained at a carrier concentration of around $10^{19}/\text{cm}^3$. However, typical carrier concentration is considered to be in a range of 10^{19} to $10^{20}/\text{cm}^3$. Cai et al. [3] and Li et al. [4] have conducted experiments with Bi_2Te_3 being alloyed with Sb for P-type and with Se for N-type as bulk materials. P-type material is doped with SiC, and N-type material is doped with SbI_3 by nano-engineering, respectively.

Results show that inclusion of suitable nanomaterial into the bulk material causes an increase in carrier concentration. This increases both Seebeck coefficient and

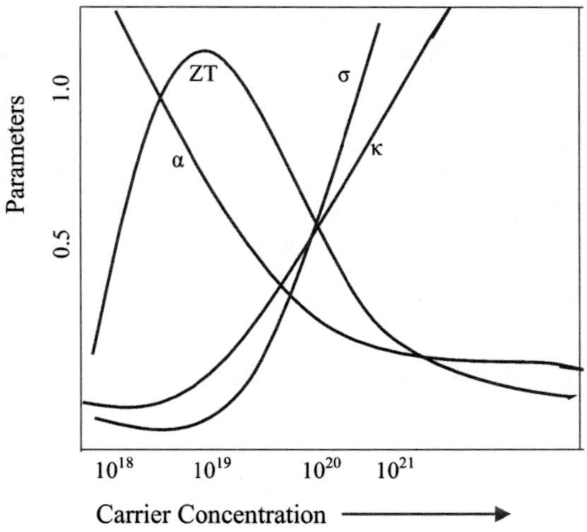

Fig. 2 Variation of TE material properties with carrier concentration

the electrical conductivity. Simultaneously, it decreases the lattice thermal conductivity though lattice thermal conductivity is independent of carrier concentration as suggested by Zhang and Zhao [5]. They have reported a case of reduction of thermal conductivity by 25% through Na doping, and inclusion of nanostructured SrTe again decreases the thermal conductivity by 55%. Thus, *ZT* increases to 1.7 at 800 K. Harman of Lincoln Laboratory has claimed a *ZT* value of 3.2 at about 300 °C for a material with nanoscale inclusions that reduced the thermal conductivity [6]. Similarly, Heremans et al. [7] and Cai et al. [8] doped thalium into PbTe. This gives an enhanced Seebeck effect due to distortion in the density of the electron states. Zhang and Zhao [5] have classified the thermoelectric materials, in bulk form, into three categories on the basis of their optimal working range of temperatures. As shown in Fig. 3 [10], Bi_2Te_3-based alloys show ZT_{max} at low temperature (<400 K), PbTe-based alloys show the peaks at middle range of temperature (600–900 K), and SiGe-based alloys show their peaks at high temperatures (>900 K). From these observations, it may be concluded that for refrigeration purpose, best suitable bulk material is Bi_2Te_3, whereas PbTe- and SiGe-based bulk materials are suitable for power generation. Yu et al. [9] experimentally verified thermoelectric properties of copper selenide and concluded that at higher temperature (at 700 K), this material changes from α-phase to β-phase, and a *ZT* value of 1.6 could be achieved at that temperature.

However, Tritt et al. [10] have opined that the 'photon-glass/electron-crystal' (PGEC) approach appears to be the best.

It is because one has to decide whether 'holey' semiconductors (materials with cages, such as skutterudites or clathrates) or 'unholey' semiconductors such as SiGe or PbTe are the best to pursue and which tuning parameters are available to improve the performance of these materials.

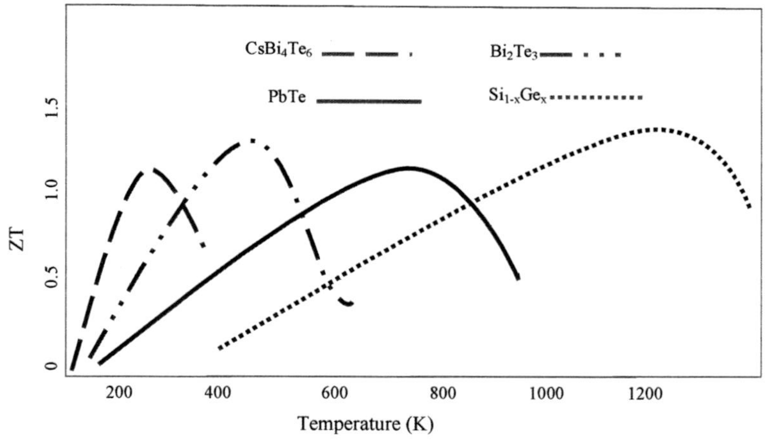

Fig. 3 Variation of *ZT* of different TE material with Temperature

4 Thermoelectric Generator and Refrigerator

4.1 Thermoelectric Generator

Thermoelectric generator works on the principle of Seebeck effect, and its efficiency increases with the increase in temperature difference between the hot end and the cold end. But thermoelectric refrigerator works on the principle of Peltier effect, and its performance increases with the decrease in that temperature difference.

Maximum efficiency of a TE generator and the maximum COP of a TE refrigerator can be expressed by Eqs. (2) and (3), respectively.

$$\eta \mathrm{max}_{TE} = \frac{T_{\mathrm{H}} - T_{\mathrm{C}}}{T_{\mathrm{H}}}\left[\frac{\sqrt{\left(1 + ZT_{\mathrm{avg}}\right)} - 1}{\sqrt{\left(1 + ZT_{\mathrm{avg}}\right)} + \frac{T_{\mathrm{C}}}{T_{\mathrm{H}}}}\right] \tag{2}$$

$$\mathrm{COPmax}_{TR} = \frac{T_{\mathrm{C}}}{T_{\mathrm{H}} - T_{\mathrm{C}}}\left[\frac{\sqrt{\left(1 + ZT_{\mathrm{avg}}\right)} - \frac{T_{\mathrm{H}}}{T_{\mathrm{C}}}}{\sqrt{\left(1 + ZT_{\mathrm{avg}}\right)} + 1}\right] \tag{3}$$

However, Kim et al. [11] have suggested more accurate formula to calculate maximum efficiency of a TE generator considering the Thompson effect and when there is a large temperature difference between hot end and cold end. They reported that it is unreliable to take T_{avg} in the case when there is a large temperature difference between two ends.

The system's overall performance can be enhanced by having maximum utilization of the energy source and reducing the irreversibility (resistance to heat and electricity flow). The heat sources may be radioisotopes, waste heat from industries or motor vehicles, solar energy, etc.

Figure 4 [12] shows the range of wavelength which carries enough energy with it to energize the electrons to that is useful for photovoltaic cell (PV cell) which is 58% as suggested by Tritt et al. [12]. The rest 42% of solar energy carried by longer wavelength (specifically infrared range) cannot be utilized by PV cells. Thus, a PV module converts around 12% of total incident solar energy to electrical energy, and rest of the radiations become available as thermal energy. Hence, this increases the temperature of the PV module, and increase in temperature of the PV module decreases its conversion efficiency. That is why a PV module needs cooling to enhance its efficiency. Again this thermal energy can be effectively utilized by thermoelectric generators to generate electricity as shown in Fig. 5 [13]. Cooling of cold end of the TEG due to flow of water (coolant) causes a large temperature difference between hot end and cold end that causes increase in TEG efficiency. Hence, it is very much beneficial to use integrated systems, which combines PV cells (solar panels) with thermoelectric generators. It has been reported by Dimri et al. [13] that by using water in place of air as the PV module cooling fluid, 18% increase in conversion efficiency

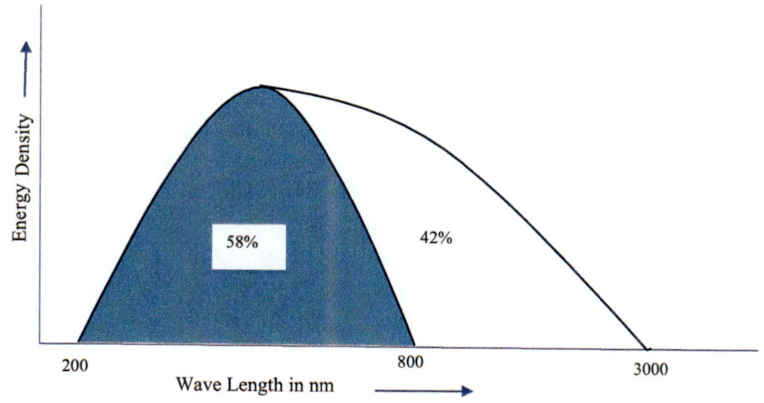

Fig. 4 Represents the Solar spectrum for PV cell and TEG

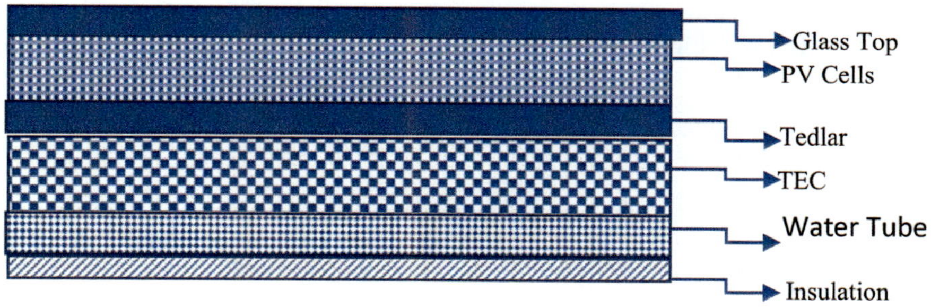

Fig. 5 A hybrid thermoelectric power generation system

is obtained for a photovoltaic thermal (PVT)-integrated thermoelectric generator. Tritt et al. [10] have suggested for a wavelength segregator through which the whole spectrum of the solar energy can be made available for electricity conversion by PV cell and TEG. Bamroongkhan et al. [14] have also studied the experimental performance of this combined system using a solar parabolic dish collector. The power generated by the PV cell is used to drive a fan which cools the PV cell, TEG as well. The efficiency of PV cell is calculated from

$$\eta_{\mathrm{PV}} = \frac{P_{\mathrm{PV}}}{A_{\mathrm{PV}}\, G} \tag{4}$$

where P_{PV} is the power generated by the PV cell, A is its surface area of exposure and G is the solar radiation. Efficiency of TEG is expressed by Eq. (3). The overall efficiency of the hybrid system is expressed as

$$\eta_{\mathrm{Overall}} = \eta \mathrm{max}_{TE} + \eta_{\mathrm{PV}} \tag{5}$$

They reported PV cell efficiency as 16.69% and TEG efficiency as 2.96% and an overall efficiency of 19.65%. The TE generator obtained an electrical power output of 2.94 W, whereas PV module obtained an output of 1.93 W.

But the experimental verification by Chavez-Urbiola et al. [15] shows that PV cell with a heat extractor gives the same efficiency as that of a sun tracking concentrating thermoelectric generator-based hybrid system. They also have a comparable cost. However, they suggest that the inclusion of TEG in PVM operating at high temperature will increase the thermal stability by reducing the losses.

4.2 Thermal Resistance

A higher value of non-dimensional Figure of Merit does not ensure a higher electricity generation. A thermoelectric module offers resistances to electric and heat flow. The schematic of these resistances are shown for one side (P-leg/N-leg) of a single TEG module as shown in Fig. 6 [17]. R_M is the electrical resistance due to metallic part

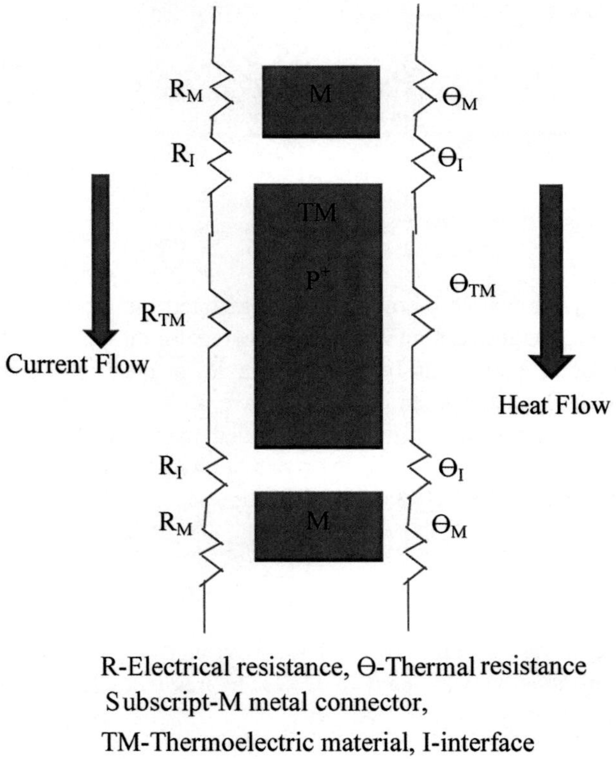

Fig. 6 Schematic representation of resistances to heat flow and current flow in P-Leg

at upper as well as lower part of the P-Leg, R_{I} is the electrical resistance due to interface of TE element and metallic part, R_{TM} is the electrical resistance due to TE element itself. In the same sequence, resistances are also offered for the flow of heat. Of these resistances, the resistances offered at the interfacial point are very high depending upon the soldering material that has been used to connect the TE module with metal connectors or electrodes. The soldering material causes an increase in interface resistances for both the electricity and heat flow due to excessive diffusion into the parent material. Increase in thermal resistance at the contact point decreases the actual temperature difference between the hot end and cold end of the TE material. Thus, it decreases the efficiency and power output. Similarly, increase in electrical resistance at the contact points increases the total electrical resistance of the system. This also decreases the reduction in efficiency and power output. This is evident from the Eq. (1). Liu and Bai [16] have conducted the experiment taking Bi_2Te_3-based TE module. They have reported that the conventionally used Ni-based soldering material is good for P-type $Bi_xSb_{2-x}Te_3$. But it is not suitable for n-type $Bi_2Te_{2.7}Se_{0.3}$. They suggested that NiFe-based alloys may be more suitable as interface or soldering material for n-type material. This gives low contact resistance. Besides this, the effect of geometrical parameters like leg length and cross-sectional area has also been studied by Wang et al. [17]. They suggested for optimal length and cross-sectional area of the legs to get maximum efficiency and maximum output power. However, for more power generation a thermoelectric module is fabricated using multiple numbers of p–n thermocouples connected thermally in parallel and electrically in series.

Besides, thermal resistances at the heat sinks, both for TEG and TEC, have a great influence over the power generation and cooling capacity, respectively. Synder et al. [18] have calculated the maximum power output when the thermal resistance of the TEM is equal to the sum of those of the heat sinks at both ends. Similarly, Lu et al. [19] have shown that the ratio of the resistances of the hot side and cold side should vary within 0.1–10, and TEM resistance should be 40–70% of the total thermal resistance so that 10.7–19.8% more cooling power can be achieved.

4.3 Portable Thermoelectric Cooler

Researches have been carried out on self-driven TEG-TEC systems, i.e. thermoelectric cooler driven by a thermoelectric generator as shown in Fig. 7a. Sometimes it is referred to as thermoelectric self-cooling (TSC) system. Lin et al. [20] reported that the cooling effect increases by 75% when a multistage TEG-TEC system for higher cooling effect is operated in individual manner. The reason is that multiple number of individual TEGs generate more electricity collectively than a multistage TEG consisting of same numbers of TEGs. In this case, TEG may be driven by waste heat from industries or motor vehicles. For space applications, radio isotopes are used as the heat source for TEG. For solar systems, combination of PV cell and the TEG modules (hybrid systems) is more beneficial. They are sometimes called as portable TEC. A portable TEC developed by Vijayarengan and reported by Pourkiaei et al.

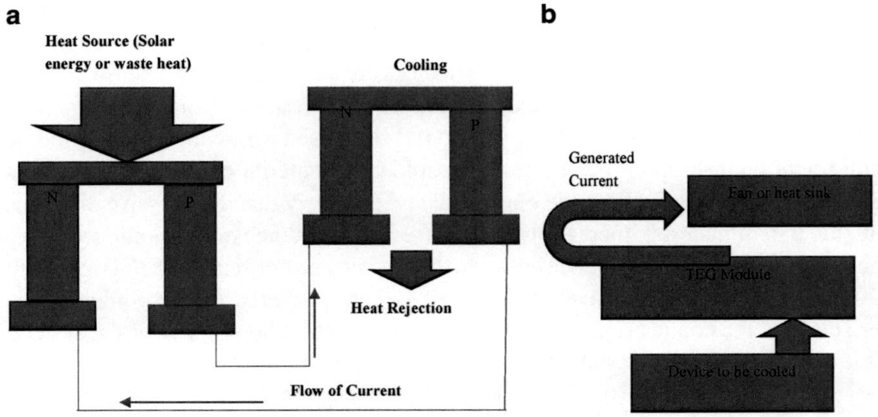

Fig. 7 **a** Schematic of a TEG operated TEC system. **b** Schematic of a thermoelectric self cooling device

[21] that this system could cool 250 ml of water by 10 °C in 2 h when the ambient was 31 °C. This shows this system has very poor performance (COP is around 0.2). Cooling of electronic devices can be done by placing the hot end of the TEG close to the device to be cooled as shown in Fig. 7b. TEG takes up the heat from the device which is normally at 55–75 °C. Current generated from the TEG is sufficient to drive a fan which will act as a heat sink. Thus, an electronic device can be self-cooled. This is in contrast to the present method of thermoelectric active cooling. These types of self-driven TEG-TEC systems are very much useful for remote areas where power grids are not available or places like aircraft, submarines, space shuttles and commodities like helmets, etc.

4.4 Integrating the Phase Change Material with Hybrid System

The performance of thermoelectric system can also be made more reliable, specifically for solar TE systems by integrating phase change material with TEG or TEC. Omer et al. [22] have reported on experimental investigation of a thermoelectric refrigeration system integrated with PCM along with thermal diode. Heat is rejected by PCM during cooling process and get solidified. Total heat rejection by PCM will be summation of its sensible heat and the latent heat of fusion. During 'Off Period', PCM absorbs latent heat and sensible heat from the refrigerated/cooling cabinet and keeps it at the desired temperature. Thermal diode is used so that heat will not be leaked to the PCM when there will be power off. Studies reveal that a salt, mixture of sodium nitrate and potassium nitrate, may be used with thermoelectric generator to act as the heat source as its melting temperature is nearly 270 °C.

5 Conclusion

The present study on thermoelectric technology gives a comprehensive idea on the potentialities of this technology for electricity generation as well as cooling. This has become more acceptable with the development of thermoelectric materials having more efficient conversion efficiency. Most prevalent TEM, alloys of Bi_2Te_3, is replaced by other alloys like $CsBi_4Te_6$ for cooling and PbTe, SiGe, CuSe, etc. for power generation, depending on their maximum *ZT* value corresponding to the temperature of application. Advanced thermoelectric systems also give higher efficiencies/COPs. Power generation can be enhanced by integrating PV cell with TEG. For intermittent heat source, specifically solar energy, TEG/TEC can be integrated with phase change material for more reliability. These systems can generate cooling of 5 mW to 500 W. In the absence of electric power, thermoelectric self-cooling system can also solve the cooling problem of electronic devices, automobile air-conditioning, storing of beverages, etc. by utilizing the waste heat or solar energy. Nowadays, this TSC system will be most useful for cooling of Internet of Things (IoT) devices.

References

1. Arora RC (2010) Refrigeration and air conditioning, text book. PHI Learning Private Limited, pp 674–687
2. Synder GJ, Toberer ES (2008) Complex thermoelectric materials. Nat Mater 7:105–114
3. Cai Y, Wang Y, Liu D, Zhao FY (2019) Thermoelectric cooling technologyapplied in the field of electronic devices: updated review on the parametric investigations and model developments. Appl Therm Eng 148:238–255
4. Li F, Zhai R, Wu Y, Xu Z, Zhao X, Zhu T (2018) Enhanced thermoelectric performance of n-type bismuth-telluride-based alloys via In alloying and hot deformation for mid temperature power generation. J Materiomics 4:208–214
5. Zhang X, Zhao LD (2015) Thermoelectric materials: conversion between heat and electricity. J Materiomics 1:92–105
6. Bell LE (2008) Cooling, heating, generating power, and recovering waste heat with thermoelectric system. Science 321:1457–1461
7. Heremans JP, Jovovic V, Toberer ES, Saramat A, Kurosaki K, Charoenphakdee A, Yamanaka S, Snyder GJ (2008) Enhancement of thermoelectric efficiency in PbTe by distortion of electronic density of states. Science 321:554–557
8. Cai B, Hu H, Zhuang HL, Li JF (2019) Promising materials for thermoelectric application. J Alloy Compd 806:471–486
9. Yu B, Liu W, Chen S, Wang H, Wang H, Chen G, Ren Z (2012) Thermoelectric properties of copper selenide with ordered selenium layerand disordered copper layer. Nano Energy 1:472–478
10. Tritt TM, Subramanian MA (2006) Guest editors: thermoelectric materials, phenomena, and applications: a bird's eye view. MRS Bull 31:188–198
11. Kim HS, Liu W, Chen G, Chu CW, Ren Z (2015) Relationship between thermoelectric figure of merit and energy conversion efficiency. PNAS 112:8205–8210
12. Tritt TM, Bottner H, Chen L (2008) Thermoelectrics: direct solar thermal energy conversion. MRS Bull 33:366–368

13. Dimri N, Tiwari A, Tiwari GN (2019) Comparative study of photovoltaic thermal (PVT) integrated thermoelectric cooler (TEC) fluid collectors. Renew Energy 134:343–356
14. Bamroongkhan P, Lertsatithanakorn C, Soponronnarit S (2019) Experimental performance study of solar parabolic dish photo voltaic-thermoelectric generator. Energy Procedia, 158, 528–533
15. Chavez-Urbeola EA, Vorobiev YV, Bulat LP (2012) Solar hybrid systems with thermoelectric generators. Sol Energy 86:369–378
16. Liu W, Bai S (2019) Thermoelectric interface materials: a perspective to the challenge of thermoelectric power generation module. J Materiom 5:321–336
17. Wang P, Wang BL, Li JE (2019) Temperature and performance modeling of thermoelectric generators. Int J Heat Mass Transf 143:118509
18. Synder GJ, Priya S, Inman DJ (eds) Thermoelectric, energy harvesting technologies. Springer, Berlin (Chapter 11)
19. Lu X, Zhao D, Ma T, Wang Q, Fan J, Yang R (2018) Thermal resistance matching for thermoelectric cooling system. Energy Convers Manag 169:186–193
20. Lin L, Zhang YF, Liu HB, Meng JH, Chen WH, Wang XD (2019) A new configuration design of thermoelectric cooler driven by thermoelectric generator. Appl Therm Eng 160:114087
21. Pourkiaei SM, Ahmadi MH, Sadeghzadeh M, Moosavi S, Pourfayaz F, Chen L, Yazdi MA, Kumar R. Thermoelectric cooler and thermoelectric generatordevices: a review of present and potential applications, modeling and materials. Elsevier, Available Online
22. Omer SA, Riffat SB, Ma X (2001) Experimental investigation of a thermoelectric refrigeration system employing a Phase Change Material integrated with thermal diode (Thermosyphons). Appl Therm Eng 21:1265–1271

Waste Heat Recovery from Walls of the Combustion Chamber of a New Portable Jaggery Plant to Dry Bagasse

A. B. Shinde and S. N. Sapali

Abstract In conventional jaggery plants, the extracted sugarcane juice is used for preparing jaggery, while the residue bagasse is used as a fuel to meet the energy requirement. Normally, bagasse contains from 48 to 52% moisture which affects the calorific value of bagasse; hence, drying of bagasse is necessary. The heat required for the evaporation is provided by burning the bagasse in a pit-type furnace. It is crucial to maintain the level of moisture in bagasse used as a fuel in order to achieve high calorific value. In conventional jaggery plant, a large amount of heat is wasted inside the wall of the combustion chamber. The performance analysis is carried out for a newly designed portable jaggery plant. The bagasse drying mechanism of this plant recovers waste heat from the combustion chamber through walls. The heat transfer from combustion gases through a brick wall, metallic wall, porous wall, etc., are simulated through ANSYS software and are tried out to validate through experimentation with certain limitations. The experimental outcome reveals that about 3.5% waste heat from the combustion chamber removes 51% moisture from wet bagasse. Jaggery preparation utilizes 23% of the combustion chamber heat. Thus, a quick and non-laborious bagasse drying process alternative to conventional sun drying process is developed.

Keywords Jaggery · Bagasse drying · Waste heat recovery

1 Introduction

In the current scenario, about 133 million t of sugar and 10 million t of jaggery are produced by 115 sugarcane cultivating countries in the world. The traditional jaggery plants utilize an underground combustion chamber, and the pan with sugar juice is arranged on the earth surface above the combustion chamber. During jaggery

A. B. Shinde · S. N. Sapali (✉)
Department of Mechanical Engineering, College of Engineering Pune, Pune, Maharashtra, India
e-mail: sns.mech@coep.ac.in

A. B. Shinde
e-mail: khodashi5@gmail.com

M. Ramgopal et al. (eds.), *Advances in Air Conditioning and Refrigeration*, Lecture Notes in Mechanical Engineering, https://doi.org/10.1007/978-981-15-6360-7_39

production, when the combustion process of fuel, i.e. dry bagasse is carried out, a large amount of heat is wasted to the surrounding area. It is necessary to carry out the drying of wet bagasse to convert into dry bagasse. However, large space and time are needed for this process. Availability and intensity of solar radiations also play a vital role in this drying process. This process reduces the moisture content and thus increases its calorific value. Different researchers have implemented various methods and techniques to improve the gross calorific value of bagasse by removing moisture content in the wet bagasse. Sharon et al. [1] have carried out work on the energy balance of conventional jaggery plant. They determined that the crushing efficiency is 60%, and the efficiency of the concentration process is 14.75%. They arrived at the conclusion that the cool water is circulated for cooling the hot jaggery can be recirculated for steam production. This makes the jaggery processing energy-efficient. Sudhakar et al. [2] have designed the bagasse dryer to eliminate the moisture content from wet bagasse up to a certain extent. The bagasse consists 51.5% of the average moisture content. After installation, about 45% of wet bagasse was routed through the dryer. Panchal et al. [3] have studied the burning characteristics of bagasse and presented in the form of charts for calorific values and bagasse composition. The efforts to improve the performance jaggery processing plant have been taken into several industries. Unused heat from the combustion chamber can be utilized for drying of bagasse and heating sugarcane juice in the preheater. Manjare et al. [4] have found about 8% improvement in thermal efficiency. Also, 1.2 kg bagasse consumption per kg jaggery production is saved due to the utilization of waste heat from exhaust gas for preheating the sugarcane juice. Jakkamputi et al. [5] worked on the utilization of solar energy using solar panels to improve the performance of jaggery plant. Anwar [6] has used a microwave oven to determine the moisture content from bagasse. The heat transfer from combustion gases through a brick wall, metallic wall, porous wall, etc., are simulated through ANSYS software and are tried out to validate through experimentation with certain limitations.

2 Methodology

The experimentation is performed with walls of the combustion chamber made up of steel plate as well as a brick wall. Dry bagasse comes into the combustion chamber from the feeder located at front face. The iron cage is having facility to hold wet bagasse around the combustion chamber. The 50 mm distance is maintained between the steel plate and wet bagasse to avoid self-ignition. The juice pan receives heat from the combustion chamber, and flue gasses are exhausted to the atmosphere. The steel plate, as well as brick wall, gets heated during the combustion. The air receives heat from walls of the combustion chamber. The hot air between steel plate and wet bagasse helps to remove the moisture from the wet bagasse. The experimental setup is shown in Fig. 1.

Fig. 1 Experimental set-up for developed portable jaggery plant

2.1 Experimentation

Case 1, the experimentation is carried out for developed portable jaggery plant with porous steel plate shown in Fig. 2. Which results in the burning of wet bagasse due to direct contact of a burning flame with the wet bagasse? Case 2, the experimentation is carried out with a porous steel plate with flame arrester. Even though by providing flame arrester, the flame travels to the wet bagasse through the porous part of the plate. Case 3, the experimentation relates with plain steel plate without porosity as walls of the combustion chamber. The direct contact of burning flame and the wet bagasse is avoided, but the temperature of the steel plate remains above the self-ignition temperature of bagasse. This results in undesired pyrolysis of bagasse. Case 4, in this case, the gap of 50 mm avoids the self-ignition of bagasse. It is observed that only the bagasse facing to combustion wall gets dried quickly. Hence in the next trial as case 5, the cage rotation by 180° is done, in order to bring the bagasse facing the atmosphere now gets facing to walls of the combustion chamber. Now the bagasse

Fig. 2 Experimental trial on a porous steel plate (Case 1)

Table 1 Experimental data for case 5 using steel plate walls with cage rotation

Sr. No.	Contents	
1	Total sugarcane used	109 kg
2	Amt. of sugarcane juice	60 kg
3	Amt. of wet bagasse	49 kg
4	Jaggery production time	2 h 15 min
5	The weight of dry bagasse	41 kg
6	Weight of moisture removed	8 kg
7	% moisture removed	33.33%

gets dried from both sides of the cage. In order to achieve quick drying of bagasse without pyrolysis, trials of case 6, case 7 and case 8 are carried out by using one-inch-thick plain brick, two-inch-thick plain brick and porous brick as walls of combustion chamber, respectively. Out of these all cases tested, experimental results of case 5 using plain steel plate as the wall of the combustion chamber. The experimental data for case 5 using steel plate walls with cage rotation is given in Table 1.

Figure 2 shows the experimentation is carried out for case 1 using porous steel plate as a wall for the combustion chamber. It shows that the burning flame comes out from the porous wall and gets in direct contact with wet bagasse results in the burning of wet bagasse. Figure 3 shows the experimentation carried out for case 5 using steel plate walls with cage rotation.

Fig. 3 Experimental trial on plain steel plate wall with cage rotation (Case 5)

3 Numerical Analysis

The temperature of the steel plate is a critical parameter for the drying of bagasse as it may cause self-ignition of bagasse as in case 3. To avoid self-ignition, a gap of 50 mm is kept between the steel plate and bagasse. The schematic representation for numerical analysis is shown in Fig. 4, which is applicable to case 4 and case 5. There is a gap of 50 mm maintained between the steel plate and the wet bagasse filled in a cage. The heat flux rate through walls of the combustion chamber is 4800 W/m^2.

The plate of 0.917 m * 0.917 m * 0.003 m size is used for numerical analysis. The boundary conditions are heat flux 4800 W/m^2, and the convective heat transfer coefficient is 2 W/m^2 K. Energy scheme of second-order upwind is used for the simulation.

ANSYS Fluent is used to plot the temperature profile along the width of the steel plate is presented in Fig. 5. The temperature of the steel plate is recorded as 976 K, while the temperature of the air between the gap varies between 300 and 800 K. The air flows over the plate due to natural convection. The similar simulation is done

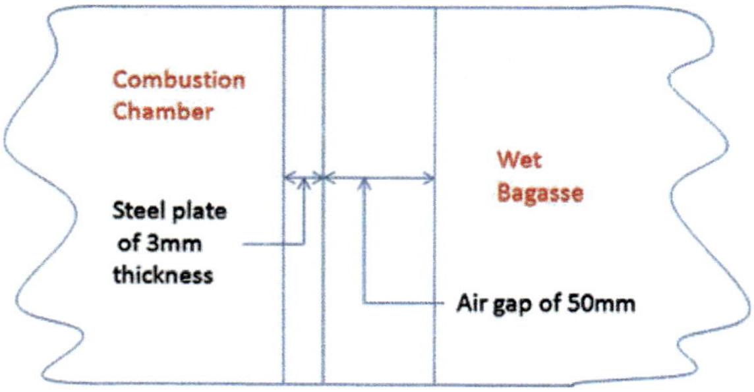

Fig. 4 Schematic representation of the combustion chamber with an air gap

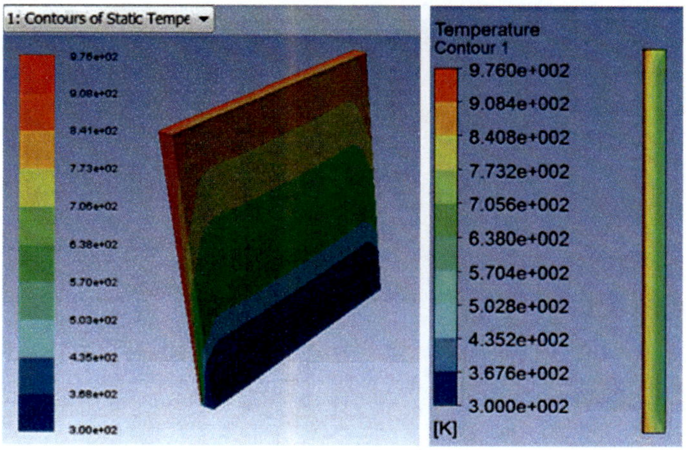

Fig. 5 Simulation results of steel plates of 3 mm thickness

for the brick wall of one- and two-inch thickness. The temperature profiles obtained through simulation in ANSYS Fluent are presented in Figs. 6 and 7. The values of the temperatures recorded are in the range from 600 to 300 K. The temperature of a brick wall is recorded as 600 K, while the temperature of the air between the gap varies between 300 and 550 K.

A plain steel plate of 3 mm thickness attains the temperature from 850 to 950 K. In case 5, in order to avoid pyrolysis one has to maintain the gap between bagasse

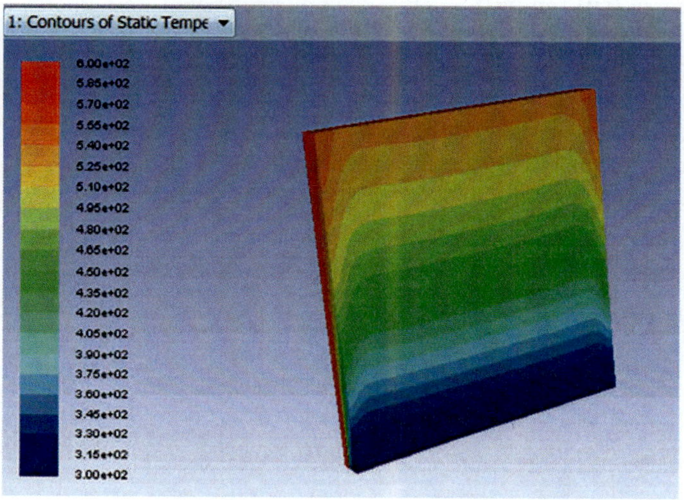

Fig. 6 One-inch brick wall simulation

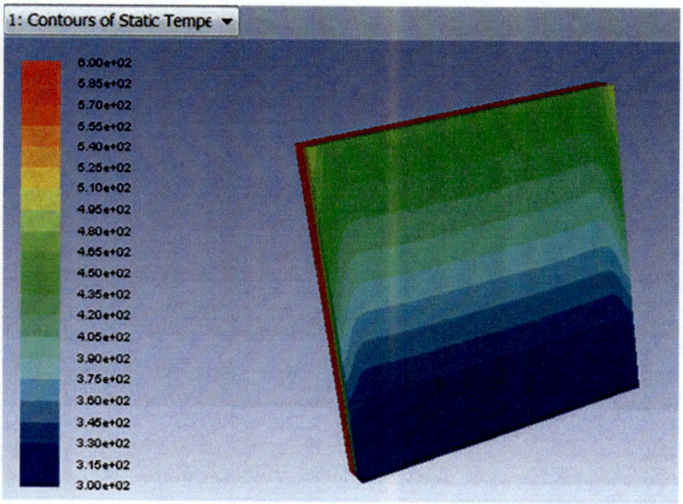

Fig. 7 Two-inch brick wall simulation

and plate. The air gap could provide the conductive resistance to heat flow; but at the same time, it enhances the convective heat transfer to the bagasse. The heated air of the gap removes the moisture of wet bagasse. The moisture removed from the combustion side is much more than the ambient side; hence, provision is required to be made to change the direction of the cage in order to dry the bagasse from both the side. The brick wall of 1-inch thickness reaches the temperature above 500 K as in case 6, and brick wall of 2-inch thickness could attain the temperature below 500 K which is below the pyrolysis temperature of bagasse, as in case 7. Increase in brick wall thickness up to 2 inches reduces the wall temperature but sufficient amount of heat is not transferred to bagasse because of higher thermal resistance of the brick wall.

4 Experimental Results

It is observed in the simulation results that the steel plate temperature is 976 K which is nearly equal to the temperature record by thermocouple during experiments that are from 800 to 900 K. Similarly, the brick wall temperature obtained through simulation is 600 K, and during the experiment, it is measured from 510 to 580 K. The heat utilization for different cases, observed during experimentation from case 1 to case 6 is presented in Figs. 8, 9, 10 and 11.

Figure 8 shows the average moisture removed during the experiments. Bagasse contains 50% moisture in it before drying. It is observed that case 5 gives the maximum moisture removal during the experiment that is 70%.

Figure 9 shows per cent heat utilized for jaggery preparation. The heat from the combustion chamber is first utilized for sensible heating of juice then for the latent heat of juice. Figure 9 shows the amount of heat utilized for jaggery preparation, which indirectly correlates with the efficiency of the portable jaggery plant.

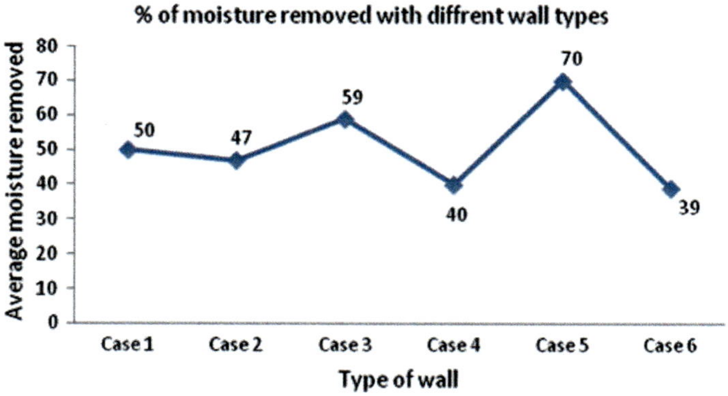

Fig. 8 Average moisture removed during experiment

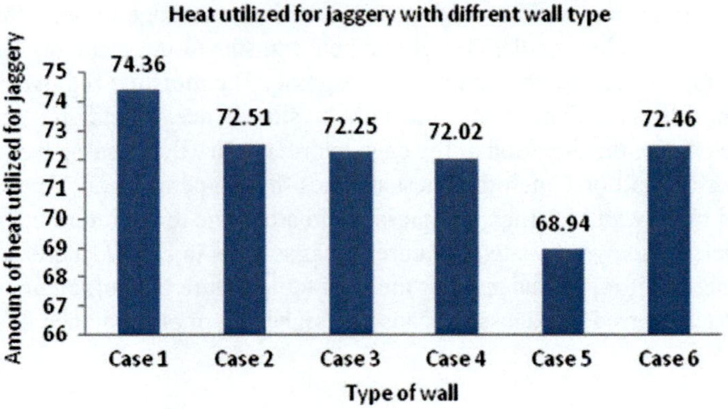

Fig. 9 Heat utilized for jaggery preparation

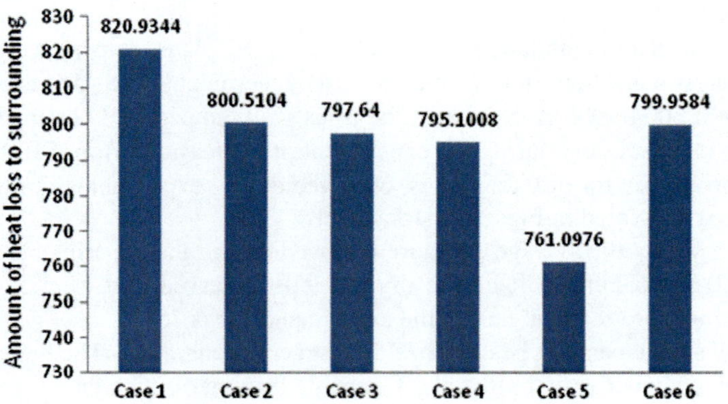

Fig. 10 Total heat loss during the experiment

Figure 10 shows total heat loss to surrounding from the combustion chamber during the experiment. Maximum heat loss observed in Case 1, that is, 74.36%, whereas the minimum heat loss is observed in case 5 that is 69%.

The heat balance is presented in the form of a bar chart as shown in Fig. 11. It shows that total heat supplied to the combustion chamber is 1104 MJ. 25% heat is utilized for jaggery making, 3.5% is utilized for moisture removal from bagasse, and 29% is heat is lost through flue gases. Also, there is 42.5% of heat loss through ash and another unaccounted form. It is observed that the utilization of heat is increasing from case 1 to case 5.

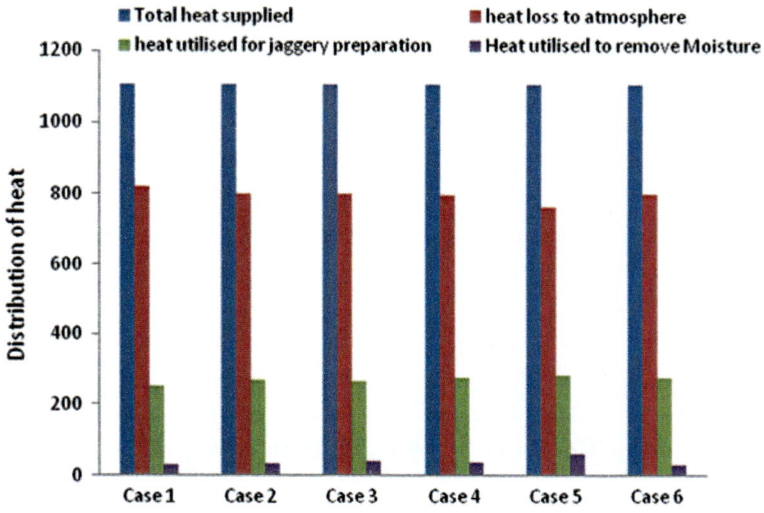

Fig. 11 Distribution of heat in jaggery preparation

5 Conclusions

The part of heat energy available through combustion is utilized for converting sugarcane juice into jaggery. The rest of heat energy is either a loss or used for moisture removal from wet bagasse packed surrounding the combustion chamber. In this research work, the analysis of heat recovery from the walls of a newly design combustion chamber of a portable jaggery plant is presented. The following conclusions are drawn from the research work.

1. Heat is recovered from the walls of the combustion chamber and utilized to remove moisture content of wet bagasse.
2. The system for quick drying of wet bagasse is developed and implemented successfully with compare to the conventional sun drying process.
3. The thermal analysis of the designed portable jaggery plant clearly indicates 70% total losses accompanied by the process.
4. The minimum energy required for jaggery processing is estimated to be only 23% of the total energy supplied.
5. The amount of heat recovered by the wet bagasse sample is observed to be around 3.5%.
6. Percentage moisture removal from the wet bagasse sample is observed to be around 69%.

References

1. Sharon MEM, Kavitha Abirami CV, Algusundaram K (2013) Energy losses in traditional jaggery processing. Indian Food Ind Mag 32
2. Panchal RJ, Shinde SM (2016) Effect of bagasse moisture on boiler performance. 2(3)
3. Manjare A, Hole J (2016) Exhaust heat recovery and performance improvement of jaggery making furnace. Int J Curr Eng Technol
4. Jakkamputi LP, Mandapathi MJK (2016) Improving the performance of jaggery making unit using solar energy. Perspect Sci
5. Anwar SI (2010) Determination of moisture content of bagasse of jaggery unit using microwave oven. J Eng Sci Technol 5(4):472–478
6. Shiralkar Kiran Y, Kancharla Sravan K, Shah Narendra G, Mahajani Sanjay M (2014) Energy improvements in jaggery making process. Energy Sustain Dev 18:36–48
7. Techno-economic analysis of jaggery production in Maharashtra. Indian Institute of Technology, Bombay (2012)
8. Vishal R, Sardeshpande DJ, Shendage, Pillai IR (2010) Thermal performance evaluation of a four pan jaggery processing furnace for improvement in energy utilization. Energy 35:4740–4747
9. Pattnayak PK, Misra MK (2004) Energetic and economics of traditional gur preparation: a case study in Ganjam district of Orissa, India. Biomass Bioenerg 26:79–88